AF800324

Tillering Behavior of Rice Plant

Pravat Kumar Mohapatra •
Ramani Kumar Sarkar •
Debabrata Panda • Ekamber Kariali

Tillering Behavior of Rice Plant

Springer

Pravat Kumar Mohapatra
School of Life Sciences
Sambalpur University
Sambalpur, Odisha, India

Debabrata Panda
School of Biodiversity and Conservation
of Natural Resources
Central University of Odisha
Koraput, Odisha, India

Ramani Kumar Sarkar
ICAR-National Rice Research Institute
Cuttack, Odisha, India

Ekamber Kariali
School of Life Sciences
Sambalpur University
Sambalpur, Odisha, India

ISBN 978-981-97-5234-8 ISBN 978-981-97-5235-5 (eBook)
https://doi.org/10.1007/978-981-97-5235-5

© The Editor(s) (if applicable) and The Author(s), under exclusive license to Springer Nature Singapore Pte Ltd. 2025
This work is subject to copyright. All rights are solely and exclusively licensed by the Publisher, whether the whole or part of the material is concerned, specifically the rights of translation, reprinting, reuse of illustrations, recitation, broadcasting, reproduction on microfilms or in any other physical way, and transmission or information storage and retrieval, electronic adaptation, computer software, or by similar or dissimilar methodology now known or hereafter developed.
The use of general descriptive names, registered names, trademarks, service marks, etc. in this publication does not imply, even in the absence of a specific statement, that such names are exempt from the relevant protective laws and regulations and therefore free for general use.
The publisher, the authors and the editors are safe to assume that the advice and information in this book are believed to be true and accurate at the date of publication. Neither the publisher nor the authors or the editors give a warranty, expressed or implied, with respect to the material contained herein or for any errors or omissions that may have been made. The publisher remains neutral with regard to jurisdictional claims in published maps and institutional affiliations.

This Springer imprint is published by the registered company Springer Nature Singapore Pte Ltd.
The registered company address is: 152 Beach Road, #21-01/04 Gateway East, Singapore 189721, Singapore

If disposing of this product, please recycle the paper.

Preface

Genus *Oryza* of the *Poaceae* family of monocot angiosperm is of high economic importance to man and animals in providing food and fodder. The genus consists of two cultivated species, namely *Oryza sativa* L. (Asian rice) and *Oryza glaberrima* Steud. (African rice), and twenty other non-domesticated species. *Oryza sativa*, a native of Asia contributes enormously to the food security of man; a minuscule of the world population in Africa survives on *O. glaberrima*. By containing decent amounts of digestible starch, protein, vitamins, and minerals, rice is the primary source of food calories for over half of the world's population, especially the poor farmers of Asia surviving on marginal landholdings. *O. sativa* originated on the river plains of the Ganges and/or Yangtze rivers and diffused and domesticated into different countries of Asia and other continents thousands of years back through human force showing the extraordinary capacity of adaptation, unparalleled for any other crop. The rice plant is semi-aquatic where the phytogeography of the plant is strongly influenced by soil water regime. The soil moisture regime, depending on the land topology, rainfall pattern, and varieties of climatic conditions has dictated the phenology, biomass, and grain yield of the rice plants growing under different environmental conditions across the world. Genotypic permissibility elicits phenotype changes of rice in the course of adaptation to a wide spectrum of ecosystems ranging from summer parched dry land to temporary or permanent water pools holding stagnant or slow running water, zero altitudes of sea level to 2600 m mountain slope of Himalaya, average crop season temperature of 33 °C in Pakistan to 17 °C temperature of Japan. The phenotypic plasticity of rice in response to the adaptation of rice to contrasting habitats world around, primarily through resilience in aberrant environments has resulted in the origin of new ecotypes, landraces, and morphological variants. Tiller dynamics has been an important indicator not only for environmental adaptation but also comes in handy for estimating primary production, reproductive growth, and grain yield. A rice genotype produces few tillers under a sub-optimal environment but becomes profuse tillering in favourable conditions. Tillers are the reproductive structures the plant generates in abundance when conditions allow and sacrifice many of them first for lack of resources and second, for survival under the exigency of exposure to unfavourable conditions. Tillering although important for yield, all tillers are not productive because of hierarchical development. Compared to early-formed basal tillers, their younger siblings arise late but coincide in maturity. The compromised longevity restricts the period for growth and yield. This book, *Tillering Behaviour of Rice Plant*, evaluates several aspects of the morphological, physiological, and molecular biology of tiller generation and hierarchy for raising rice grain yield ceiling. There are ten chapters in this book. Rice facts and fundamental information on tiller production in rice ecotypes dwelling in contrasting habitats are described in the preliminary chapters before moving to the molecular biology of tiller dynamics in the subsequent chapters enunciating a new hypothesis for designer rice.

Sambalpur, Odisha, India — Pravat Kumar Mohapatra
Cuttack, Odisha, India — Ramani Kumar Sarkar
Koraput, Odisha, India — Debabrata Panda
Sambalpur, Odisha, India — Ekamber Kariali

Contents

About the Authors

Pravat Kumar Mohapatra (Corresponding author) obtained his Ph.D. (Plant Physiology) from University of Adelaide. He joined the School of Life Sciences, Sambalpur University, and superannuated as Professor in 2008. Post-superannuation, he served as CSIR and UGC Emeritus Scientist. His epoch-making research on rice yield potential earned him distinctions in India, Australia, China, Japan, and IRRI, Philippines.

Ramani Kumar Sarkar obtained his Ph.D. (Botany) from University of Calcutta. He joined Agricultural Research Service, ICAR, and superannuated as Principal Scientist, ICAR-NRRI, Cuttack. Post-superannuation, he was Emeritus Scientist till 2020. He achieved reputations in India, CIMMYT, Mexico, and IRRI, Philippines for research on drought, salinity, and submergence tolerance.

Debabrata Panda was Scientist-B at Rubber Research Institute of India before joining faculty at Central University of Odisha, Koraput, in 2012. His research interests include physiology of abiotic stress tolerance in plants. He contributed to collection, evaluation, and popularization of indigenous wild crops strengthening tribal food security.

Ekamber Kariali joined the School of Life Sciences, Sambalpur University, as Lecturer in 1997 and obtained his Ph.D. in Life Sciences (Plant Physiology). He became Professor in 2017. He specializes in rice tiller dynamics. He made significant contributions to biodiversity of indigenous rice cultivars and their utility for mankind.

Origin and Evolution of Rice as Domesticated Staple Food Crop

1

Abstract

Rice (*Oryza sativa* L.) plays a pivotal role in the economy of developing nations of Asia as the staple food for citizens of countries bereft of any substitute crop of sustenance. It is an ancient crop older than 8000 years, but its ancestry is shrouded with mystery. Domesticated rice evolved with man and nature adapting to widely contrasting ecological niches most prominently present in the undulated river plains of the Ganges in India and the Yangtze in China. Genus *Oryza* has 27 species belonging to 11 genome groups: *Oryza sativa* L. (Asian rice) and *Oryza glaberrima* Steud. (African rice) are cultivated and the rest are wild. Geographical diversity is more in Asian rice than in African rice because of its diffusion to various ecological conditions. To meet the demand of an ever-growing number of rice consumers, the Green Revolution was introduced into rice cultivation in the late 1960s through the adoption of semi-dwarf high-yielding lodging-resistant traits. This remarkable achievement breached the yield potential of rice and increased grain production, but the success of bountiful production vanished in the 1980s owing to a significant rise in population growth. As of now, population growth continues unabated. Matching with the level of population growth the demand for production is expected to rise in the forthcoming decades. Bereft of discoveries, our food security will be at stake.

Keywords

Oryza sativa · Antiquity · Evolution · Yield potential · Green revolution · Domestication · Population growth

© The Author(s), under exclusive license to Springer Nature Singapore Pte Ltd. 2025

P. K. Mohapatra et al., *Tillering Behavior of Rice Plant*,
https://doi.org/10.1007/978-981-97-5235-5_1

1.1 Origin of Rice

Rice is a food crop grown in the tropical conditions of the world, mostly in Asian countries. It flourishes in hot humid tropical or semi-tropical climates. It is the only crop in rain-fed areas that receives heavy annual seasonal rainfall. The crop has been domesticated by mankind for more than 8000 years. The plant evolved along with man and nature adapting to a wide diversity of ecologies significantly different from one another in physicochemical and biological factors. Although domestication is akin to the progress of human civilization, rice antiquity has remained very complex and mysterious. The exact time and place of origin of rice have been debated. According to Chang (2003), its early progenitor was a grass that was embellished into different forms in contrasting humid regions of the southern landmass called the Gondwana supercontinent more than 130 million years ago. After the splitting up of the continent and subsequent continental drift into land fragments, rice progenitors occupied disjunct positions on the present landmasses. The ancestral form positioned in West Africa gave rise to cultigens *Oryza glaberrima* Steud., whereas the ancestral form occupying South and Southeast Asia generated cultigens *Oryza sativa* L. in the course of evolution passing through a series of intermediaries. The centre of origin for any crop is identified based on specific criteria, including genetic diversity and distribution of the species, antiquity, diverse uses in a region, tradition, and folklore as well as archaeological findings. Considering these criteria it is postulated that Asian rice might have originated in any place from India to China. The Yangtze and Mekong River deltas were probably the primary centre of origin of Asian cultivated species *O. sativa*, whereas the river delta of Niger was the primary centre of origin of African rice *O. glaberrima*. Besides, it is widely acknowledged that the Jeypore tract of Odisha, India, serves as a secondary centre of origin for rice. *Indica* rice as indicated independently, originated in India (Gross and Zhao 2013). Later on, it spread throughout the world. 'Champa Kingdom' of Central Vietnam grew a cultivar known as 'Champa', which is supposed to have originated anywhere in the eastern part of India, Bangladesh or Myanmar (Barker 2012). Unlike the varieties that originated in the river delta, it was a short-duration variety, maturing within 100 days and avoiding the vagaries of drought/flood. Cultivar 'Champa' was widely grown from India to China and other Southeast Asian countries. The genus has evolved, diversified, and diffused into various geographic habitats in the tropics through time. The place of origin and drifting of *indica*, *japonica*, *javanica* and wild species are depicted in Fig. 1.1.

1.1.1 Rice Antiquity

The antiquity of Asian rice has been a topic of prolonged debate and discussion. In addition to the two cultigens, the genus *Oryza* has several well-organized wild species. Along with geographic distribution, these species differ in chromosome

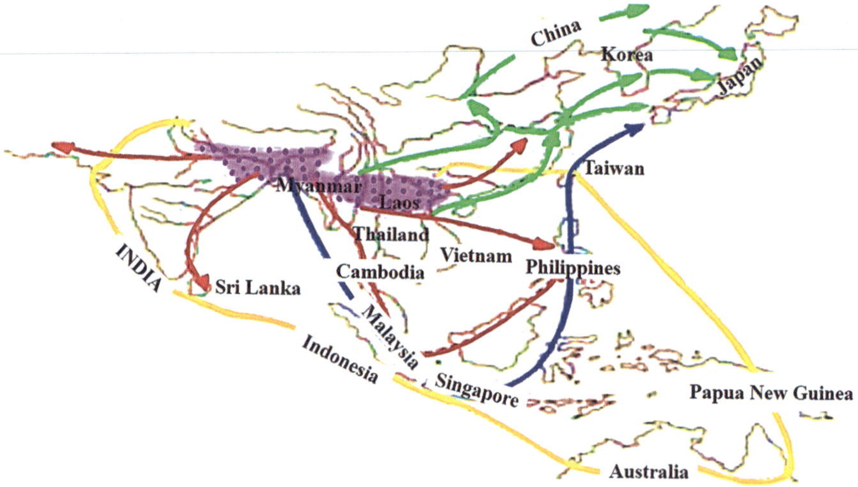

Fig. 1.1 World map of origin of rice. (Source: Jayasinghe 2014)

number and genome symbol. There are altogether 27 species of *Oryza* grown in different parts of the world. These 27 species belong to 11 types of genome among which 6 are diploid, i.e. $n = 12$ (AA, BB, CC, EE, FF, and GG), and 5 are polyploid, i.e. $n = 24$ (BBCC, CCDD, HHJJ, HHKK, and KKLL). Cultivated species, i.e. *O. sativa* and *O. glaberrima* belong to the genome AA. Two wild species, namely *Oryza australiensis* and *Oryza brachyantha* have no close relations and are placed independently out of the two sections. The 11 genome groups were labeled from AA to LL. The species of *Oryza*, genome symbols and geographical distributions are given (Table 1.1).

1.1.2 History of Rice Farming

The history of rice farming is synonymous with the evolution of human civilization. Its domestication initiated a changeover in the life of early humans. The primitive man shifted from hunting-gathering to agriculture for food. The human lifestyle changed, and colonies of settlements were established preferably on river banks for an easy source of irrigation water to tend crops. Primitive men dug up the ground, put the wild grains into the field, and learned to irrigate and raise them on the farm. This incredible discovery was made by the early nomadic humans probably around 12,000 years ago (National Geographic 2019) in the Neolithic revolution (First Agriculture revolution). It could be the rice plant that was used in the switch over to the agricultural revolution 12,000 years ago by primitive man. Archaeological pieces of evidence of plant remains dating back to 10,000 B.C. have been discovered in Thailand (Rice Almanac 2013). In India, rice domestication probably

Table 1.1 Different species of *Oryza* with their genomic composition, distribution, and habitat

Taxa	Genome group	Distribution	General habitat
Section *O. sativa*			
O. sativa L.	AA	All continents except Antarctica	All rice ecologies
O. glaberrima Steud	AA	West Africa	All rice ecologies
O. rufipogon Griff.	AA	Tropical Asia to northern Australia	Deepwater, wet ecology
O. nivara Sharma et Shastry	AA	Tropical Asia	Upland, dry ecology
O. barthii Chev. et Roehr	AA	Africa	Upland, dry ecology
O. longistaminata Chev. et Roehr	AA	Africa	Dry to deep-water ecologies
O. meridionalis Ng	AA	Northern Australia	Upland, dry ecology
O. glumaepatula Steud.	AA	South America	Rainfed lowland
Section *O. officinalis*			
O. officinalis Wall ex. Watt	CC	Tropical Asia to Papua New Guinea	Upland, dry ecology
O. minuta J.S. Presl. ex C.B. Presl.	BBCC	Papua New Guinea, Philippines	Along the stream sides; low light
O. malapuzhaensis Krishnaswamy and Chandrasekaran	BBCC	India	Seasonally dry near forest pool, low light
O. rhizomatis Vaughan	CC	Sri Lanka	Seasonally dry, open
O. eichingeri Peter	CC	Sri Lanka, central, east, and West Africa	Seasonally dry near forest pool, low light
O. punctata Kotschy ex Steud.	BB	Africa	Seasonally dry, open
O. punctata (Syn: *O. schweinfurthiana*)	BBCC	Africa	Seasonally dry near forest pool, low light
O. alta Swallen	CCDD	Central and South America	Occasionally flooded
O. latifolia Desv.	CCDD	Central and South America	Upland, dry ecology
O. grandiglumis Prod.	CCDD	South America	Occasionally flooded
Section *Australiensis*			
O. australiensis Domin	EE	Australia	Occasionally dry
Section *Brachyantha*			
O. brachyantha Chev. et Roehr	FF	Africa	Shallow pools on ironstone
Section *Ridleyanae*			
O. longiglumis Jansen	HHJJ	Indonesia, Papua New Guinea	Swampy forest floor, low light
O. ridleyi Hook	HHJJ	Papua New Guinea, Southeast Asia	Swampy forest floor, low light
Section *Meyerianae*			
O. granulata Nees et Arn. ex. Watt	GG	South and Southeast Asia	Shade, forest floor

Table 1.1 (continued)

Taxa	Genome group	Distribution	General habitat
O. meyeriana Baill	GG	South and Southeast Asia	Shade, forest floor
O. neocaledonica Morat	GG	New Caledonia	Upland, dry ecology
Section primitive *Oryza*			
O. schlechteri Pilger	HHKK	Indonesia, Papua New Guinea	Streamside, river bank
O. coarctata Roxb	KKLL	Myanmar, South Asia	Halophyte, saline ecology

matched with the development of the Indus Valley civilization in c. 3000–1500 B.C. (Possehl 2002). Discoveries of rice grains and early pottery found in the excavation site of Lahuradewa, in Uttar Pradesh located in the middle reaches of Gangetic plains, India evidences the domestication of rice dating to c.6409 B.C. (Fuller et al. 2007; Tewari et al. 2008). Similarly, archaeological evidence indicates the first domestication of rice in the middle reaches of the Yangtze and Upper Huai rivers of China dating back to the year 8000 B.C. In the Neolithic period, it was conjectured that the cultivation of early rice took place in a cleared area within a forest using a shifting cultivation method. During this cultivation, the rice crop was directly seeded into the soil without the use of standing water. A changeover in field preparations occurred later through deep ploughing, puddling the soil to mud, transplanting of young seedlings, and management of standing water. The farmers also learned to raise the crop with supplementary provisions of organic materials. From South and Southeast Asia, the crop diffused in different directions with human forces. Save Antarctica, it is now cultivated in all continents of the globe.

1.1.3 Diffusion and Diversification of Rice

As discussed, the sequence of evolution in which plant species become crops goes from food gathering to cultivation and then to domestication. Rice evolution leading to the development of two cultigens, *O. sativa,* and *O. glaberrima* reveals the wide and disjunctive distribution of *Oryza* species spanning from West Africa across tropical Asia.

The importance of this pattern of distribution is not comprehensively grasped, primarily due to the lack of clear evidence regarding the geographical origins and subsequent fragmentation of the *Oryza* genus, which remains a puzzling aspect. It is believed that the evolutionary pathways of the two widely located cultigens run parallel to each other. The cultigens underwent distinct differentiations and diversifications in various agricultural systems, driven by human populations with varying levels of cultural and technological advancement. This crop relied on human intervention for its continued existence, with grains being transported back to homes to support the food supply during the time between successive harvests and to serve as seeds for the upcoming generation. Both rice cultigens descended from a common ancestor of *Oryza* stock. Several researchers have proposed evolutionary pathways of rice cultigens from the common ancestor, i.e. *O. perennis* (Sampath 1962; Tateoka 1964; Sharma and Shastry 1965; Chang 1976). The species are widely distributed all over the tropical countries of the world. *O. perennis* Moench is not a homogeneous species composition rather is composed of highly diverse members of different propagation habits with annual, perennial, and intermediate life cycles (Oka and Morishima 1967; Agnoun et al. 2012; Barrera-Figueroa and Peña-Castro 2018). *O. perennis* Moench can be divided into four main races depending on the prevalence in the four geographical regions such as Asia, Africa, America, and Oceania. It is believed that the Asian form of *O. perennis* Moench evolved into the cultivated form *O. sativa* whereas the African form into *O. glaberrima*. In due course of time, several changes occurred in *O. sativa* species and it is now considered based on different molecular tools that *O. sativa* contains at least five genetically distinct groups such as *indica*, *aus*, temperate *japonica*, *javanica* (tropical *japonica*,), and aromatic (aromatic *japonica*) (Garris et al. 2005; Sweeney and McCouch 2007; Wang et al. 2013). Traditional thought was both *aus* and aromatic basmati-type rice belonged to *indica* group. The *aus* sub-population is close to the *indica* type whereas the rest are very close to the *japonica* type.

O. sativa's evolutionary lineage can be traced back to a wild, perennial plant similar to *Oryza rufipogon*, which thrived in aquatic ditches within undulating landmasses. It migrated to parched dryland and became annual in the form of *O. nivara*. Wild rice *O. nivara* propagated by seed, unlike propagation by stolons in the perennial *O. rufipogon*, could have been the predecessor of the annual cultigen *O. sativa* through its annual form *O. nivara* (Fig. 1.2). Genetic factors apart, the great power and efficacy of human selection could be the driving force behind the evolution of primitive agriculture. Bereft of indigenous agriculture, wild *Oryza* species do not express evolution or differentiation into cultigens. The absence of indigenous agriculture in Australia and of wetland agronomy in South America nullified the driving forces for rice evolution and the genesis of new cultigens (Chang 2003).

1.1.4 Spread of Rice

The origin of rice cultivation and the subsequent domestication process can be attributed to the introduction of puddling and transplanting techniques. The *indica* sub-species of Asian rice *O. sativa* have a great diversity in Asia. It served as the primary source of variation of other exogenetic races. The divergence is greater in Asian rice compared to its West African counterpart. According to Chang (2003), the process of differentiation of Asian rice continued for 10,000 years in China. The history of rice crops in the river valley or low-lying area is older than the dryland habitat (Huke and Hume 1990). Conversely, rice was initially cultivated in arid, upland regions of South Asia and was subsequently moved to river deltas at a later stage. Migrating populations from China and/or Vietnam brought rice to the

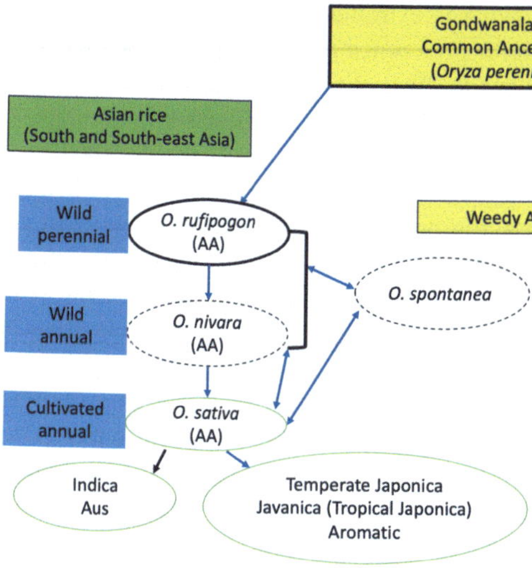

Fig. 1.2 Origin and evolution of rice *Oryza*. Wild perennial, solid black oval line; Wild annual, broken oval line; Cultivated annual, solid green oval line; ⟶, indicates direct progenies; ⟷, indicates hybridization; —┤ ├— indicates indirect progenies

Philippines for cultivation in wetland areas, while Malays later introduced it to Indonesia. In another route, the movement of rice to India might have been accomplished very early in the era of 2500 B.C. and historical linkages with India could have facilitated rice spreading to Sri Lanka and its adoption in 1000 B.C. It is possible that rice was introduced to Greece and neighbouring Mediterranean regions during the period of Alexander the Great's invasion, between 344 and 324 B.C., as his troops from India might have carried and disseminated this crop. European settlers introduced rice cultivation into several parts of the South and North Americas in pursuance of their zeal for exploration of a new land. Human forces working relentlessly in crop management and improvement have extended its cultivation to over 100 countries worldwide, encompassing a broad geographical range from approximately 43°S to 53°N latitude, spanning from the southern to the northern hemisphere.

1.2 Domestication of Rice

Plants evolve subject to natural selection. For rice, a common trend for selection following cultivation could be both natural and human forces. It could be more of the latter than the former, because of the preference of human desire in the selection of favourable characters against undesirable traits, which is very crucial for the exploitation of high yield. The power and efficacy of human selection have been the primary factors responsible for tremendous progress made in plant breeding according to the development of new variant cultivars although not new species.

Because the genetic diversity of *O. sativa* is extremely high, the cultigens might have incurred greater attention of man for preference in cultivation and subsequent adoption in domestication leading to wide diversification. In addition to human force, it could be the genetic compatibility that permitted the extraordinary capacity of rice to adapt to a large variety of contrasting habitats. It is construed that spontaneous gene mutation fuelled the beginning of rice, the Asian rice (Chang 2003). It could have provided the opportunity for individual plants growing in an environment for manifestation of their adaptive traits and enable human selection. The trend in human selection was always a preference for more grains at the cost of reducing biomass accumulation in vegetative parts of the plant. Bereft of such advantages in genetic mutation, geographical dispersal, land topology variation of habitats, and absence of enthusiasm for food production and consumption possibly could be the factors responsible for the restricted diversity of African rice. The effects of geographical dispersion and the power of human choice for the adoption of introduced rice have contributed immensely to rice diversification. It is best exemplified in the ancient rice culture of China, where, based on grain quality, rice was divided into two types: the drier cooking variety of rice is referred to as *shien* or *sen*, while the stickier type is known as *keng* (Ting 1949). In plant characteristics, the former corresponds with those plants of tropical *indica*; the latter possesses high yield potential and the advantage of the growth in cooler temperate climates corresponds to *japonica* type. Accordingly, the races were named *indica* and *japonica*, respectively. A third ecogeographic race was identified, characterized by larger plant size and larger, more robust grains, and it was subsequently named *javanica* (Morinaga and Kuriyama 1958).

Morphological and physiological attributes apart, there exists a genetic incompatibility in the hybrid offspring of these varieties. With ample support from human forces, the races dispersed into different locations and climatic conditions of Asia, showing the extraordinary capacity for domestication. Genetic data along with geographic information provides solid proof that multiple autonomous domestications of *O. sativa* (*japonica*) and *O. sativa* (*indica*) (Garris et al. 2005). From ecogeographical points, *japonica* rice cultivars are primarily known as upland rice that is prevalent in temperate East Asia to upland areas of South and Southeast Asia, while the *indica* species are primarily lowland rice (Londo et al. 2006). However, the subsequent intervention of humans and the free flow of *japonica* and *indica* subpopulations started to grow under different hydro-ecological and environmental conditions. Practically all types of sub-populations can be noticed at a specific season or hydrological conditions, e.g. deepwater rice of Bangladesh where one can notice the presence of *indica*, *japonica*, and *aus* types of rice (Travis et al. 2015). Likely, aromatic Jasmine rice is an *indica* whereas aromatic Basmati rice is too close to *japonica* than *indica* (Kovach et al. 2009). Depending on the season and hydro-ecological factors different rice sub-populations are named as same ecotypes. With time, genetic modification occurred among different sub-populations and new genes were developed so that they could adopt a specific environmental condition. Among the different sub-populations and seasonality of cultivation, greater genetic diversity happened in the *aus* sub-population and winter (boro of Bangladesh) ecotype (Ali et al. 2011). Different important genes were discovered especially from the *aus* ecotype and were introgressed in the high-yielding background, which resulted from the development of climate resilient rice, e.g. submergence tolerant gene *SUB1* from FR13A (Xu et al. 2006), efficient phosphorous uptake under P-deficient condition gene *Pstol1* from Kasalath (Gamuyao et al. 2012); drought-tolerant gene, which is associated with higher root density and greater root length from Dular (Henry et al. 2011), heat tolerant gene from N22 (Jagadish et al. 2008), high seedling vigour characteristics from Black Gora (Redoña and Mackill 1996), etc. Variations in cultural regimes and morphological and physiological attributes of epigenetic races of *O. sativa* have been shown in Table 1.2.

Table 1.2 Exogenetic races of *Oryza sativa* and their morpho-physiological characteristics

Indica	Aus ecotype	Temperate japonica	Javanica (tropical japonica)	Aromatic japonica
Tropics to sub-tropics	Tropics	Temperate to cooler zones of sub-tropics	Tropics	Tropics to sub-tropics
Tall except semi-dwarf	Tall	Short to intermediate height	Tall	Tall except semi-dwarf
Narrow to broad, light green leaves	Narrow to broad, light green leaves	Narrow, dark green leaves	Broad and stiff, light green leaves	Narrow, light green leaves
Profuse tillering	Low to medium tillering	Medium tillering	Low tillering	Medium tillering
Mostly awnless	Awnless to long-awned	Awn-less to long-awned	Awnless to long-awned	Mostly awnless
Grain short to long, slender to bold	Short, long, bold grains	Short, roundish grains	Long, broad, thick grains	Short, slender grains
Easy shattering	Low to no shattering	Low shattering	Low shattering	Easy to low-shattering
Soft plant tissue	Soft to hard plant tissue	Hard plant tissue	Hard plant tissue	Soft plant tissue
Low to strong photoperiod sensitivity	Low sensitivity to photoperiod	Low sensitivity to photoperiod	Low sensitivity to photoperiod	Low to strong photoperiod sensitivity
Amylose 23–31%	Amylose 22–27%	Amylose 0–20%	Amylose 0–25%	Amylose 18–24%
Protein 5–8%	Protein 8–12%	Protein 5–7%	Protein 8–14%	Protein 5–7%
Low to high gelatinization temperature	Low to high gelatinization temperature	Low gelatinization temperature	Low gelatinization temperature	Low gelatinization temperature

1.3 Economic Importance

Rice (*O. sativa L.*) is intimately involved economically, in foodways, cultural, and religious rituals of a large population on earth. In many countries of Asia, rice is an indispensable part of many cultural folklore. Rice is given to mankind by Lord Vishnu and Lord Indra, the God of rains taught mankind to culture its field and nurture the plants. Rice grains are used on all auspicious occasions of life and other rituals of God worship and festival celebrations for invoking the blessings of God. Decoration of paddy rice on the doorstep of households signifies the abode of the Goddess of Wealth (*Lakshmi*) in Hindu culture and traditions. According to Indian traditions, rice grains hold great spiritual and ritual significance. Grains are considered a potent symbol of good fortune, prosperity, and fertility because of their life-sustaining properties. There is no celebration of any auspicious carnival bereft of rice grains, which is considered most sacred. Rice grains are a gift of God immensely bestowed on man, and are always treated with reverence now. So pleased with his child friend Sudama's gift of a handful of roasted rice, Lord Krishna gave him the earth and heavens in exchange. Age-old social customs and celebrations of various festivals in India, especially in Odisha, have a strong linkage with the different developmental phases of the cultivated rice plant. Rice cultivation begins with the celebration of *Akhyayatrutia* during May and June when the farmers initiate the sowing of seeds on their land whereas the other festivals like *Raja Sankranti* in mid-June mark the completion of sowing the seeds in the field (Das 2012). *Garbhana Sankranti* is another festival in October, which coincides with the booting stage of rice and represents the beginning of the reproductive phase of crop growth whereas the *Karama* festival symbolizes both vegetative and reproductive success, especially against insects and pests during flowering and milking stages of the crop. Two other important festivals *Nuakhai* and *Laxmipuja* correspond to harvest of upland and lowland rice, respectively (Das 2012). *Makar Sankranti* in mid-January is also celebrated with great joy especially by the people of Odisha as *Chaitaparab* to mark the occasion for threshing and bringing the harvested seeds to a granary for storage. *Pongala*, another important Tamil and South Indian festival is celebrated during mid-January to mark the closure of the rice harvest season where freshly cooked rice dishes are offered to the Sun God (*Suryadeva*) to express gratitude for the completion of the successful harvest of the crop and to obtain blessings for prosperity and disease-free life.

Supplement to Hindu myth, Chinese myth and folklore consider rice as a gift of God, Goddess, and sacred animals rather than only God. According to their mythology, bunches of long and yellow seeds were hanging to the dog's tail and these seeds generated rice cultivation to provide food for distressed people who passed through a period of calamity caused by devastating floodwaters. Gods and Goddesses taught the ancient population how to grow it and harness grains for food. Religious uses of rice also take place in Myanmar, Indonesia, Thailand, Sri Lanka, and Malaysia since ancient times. According to legends of Indonesia, rice is intricately intertwined with the religion, tradition, and culture of many ethnic groups of the country. Rice cultivation has had a long history in the Malay Archipelago. Scholars believe that the name Java Island was derived from the Sanskrit language meaning the island of rice. In Japanese folklore, the Emperor of the country is a living legend of *Ninigo-no-Mikoto*, meaning the God of ripened rice plants. Similarly, there are many original myths and much folklore on rice in different parts of Asian countries since ancient times because of its very old history of cultivation and support provided to mankind as the primary source of food calories.

1.4 Trends in Rice Production

Rice serves as the primary food source for over 60% of the world's population. This crop offers livelihood security to the majority of rural populations in Southeast Asian countries. The total land area dedicated to rice cultivation worldwide is approximately 158 million hectares. Over 3.5 billion people around the world heavily rely on rice, which provides roughly 20% of their daily calorie intake. In 2020, rice production for India was 178 million tonnes (Fig. 1.3). Rice or paddy production in India increased from 20.58 million tonnes in 1951 to 178 million tonnes in 2020 growing at an average annual rate of 2.68% (DES-GOI 2021). India and China are the top rice-producing countries, collectively contributing 640 million tons of rice from Asia, which accounts for nearly 90% of the world's total rice production.

China has produced more than 211 million metric tons of milled rice, surpassing any other country in terms of rice production volume (Fig. 1.4). In that particular crop year, India secured the second position, producing 178 million metric tons of milled rice (Shahbandeh 2022). Rice holds the utmost significance as a food crop in the developing countries of Asia. Rice serves as the staple food for over half of the world's population now. For man, carbohydrates are the main fuel source and rice

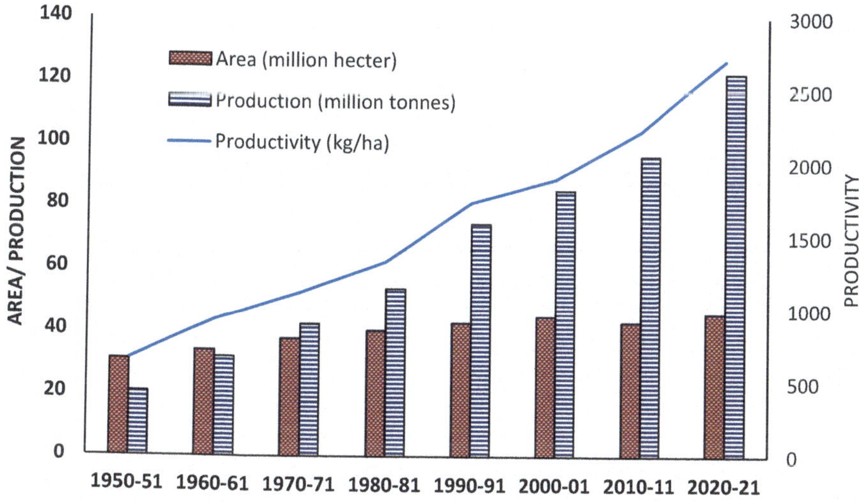

Fig. 1.3 Area, production, and productivity of rice from 1951 to 2021 in India (DES-GOI 2021)

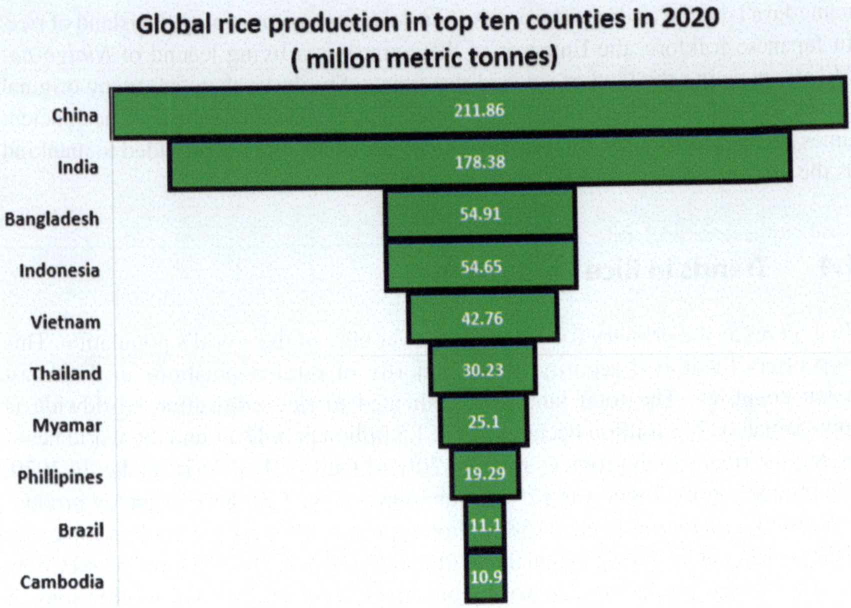

Fig. 1.4 Global rice production during 2020–21 in the top ten countries (DES-GOI 2021)

grains have them plenty because of the presence of starchy endosperm. In the world, there are three major cereal grain crops, such as rice, wheat, and maize. They directly contribute to supplying more than 50% of the total calories consumed by the entire global human population. For rice, most of the crop is grown for human consumption, whereas a small part of wheat and a major portion of maize crops are used for livestock feeding (Huke and Hume 1990). As reported by the Rice Knowledge Bank at IRRI (International Rice Research Institute), human consumption makes up 85% of the total rice production, whereas it accounts for 72% of wheat and 19% of maize. More than 3.5 billion people, nearly half of the world's population, depend on rice for daily calories (Ricepedia; https://ricepedia.org › the-global-staple-rice-consumers). In general, rice provides 20% per capita human food energy and 15% per capita protein. Asia is home to 59% of the global population, and 95% of rice production comes from this continent only. For a vast majority of poor and marginal farmers living in underdeveloped Asian nations, rice becomes the cheapest source of food energy and protein. Rice is typically grown and consumed locally on farms, with limited commercial trade occurring outside of the farm's immediate vicinity. In humid and sub-humid Asia, options for crops other than rice are remote. Water supply for cultivation is limited to the rainy season of the year in the absence of large-scale infrastructure for storage of runoff water in reservoirs and its follow-up utilization for crop irrigation in the lean season. Hence, agrarian economy as well as community food production sustained rice cultivation. Additionally, rice production has another agronomic advantage over other cereals. In comparison to wheat and maize, rice produces a higher grain yield on an average basis. Rice cultivation can provide food security for a greater number of people per hectare of land compared to many other crops. Therefore, the rapid increase in the human population often relies on intensive rice cultivation to sustain itself (Hanks 1972; Lu and Chang 1980; Chang 1987).

1.5 Demand for Rice

The world's population was projected to rise from the level of 7.21 billion in 2015 to 8.212 billion in 2030. It indicates a corresponding increase in demand for rice from 680 million tonnes to 771 tonnes in the same time interval to avoid food scarcity and malnutrition. As rice is the staple food for more than half of the world's population, they depend on rice for 20% of their daily calories. According to Ricepedia, in 2009, rice contributed to 19% of the per capita energy intake and 13% of the per capita protein intake for humans. Asia significantly dominates global rice consumption, accounting for approximately 90% of the world's total rice consumption, and this demand continues to increase in tandem with population growth in the region. Driven by both population and economic growth, rice consumption remains strong, making rice the fastest-growing food staple in several Asian and African nations and Latin America (Fig.1.4). Besides, rice production and consumption are associated with low income and poverty very often. The incidence of large-scale poverty and the rapid growth of population in South Asia therefore may increase the food need for rice beyond the projected levels. The expectations for food security in these countries can be ensured by rice only because of the friendly production climate, food values, safety, and ease of storage extending over 2–3 years. The long storage period assures food safety around the year. In contrast to wheat or maize, rice seed caryopsis, is protected by a hard covering named hull (husk), which precludes intrusion of inclement external biological and physical factors potentially causing damage to the seed. Compared to nationals of other countries of the world Asians consume more rice. Throughout Asia, rice is synonymous with food. An average Asian consumes 100 kg of rice per annum. Poor and marginal farmers in the Asian countryside derive more than 50% of their caloric intake from rice. China holds the distinction of being the world's most heavily populated nation and it consumed 149 million metric tonnes of rice in 2020–21 followed by India with a figure of 106 million metric tonnes (https://www.statista.com › Agriculture › Farming). The per capita annual consumption is 103 kg in India. On average a Chinese eats 315 g of rice per day amounting to 126 kg per annum. Conversely, in some Asian nations, per capita, annual rice consumption is very high amounting to 306 kg in Myanmar, followed by Vietnam at 285 kg, Thailand at 233 kg, Bangladesh at 229 kg, and Indonesia at 210 kg.

1.6 Parts of Rice Grain and Their Utility

The unprocessed rice harvested from the field is known as paddy rice. Paddy also means the flooded field where rice is grown. Rice seed is a fruit caryopsis. During development, the ovary wall (pericarp) joins with the seed in a grain comprising a seed coat, nucellus, endosperm, and germ. Caryopsis is a typical fruit in the Family *Poaceae*. The caryopsis is enclosed rigidly by an outer surrounding cover named a hull or husk. The dehulled caryopsis or brown rice is composed of the pericarp, seed coat, nucellus, endosperm, and germ. Intensive milling processes involve the removal of the outer layers, including the pericarp, aleurone, and germ (embryo) to refine the grain and produce white rice.

1.6.1 Hull (Husk)

The hull of indica rice is comprised of the palea, lemma, and rachilla collectively (Fig. 1.5). In the case of *japonica* rice, the hull consists not only of rudimentary glumes but also includes a portion of the pedicel. Rice hull is not edible but is used as animal feed occasionally. It is produced as a waste product during milling and represents 20% of grain weight. The constituents of the rice hull encompass cellulose, lignin, and mineral ash. The mineral ash consisting of amorphous silica accounts for 15–20% of the weight of rice hull. For various purposes, the rice hull is used as a complex renewable material resource. Glusankova et al. (2018) have suggested the technology for producing construction materials made from rice hulls as a sustainable resource involves creating cellular glass silica materials.

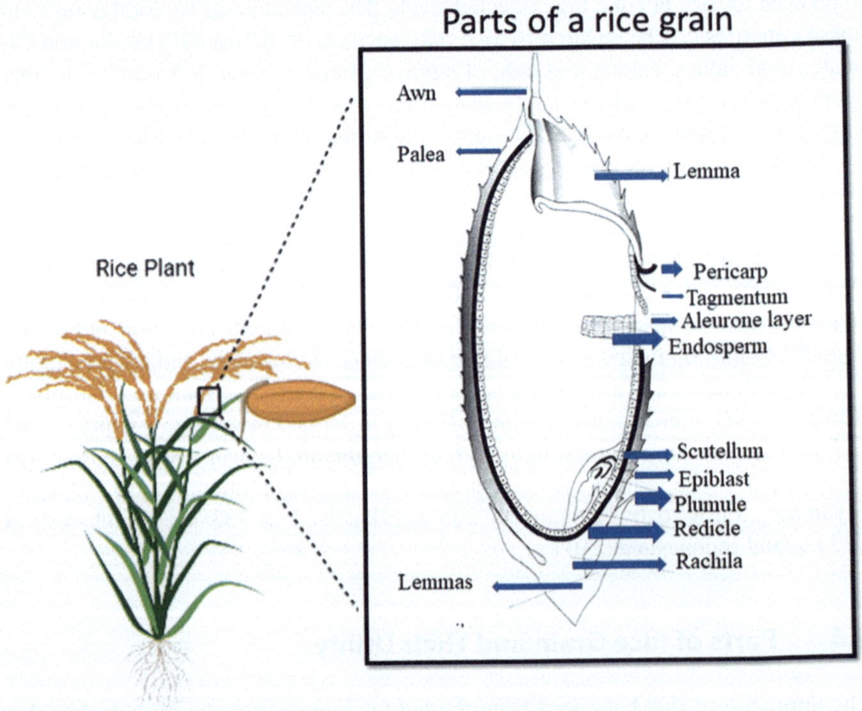

Fig. 1.5 Photographs showing different parts of the rice grain

1.7 Brown and White Rice

Brown rice is the whole grain caryopsis, which contains three integral parts; the fibrous bran, germ, and carbohydrate-rich endosperm. The germ and bran are not present in white rice because of extensive polishing. White rice is consumed by rice eaters normally, but brown rice is a better option for nutrition. The difference in food values of the two types is listed in Table 1.3 concerning McDonell (2016), Megan Ware (2019), and Kay (2020).

1. Brown rice has more fibre, antioxidants, vitamins, and minerals. The table below provides differences in the nutritional properties of brown and white rice.
2. Brown rice contains anti-nutrients: The bran and germ parts of the brown rice contain anti-nutrients like phytate or phytic acid. The chemical gives protection to the seed against external invasion by pathogens. The phytates bind to minerals like calcium, zinc, and magnesium and reduce the body's ability to digest them. The presence of arsenic in brown rice also causes problems for rice consumers.
3. The presence of magnesium and fibres helps the control of blood sugar in men. Compared to brown rice, white rice has a high glycaemic index score that promptly increases blood sugar levels increasing the risk of type 2 diabetes.
4. Brown rice has lignans that reduce the amount of fat in the blood, preclude the incidence of lower blood pressure, decrease inflammation of arteries, and help reduce risk factors for heart diseases.
5. The presence of antioxidants in brown rice gives protection against cancer, heart disease, and type 2 diabetes. It also helps preclusion of obesity.

Table 1.3 Comparison of nutritional values of brown and white rice (RDI recommended dietary intake)

Vitamins/minerals	White rice (RDI)	Brown rice (RDI)
Thiamine	1%	6%
Niacin	2%	8%
Vitamin B6	5%	7%
Manganese	24%	45%
Magnesium	3%	11%
Phosphorus	4%	8%
Iron	1%	2%
Zinc	3%	4%
Other nutrients of rice per serving cup		
Fibre/100 g cooked rice	0.4 g	1.8 g
Glycaemic index score	72	50
Fat/cup of cooked rice	0.4 g	2.0 g
Protein/cup of serving rice	4.3 g	5.5 g
Carbohydrates/cup serving rice	45 g	52 g
Food calories/cup serving of rice	205	248
Selenium/per serving cup	14 μg	11 μg
Copper/serving cup	0.1 mg	0.2 mg
Folate/serving cup	108 μg	18.2 μg

References

Agnoun Y, Biaou SHS, Sié M, Vodouhè RS, Ahanchédé A (2012) The African Rice *Oryza glaberrima* Steud: knowledge distribution and prospects. Int J Biol 4:158–180

Ali ML, McClung AM, Jia MH, Kimball JA, McCouch SR, Eizenga GC (2011) A rice diversity panel evaluated for genetic and agro-morphological diversity between sub-populations and their geographic distribution. Crop Sci 51:2021–2035

Barker R (2012) The origin and spread of early-ripening Champa Rice: its impact on song dynasty China. Rice 4:184–186

Barrera-Figueroa BE, Peña-Castro JM (2018) *Oryza granulate* Nees et Arn. Ex watt. In: Monda TK, Henry RJ (eds) The wild *Oryza* genome. Springer Nature, Cham, pp 145–150

Chang TT (1976) Manual on genetic conservation of rice germ plasm for evaluation and utilization. International Rice Research Institute, Manila

Chang TT (1987) The impact of rice on human civilization and population expansion. Interdiscip Sci Rev 12:63–69

Chang TT (2003) Origin, history, and domestication. In: Smith CW, Dilday RH (eds) Rice: origin, history, technology, and production. Wiley, Hoboken, NJ, pp 3–26

Das SR (2012) Rice in Odisha. Technical bulletin, no.16. International Rice Research Institute, Manila

DES-GOI (2021) Agricultural statistics at a glance 2020. Directorate of Economics and Statistics; Ministry of Agriculture, Government of India, New Delhi

Fuller DQ, Qin L, Harvey E (2007) Evidence for a late onset of agriculture in the lower Yangtze region and challenges for archaeobotany rice. In: Sanchez-Mazas A, Blench R, Ross MD, Peiros I, Lin M (eds) Past human migrations in East Asia, matching archaeology, linguistics, and genetics. Routledge, London, pp 40–83

Gamuyao R, Chin JH, Pariasca-Tanaka J, Pesaresi P, Catausan S, Dalid C, Slamet-Loedin I, Tecson-Mendoza EM, Wissuwa M, Heuer S (2012) The protein kinase *Pstol1* from traditional rice confers tolerance of phosphorus deficiency. Nature 488:535–539

Garris AJ, Tai TH, Coburn J, Kresovich S, McCouch S (2005) Genetic structure and diversity in *Oryza sativa* L. Genetics 169:1631–1638. https://doi.org/10.1534/genetics.104.035642

Glusankova I, Ketov A, Krasanovskikh M, Rudakova L, Vaisman I (2018) Rice hull as a renewable complex material resource. Resources 7(2):31

Gross BL, Zhao Z (2013) Archaeological and genetic insights into the origins of domesticated rice. PNAS 111:6190–6197

Hanks LM (1972) Rice and man. Aldine-Atherton, Chicago

Henry A, Gowda VRP, Torres RO, McNally KL, Serraj R (2011) Variation in root system architecture and drought response in rice (*Oryza sativa* L): phenotyping of the *OryzaSNP* panel in rainfed lowland fields. Field Crop Res 120:205–214

Huke RE, Hume EH (1990) Rice: then and now. International Rice Research Institute, Manila

Jagadish SVK, Craufurd PQ, Wheeler TR (2008) Phenotyping parents of mapping rice populations for heat tolerance during anthesis. Crop Sci 48:1140–1146

Jayasinghe R (2014) Rice origin and distribution. A Baurs and Com, Colombo

Kay S (2020) Brown rice versus white rice: which is better? https://kaynutrition.com/. Accessed 28 Jan 2020

Kovach MJ, Calingacion MN, Fitzgerald MA, McCouch SR (2009) The origin and evolution of fragrance in rice (*Oryza sativa* L.). PNAS 106:14444–14449

Londo JP, Chiang YC, Hung KH, Chiang TY, Schaal BA (2006) Phylogeography of Asian wild rice, *Oryza rufipogon*, reveals multiple independent domestications of cultivated rice, *Oryza sativa*. PNAS 103:9578–9583

Lu JJ, Chang TT (1980) Rice in its temporal and spatial perspectives. In: Luh BS (ed) Rice production and utilization. AVI, Westport, CT, pp 1–74

McDonell RDK (2016) Brown versus white rice: which is better for your health? Healthline 31:2016

Megan Ware RDN, LD (2019) Is brown rice or white rice better for health? Medical New Toda, November 5, 2019

Morinaga T, Kuriyama H (1958) Intermediate type of rice in the subcontinent of India and Java. Japan J Breed 7:253–259

National Geographic (2019) Hunter-gatherer culture. In: Subjects: anthropology, sociology, and world history. Grades, pp 6–8

Oka H-I, Morishima H (1967) Variations in breeding system of wild rice, *Oryza perennis*. Evolution 21:249–258

Possehl GL (2002) The Indus civilization: a contemporary perspective. Alta Mira Press, Walnut Creek, CA

Redoña ED, Mackill DJ (1996) Genetic variation for seedling vigour traits in rice. Crop Sci 36:285–290. https://doi.org/10.2135/cropsci1996.0011183X003600020012x

Rice Almanac (2013) IRRI Books, 4th edn. International Rice Research Institute (IRRI), Los Banos

Sampath S (1962) The genus *Oryza*: its taxonomy and species relationships. Oryza 1:1–29

Shahbandeh M (2022) Top countries based on the production of milled rice 2020/21. https://www.statista.com/statistics/255945/top-countries-of-destination-for-us-rice-exports-2011

Sharma SD, Shastry SVS (1965) Taxonomic studies in genus *Oryza* L. ?. *O. Rufipogon* Griff. Sensustricto and *O. nivara* Sharma et.Shastry nom nov. Indian J Plant Genet Breed 25:157–167

Sweeney M, McCouch S (2007) The complex history of the domestication of rice. Ann Bot 100:951–957

Tateoka T (1964) Taxonomic studies of the genus *Oryza*. In: Rice genetics and cytogenetics. IRRI–Elsevier, Amsterdam, pp 15–21

Tewari R, Srivastava RK, Saraswat KS, Singh IB, Singh KK (2008) Early farming at Lahuradewa. Pragdhara 18:347–373

Ting Y (1949) The origin of rice cultivation in China. In: Agron Bull, Sun Yatsen University, Ser III. No. 7, p 18

Travis AJ, Norton GJ, Datta S, Sarma R, Dasgupta T, Savio FL, Macaulay M, Hedley PE, McNally KL, Sumon MH, Islam MR, Price AH (2015) Assessing the genetic diversity of rice originating from Bangladesh, Assam and West Bengal. Rice 8:35

Wang M, Zhu ZF, Tan L, Liu FX, Fu YC, Sun CQ, Cai HW (2013) Complexity of *indica-japonica* varietal differentiation in Bangladesh rice landraces revealed by microsatellite markers. Breed Sci 63:227–232

Xu K, Xu X, Fukao T, Canlas P, Maghirang-Rodriguez R, Heuer S, Ismail AM, Bailey-Serres J, Ronald PC, Mackill DJ (2006) Sub1A is an ethylene-response-factor-like gene that confers submergence tolerance to rice. Nature 442:705–708

2 Ecology and Diversity of Rice Habitats

Abstract

Globally, one-fifth of the Earth's land area of more than two billion hectares has been degraded, representing more than half of all agricultural land. Unless we change how we manage soil, cent more than 90% could become degraded by 2050, the UN said. Land degradation negatively impacts one-fifth of the planet's land area and livelihoods, 3.2 billion people and 40% of the global population. It accelerates climate change and biodiversity loss and contributes to droughts, wildfires, involuntary migration, and the emergence of zoonotic infectious diseases. Rice is the premier crop that is affected by these erratic conditions more than other crops. Rice is grown in agroecosystems that are hydrologically marginal for agronomic plants. Rice fields are unique biotypes characterized by temporary wetland agroecosystems. In terms of the physical environment dictating water control, five rice ecosystems grow *indica* and *japonica* sub-species of rice in Asia. Each agroecosystem has specific production problems, as they differ considerably concerning their soil water regime, particularly the water depth of their habitat. These habitats harbour different rice cultivation systems, namely, dryland, shallow and medium rainfed, deep-water rainfed, irrigated, and very deep water. *Japonica* rice, the other ecogeographic race of *Oryza sativa,* is grown extensively in the cooler zone of the subtropics and temperate zone under a relatively stable environment. Rice plants are extremely sturdy, but their phenotype changes under drought, flooding, alkaline conditions, and other stresses.

Keywords

Compatible solute · Drought escape · Ethylene signaling · Flood-prone rice · Irrigated rice · Osmoregulators · Quiescent strategy · Rain-fed rice · ROS scavenging · Salt sensing · Stagnant flooding · Upland rice

© The Author(s), under exclusive license to Springer Nature Singapore Pte Ltd. 2025

P. K. Mohapatra et al., *Tillering Behavior of Rice Plant*,
https://doi.org/10.1007/978-981-97-5235-5_2

2.1 Diversity of Rice Habitats

Rice habitat degradation threatens the food security of rice consumers and has been a major issue for humans. According to United Nations estimates, more than two billion hectares of land are currently degraded, this includes more than half of all agricultural land. In the absence of appropriate soil conservation methods, the progression of land degradation will continue to increase to 90% degradation by the year 2050. Such devastation will accelerate climate change and biodiversity loss and thereby jeopardize the livelihood of 40% of the global population. Considering the fragility and instability of the ecological niche, the effect of degradation will be greater on rice crops. Modern farm management notwithstanding, rice consumers will be affected more than consumers in other sections of the world population using wheat, barley or maize as primary staple foods. Rice is the primary food in that part of the world; where population density is high, per capita crop area is severely constrained, and climatic conditions do not augur well for any other major food crop. It is cultivated in diverse environments, including Myanmar's Arakan Coast, where the growing season receives an average of more than 5100 mm of rainfall, and in places such as the Al Hasa Oasis in Saudi Arabia, where the annual rainfall is less than 100 mm. The temperature suitable for the cultivation of rice varies significantly. For example, in Upper Sind, Pakistan, the average temperature during the rice-growing period is approximately 33 °C, while in Otaru, Japan, the average temperature during the growing season is approximately 17 °C. Rice plants grow at different elevations, ranging from sea level on coastal plains and delta regions across Asia to elevations of up to 2600 m on the slopes of Nepal's mountains. Geographical diversity affects solar radiation levels for rice cultivation, ranging from 25% of the potential during the primary rice season in certain areas of Myanmar, Thailand, and India's Assam State to approximately 95% of the potential in southern Egypt and Sudan. Rice cultivation is widespread in Southeast Asia and is primarily sustained by seasonal rainwater in paddy fields. The weather in this region is unpredictable, leading to an alternating wet and dry seasonal cycle in crop patterns. Additionally, this region is characterized by major rivers with extensive deltas, and the flat, low-lying agricultural land along these riverside areas is inundated annually during and shortly after the rainy season.

2.2 Ecology of Rice Environments

Following domestication, the common trend in crop selection is for plants to be developed to conform further to human aspirations relative to various phenotypic attributes, such as growth duration, plant height, grain yield, photoperiod, and temperature sensitivity, flood and drought resiliency, grain morphology and quality, plant height, tillering, and resistance to insect and pest infestations prevalent in the locality. Humans can control all these attributes effectively by cleverly manipulating agronomic practices, whereas plants growing in wildernesses fend for themselves in natural selection (Chang 2003). Agronomic practices, thanks to human interventions, modify the physical, chemical, and biological conditions within the rice ecosystem, creating a favourable environment for certain soil organisms while rendering it unsustainable for others. In addition to human activities, rice habitats are influenced by climatic factors such as solar radiation, ambient temperature, wind velocity, and relative humidity. Atmospheric precipitation regulates the soil moisture regime; the incidence of heavy rainfall floods fields and has a tremendous impact on soil organisms and rice plants. As discussed, rice farming is performed in several agroecological zones worldwide, which sharply differ from one another. Farming predominantly occurs in warm/cool humid subtropical regions, warm humid tropical areas, and warm sub-humid tropical regions. Irrespective of differences in agroecological zones, rice ecosystems are composed of surface water, a ploughed soil layer, and subsoil; the soil layer has two divisions, the oxidized soil layer and the reduced soil layer.

Rice farming involves a monoculture agroecosystem in which rice plants are cultivated in a temporary aquatic environment influenced and maintained by humans. The habitat undergoes three major ecological phases sequentially in every single paddy cultivation cycle, namely, the terrestrial dry, semiaquatic, and aquatic phases (Fernando 1995). During the aquatic phase, the water depth typically ranges from 5 cm to 30 cm, with its duration varying from one site to another. This phase encompasses both the vegetative and reproductive stages of rice plant growth. The grain starts ripening in the semiaquatic phase, followed by maturity and harvest in the terrestrial phase. Thus, paddy fields are inundated with water during cultivation and left to dry after harvest in alternate cycles every year. Together with their contiguous flooding and drying features, rice habitats comprise a mosaic of rapidly changing ecotones allowing congenial life support for a rich variety of flora and fauna over several millennia (Heckman 1979). These organisms establish colonies in rice fields by rapid multiplication and increase soil fertility after death and decomposition.

2.3 Composition of Rice Soils

Because rice is a typical semiaquatic plant, the characteristics of rice soils are unique compared to those of other upland crop soils. The rice ecosystem has two different habitats: a rectangular-shaped field that harbours rice plants and surrounding bunds (levees) infested with different types of weeds and grasses. The levees retain water that is valuable for rice growth in semiaquatic environments. Ecosystems vary with land topology, where soil textures differ significantly. In general, silt clay, silt clay loam, and clay soils are most appropriate for rice farming. The riverine alluvial soil is very fertile and immensely supports rice crops. Although a neutral soil pH is preferred, crops can grow within a pH range of 5–8. The physicochemical properties of the soil in various rice environments are frequently disturbed by agricultural practices, such as irrigation, tillage, crop establishment, mineral and organic fertilization, and weeding. Additionally, natural phenomena such as low or excessive rainfall also frequently destabilize crop habitats. Depending on their respective situations, ecosystems are influenced by one or more abiotic stresses, such as salinity, drought, high or low temperatures, stagnant or flash flooding, nutrient deficiency or cyclonic winds.

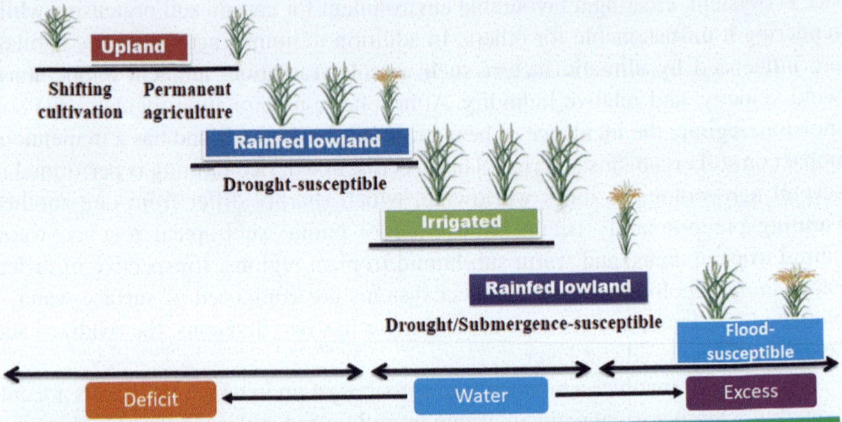

Upland	Rainfed lowland	Irrigated	Flood-susceptible
• Aerobic soil • Sloping fields • Rarely flooding • Direct seeding on plowed soil • Non-puddled soil condition	• Bunded field • Dis-continuous flooding • Submergence < 50cm • Puddle soil condition • Wet and dry condition	• Transplanted or direct seeded • Bunded fields with water control • Shallow flooding	• Medium to deep flooding • Water depth 50 to > 300cm • Transplanting or direct seeded • Aerobic or anaerobic soil

Fig. 2.1 Ecology of different rice ecosystems. (Adapted from Halwart and Gupta 2004)

Rice plants are grown in several agro-ecological zones worldwide (Fig. 2.1). Nonetheless, rice cropping is primarily practised in agro-ecological zones characterized by warm/cool humid subtropical climates, warm humid tropical climates, and warm sub-humid tropical climates. A paddy field is a successor of a lowland or temporary marshland area supplied with adequate water to maintain the growth of rice, a semiaquatic plant by nature. Irrespective of the variation in agro-ecological zones, the water depth in paddy fields is the level at which the crop phenotype is most conspicuously and prominently determined in various geographic locations. According to Halwart and Gupta (2004), the IRRI (1993) has classified rice farmland ecosystems into five types (Fig. 2.1) in terms of the physical environment, especially water control. Rice plants have undergone distinctive culture modes under these conditions, with changes occurring distinctively in terms of plant phenotype and grain yield.

2.4 Growing Habitats of Rice

2.4.1 Upland Rice

Upland rice is grown on riverbanks as well as steeply sloping fields on mountain terraces as water recedes at the end of the rainy season. The crop depends on rainfall for moisture but is rarely flooded. With no standing water, the soil is often termed aerobic soil. The seeds are directly sown on ploughed soil but not on puddling. Soil conditions limit many essential elements, such as P, Si, Fe, and Mn, and thereby adversely effects on yield. Compared to wetland situations, weed populations are more common under dryland conditions which negatively influence grain yield. Weeds compete with rice for nutrients, light, and water. The crop is often subjected to episodes of drought when the monsoon becomes erratic and fickle. The plant has a short life cycle and produces low amounts of biomass and grain yield, below 1 t/ha. The plants are short in stature and have fewer ear-bearing tillers.

2.4.2 Rainfed Lowland

Rainfed lowland rice areas exhibit diverse growing conditions influenced by hydrological factors, encompassing rainfall quantity, duration, standing water depth and extent, flooding frequency, timing of floods during the growing season, soil composition, and topography. These hydrological characteristics lead to the classification of rainfed lowlands into five main categories, as outlined by Sarkar et al. (2009). The developmental stages of the crop are adapted based on the periods of water abundance.

2.4.2.1 Shallow and Favourable Rainfed Lowlands

In these areas, the rainfall and water control are generally sufficient. Although short episodes of drought stress or mild submergence may occur occasionally, they are not significant hindrances. Supplementary irrigation might be accessible to manage such conditions.

2.4.2.2 Shallow and Drought-Prone Rainfed Lowlands

The conditions in these areas range from resembling upland areas with no standing water to lowland areas where water may collect in the field for a part of the season. The rainy period usually lasts for approximately 90 to 110 days, and water deficits can occur at any stage of crop growth.

2.4.2.3 Shallow Drought- and submergence-Prone Rainfed Lowlands

In shallow, drought-prone and submergence-prone rainfed lowlands, growing conditions are challenging. Complete submergence often occurs due to heavy rains and overflow from nearby rivers and streams. Extended periods of no rainfall can lead to water deficits during the growing season. Soils in these areas typically have a light texture and low fertility, adding to the difficulties faced by farmers.

2.4.2.4 Shallow Submergence-Prone Rainfed Lowlands

In shallow, submergence-prone rainfed lowlands, floodwater typically remains at shallow depths. However, there is a possibility of complete submergence lasting for up to 10 days or more during periods of heavy rainfall. This presents a specific challenge for rice cultivation in these areas.

2.4.2.5 Medium-Deep, Waterlogged Rainfed Lowlands

In medium-deep, waterlogged rainfed lowlands, water tends to accumulate and stagnate due to restricted drainage, sometimes persisting for a period of 1–4 months. The water depth in these areas can range from 25 to 50 cm.

2.4.3 Irrigated Rice

Irrigated rice is cultivated in fields that have been levelled and bunded, allowing for effective water control (Fig. 2.2). Irrigation becomes the main source of water in the lean season. Supplementary water is provided in the field when monsoonal rains are rare. Irrigation water comes from large and small reservoirs and pumps taping underground water tables with strenuous human interventions. The seedlings are subsequently transplanted to puddle soil. The culture type is highly productive because of the stability of the agroecosystem. Irrigated fields are the preferred choice for the successful cultivation of modern high-yielding semi-dwarf rice varieties, primarily because they help prevent drought-related challenges throughout the cultivation period. Supplementary water can be added in small quantities either during the early wet season or amid a dry period within the growing season, particularly when the crop is at risk due to irregular precipitation patterns. Rainfall is low, and cloud cover is rare in the dry season, during which no rice is produced in the absence of irrigated water. Cloudless skies and high ambient temperatures enhance evapotranspiration, imposing greater demands on water needs. Unperturbed photosynthesis under clear-sky conditions sustains biomass and yield gains better than the wet-season template. The combination of adequate fertilizer input, irrigation water, solar radiation and low incidence of pests contributes to successful yield gains. Irrigated rice fields are prominent sources of methane production. It is a potent greenhouse gas emitted from submerged rice fields by bacteria in a reducing environment in large quantities. Gas is released from flooded soil by diffusion and ebullition and through the root and stem aerenchyma of rice plants to the atmosphere (Gogoi et al. 2005) (Fig. 2.3). The soil microbes of rice fields also produce nitrous oxide.

Fig. 2.2 Irrigated high-yielding rice cultivation with drainage and irrigation facility (Source: Photo gallery of P.K. Senapati and E. Kariali)

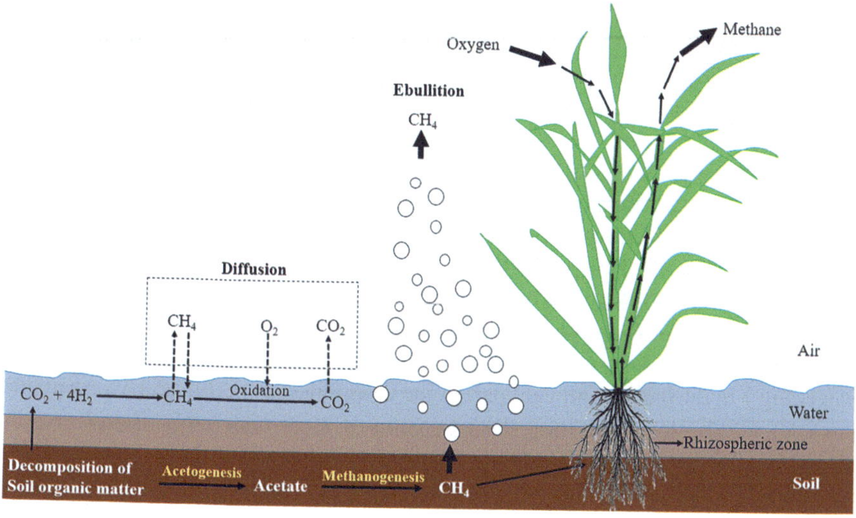

Fig. 2.3 Model showing methane emission from rice field

Due to the limitations of water, a new system of crop cultivation has been developed. The cultivation method is comparable to that used for upland rice. Here, ponding of water is avoided. Instead, surface irrigation is provided when water is needed for the rice crop. This method of rice cultivation is known as the aerobic rice cultivation system. Rice plants are grown through direct-seeded or transplanted methods depending on the soil type and availability of water. Aerobic rice saves water input and reduces the need for farming implements as well as controls greenhouse gas emissions. Aerobic rice can be cultivated in favourable upland areas characterized by flat terrain and sufficient rainfall to maintain soil moisture levels close to field capacity throughout the entire rice growing season. Aerobic rice is also suitable for irrigated lowlands where water shortages frequently occur. The adoption of an aerobic rice system can help create areas suitable for rice cultivation during the dry season, where water availability is strongly limited. Rice yield was comparable between the aerobic system and the irrigated system. However, it has been observed that rice yield decreases under such conditions if the system is followed continuously for a few seasons. To address this problem, a specialized package of practices has been proposed for aerobic rice cultivation.

2.4.4 Flood-Prone Rice

Deepwater rice cultivars are cultivated in paddy fields with water stagnation greater than 50 cm. For convenience, Huke and Huke (1990) divided this culture into two sections, i.e. deep-water (water depth 50–100 cm) rainfed rice and very deep (water depth > 100 cm) water rice. There is another category, i.e. tidal wetland rice, which is mainly confounded nearer to the sea coast and has encountered submergence even with salty water. In rainfed deep water, rice culture is practised on riverbanks or in deltas in southern and southern East Asia, where floodwater often intrudes into rice fields during different periods of crop growth. The flooding duration sometimes exceeds several weeks. As such, flood-prone ecosystems are highly diverse. The adoption of modern rice varieties is lower in flood-prone areas because high-yielding semi-dwarf rice varieties often cannot withstand prolonged water stagnation, which is common in such regions. Farmers continue to cultivate various traditional rice genotypes that exhibit tolerance to diverse flooding conditions. Although these traditional rice varieties have lower yields than modern varieties, they possess adaptive traits necessary for survival in their specific ecosystems and contribute to overall plant productivity. Traditional tall rice varieties have longer stem internodes and taller plant heights, and larger leaves are more suitable for these systems than standard semi-dwarf rice varieties.

Floating or very deep-water rice plants grow in paddy fields with water depths exceeding 1 m during a significant part of the cropping season (Fig. 2.4A). An abundance of floodwater accelerates the extension of stem internodes. Rapid growth allows the plant to maintain some of its foliage above water and preclude death by drowning. The water depth may increase as high as 5 m, which is observed in the river deltas of Vietnam and Bangladesh. This mechanism of adoption developed by rice plants in deep-water situations is called the 'escape' strategy. Often, many traditional rice plants remain submerged completely inside water without loss of life and can grow comfortably after the water level recedes; such a strategy is called the 'quiescent' mechanism of submergence tolerance. Several traditional landraces have been cultivated by the local indigenous community upstream of the famous Hirakud dam (India) reservoir for a long time (Fig. 2.4B). Rice fields are seldom bounded by dikes. As the monsoon season begins, rice plants are broadcast into unpuddled soil. As the water level rises briskly, the field is left uncared until the end of the season. The plants are mono-tillering, tall and leafy, and float on the surface of water. Being strongly photosensitive, the plant matches flowering time and harvest

Fig. 2.4 Photograph of flood-prone rice (**A**). (Source: Photo gallery of R. K. Sarkar) and Flood-prone rice cultivated in Hirakud dam (India) reservoir. (**B**) (Source: Photo gallery of E. Kariali)

towards the end of the rainy season. A long delay in reproductive growth increases vegetative biomass and leads to a decrease in the harvest index.

2.5 Biodiversity of Organisms Associated with Rice Field Ecosystems

The aquatic environment in rice fields is characterized by a soil moisture regime. Because of shallowness, the water level of the habitat is strongly influenced by weather conditions such as incident solar radiation, rainfall, wind velocity, and ambient temperature. The system generally lacks stability. It is characterized by variations in water depth and turbidity and extensive fluctuations in temperature, soil water pH, and dissolved oxygen. Under these circumstances, rice environments invite biota colonies to act as temporary guests for a period matching the duration of crop growth. As the plant ripens, the crop needs very little water, and field water is drained. The microbiota composition changes drastically in dry rice fields.

The biodiversity of rice fields is related to agricultural practices and man-made ecosystems and includes the diversity of plants and animal species. Some species, such as soil microbes, earthworms, blue-green algae, and pest-controlling animals, contribute immensely to agricultural production, whereas others, such as pests, damage the system. The aquatic phase of rice cultivation provides a habitat for a variety of animals that live on vegetation, primarily arthropod insects and spiders. Furthermore, numerous species of birds, reptiles, amphibians, pests, and mammals from the surrounding areas periodically visit rice fields in search of food. In the opinion of Heckman (1979), rice cultivation on earth over a long duration of several thousand years has enabled many of these organisms to adapt and establish themselves in the rice field ecosystem. These organisms colonize and survive in niches where they are resilient to drastic fluctuations in environmental conditions (Fernando 1996). According to Bambardeniya and Amerasinghe (2003), most of the biota of rice fields are opportunists with high resilience to react physiologically and behaviourally to widely fluctuating conditions in temporary wetland paddy fields.

2.5.1 Flora of Rice Fields

The aquatic flora in rice fields can be divided into two types, namely, microphytes and macrophytes. Microphytes include various floating algae, blue-green algae, and macrophytes, which constitute a range of weeds that confound allelopathic competition with rice plants for space and nutrition.

2.5.1.1 Macrophytes

The macrophyte flora of rice fields includes several species of grasses, sedges and broad-leaf dicot plants rooted in the soil media, as do those of rice plants. In general, weeds compete with rice for solar radiation, space, and nutrients (Fig. 2.5). Several types of weeds differ in their habitats. Weeds in rice fields come in various forms,

Fig. 2.5 Weeds associated with rice field ecosystems. (Source: Photo gallery of R. K. Sarkar)

including aquatic, wetland, and dryland weeds. The deep-water rice field remains flooded throughout the year, and the plants harbour many free-floating aquatic weeds. The aquatic environment of rice fields is a stable habitat for many algal species. Terrestrial weeds colonize and become conspicuous when rice fields are drained for the final harvest. Weed communities colonizing bunds are perfectly terrestrial. In rice fields of Asia, grass species (*Poaceae*) are the most common weed species, followed by *Cyperaceae* (sedges). Barnyard grass (*Echinochloacrus grandis*) is a severe problem for rice cultivation. Other grasses, such as *Eleusine indica* and *Ischaemum rugosum*, *Echinochloa crus-galli*, *Echinochloa glabrescens*, *Leptochloa chinensis*, *Echinochloa colona*, *Pisticum distichum*, *Digitaria setigera;* sedges *such as Cyperus difformis*, *Cyperus rotundus*, *Cyperus iria*, and *Fimbristylis miliacea;* and the broad leaf plant *Monochoria vaginalis* and *Sphenoclea zeylanica*, compete with rice for growth in paddy habitats (Smith 1983, IRRI Knowledge Bank; Ashraf et al. 2018). In addition, some other plants belonging to the macrophyte group are beneficial for rice crop production.

1. The water fern *Azolla* floats on rice paddies in Asia. In association with the blue-green algae *Anabaena, Azolla* fixes nitrogen in rice fields and improves productivity. *Azolla* is used as a green manure for rice crops to increase rice productivity two times. Under an ideal environment, *Azolla* and *Anabaena* symbiosis enables vigorous growth. The fern can double its biomass within 2–5 days. The popula-

Fig. 2.6 *Azolla* with rice field ecosystems. (Source: Photo gallery of D. Panda)

tion proliferates into a mat of floating plants to increase the fertility of the rice field (Fig. 2.6).

2. *Duckweeds:* Duckweeds are small, fragile, and free-floating aquatic flowering plants. When water, temperature, light, pH, and nutrients are ideal, plants photosynthesize vigorously, and the biomass increases two times within 2 days. Duckweed plants are used to feed ducks and fish but are also utilized for enriching paddy field fertility. The free-floating aquatic duckweed fern *Lemna minor* L. reduces ammonia losses from urea-N applications and improves N-use efficiency in rice fields (Li et al. 2009).

2.5.1.2 Microphytes

Therophyte flora includes a variety of green algae, diatoms, desmids and blue-green algae. Among them, the contribution of blue-green algae has been most pertinent to rice cultivation for economic consideration. They are capable of fixing atmospheric nitrogen and enriching soil fertility. Cyanobacteria (blue-green algae) are an important component of the nitrogen-fixing biomass of paddy fields. They fix nitrogen and release it to the field to enhance rice productivity. The soil conditions in rice fields are conducive to the growth of cyanobacteria. These microbes are by far the largest group of photosynthetic prokaryotes on earth adjudged for their widespread occurrence, stress resistance, morphological diversity, frequency, and abundance. They can inhabit almost any form of the environment without sufficient energy input, as needed for photosynthetic microbes. (Whitton 2007). They work as biofertilizers, and farmers often apply them as inoculants to increase soil fertility. The majority of nitrogen-fixing blue-green algae (cyanobacteria) are indeed heterocystous, meaning that they possess specialized cells called heterocysts that are responsible for nitrogen fixation. The dominant blue-green algae of rice fields are *Merismopedia*, *Anabaena*, *Gloeocapsa*, and *Nostoc,* although several other cyanobacteria, such as *Calothrix*, *Aulosira*, *Cylindrospermum*, *Fischerella*, *Lyngbya*, *Nostoc*, *Plectonema*, and *Stigonema,* also fix nitrogen (Whitton 2007). In rice, nitrogen enrichment is successful when local species are used as inoculums. According to Paudel et al. (2012), mixed inocula of cyanobacteria such as *Nostoc, Anabaena, Westiellopsis, Aulosira*, and *Scytonema* can serve as dependable and sustainable sources of nitrogen to increase rice productivity. These cyanobacteria are known for their ability to fix atmospheric nitrogen and contribute to soil fertility, which can ultimately benefit rice plants.

2.5.1.3 Soil Fauna of Rice Fields

The biodiversity associated with the rice agroecosystem is very high in any tropical rainfed system. The paddy rice ecosystem contains a variety of wild species, including fish, molluscs, crabs, prawns, snakes, insects, and water birds. These organisms play significant roles in organic matter decomposition and soil mineralization. The fauna is classified into four major categories.

2.5.1.4 Aquatic Invertebrate Fauna

In rice farming, the agroecosystem undergoes a transformation from an open littoral environment to a vegetative littoral system as the rice plants grow and interact with associated aquatic organisms (Heckman 1979), and the biota colonizes the rice field successively with the change in habitat from the dry phase to the aquatic phase (Fernando et al. 1979). The rice ecosystem hosts a wide array of organisms that have adapted to dynamic and managed environments and are periodically disrupted by various agricultural practices and interventions (Fernando 1995). Bambardeniya and Amerasinghe (2003) classified aquatic invertebrates into four groups: (a) neuston-surface dwelling insects; (b) zooplankton-protozoans, rotifers and microcrustaceans; (c) nektons-aquatic insects and their larvae; and (d) benthos-bottom dwelling annelid worms, molluscs, and nematodes. Among the invertebrates, insects are the dominant species (Heckman 1979). In another study, Bambardeniya (2000) identified 179 species of aquatic invertebrates in a typical irrigated rice field in Sri Lanka; these species span 96 families and 10 phyla and include 96 arthropods and 65 insect species. The Diptera order of insects, specifically the *Culicidae* family, was the predominant group in terms of insects, while the majority of the arthropods were Oligochaetes. The residual invertebrates consisted of Rotifera, protozoa, nematodes, molluscs, platyhelminths, etc. The zooplankton of the rice fields consist largely of rotifers and microcrustaceans. The population varies with temporal and field site changes (Lim et al. 1984; Simpson et al. 1994; Taniguchi et al. 1997). The species compositions of rotifers and microcrustaceans differ significantly between the rice environments of Asian countries, such as Japan, Sri Lanka, Thailand, Indonesia, and Malaysia (Bambardeniya and Amerasinghe 2003).

The aquatic environment of wetland rice fields constitutes a favourable habitat for mosquito breeding. Lacey and Lacey (1990) listed 137 mosquito species that breed in rice fields around the world. Other workers reported a high density of mosquito species in rice fields in Japan (Takagi et al. 1995, 1996) and Sri Lanka (Amersinghe 1993). There are also reports on aquatic insects living in rice fields other than mosquitoes. The dominant insect species belong to the families *Chironomidae, Dytiscidae, Corixidae,* and *Belostomatidae*; most of the insects are predatory (Rozilah and Ali 1998). Alongside these insects, oligochaetes, chironomid dipteran larvae, and freshwater crabs are commonly found as components of the benthos in rice fields. Prot and Rahman (1994) also documented parasitic nematodes associated with the rice ecosystem in Southeast Asia. Molluscs are very commonly observed in the freshwater habitats of Asian rice fields. Golden apple snails are reported to invade and grow profusely in rice fields (Naylor 1996).

2.5.2 Terrestrial Invertebrates in Rice Fields

In addition to aquatic invertebrates, terrestrial invertebrates, mainly arthropods, grow and multiply near rice fields. The arthropod community is composed of insects and spiders. They live on the soil surface and survive on plants and weeds. This group of organisms includes rice pests and their enemies, both predators and parasitoids. A diverse array of predators and parasitoids serve as natural biological control agents. The predator group consists of a variety of spiders and insects, such as carabid beetles, dragonflies, and aquatic and terrestrial predatory bugs. However, the parasitoid group included numerous species of hymenopteran wasps and dipteran flies. The dynamics of arthropod communities change with the growth of rice plants. Abdullah et al. (1998) detected 36 families of insects and arachnid arthropods in the Muda irrigation scheme in Malaysia. In Sri Lankan rice fields, 280 species of insects and 60 species of arachnids were identified by Bambardeniya (2000). The majority of the insects were classified under the order Hymenoptera, with the order Lepidoptera following in terms of abundance.

2.5.3 Vertebrate Fauna of Rice Fields

Several vertebrates visit rice fields for food and sustenance. Rice plants can be emptied from field sites for feeding purposes. These include rats, mice, bandicoots living near the fields and several types of migratory birds occasionally coming to the field in search of food. Aquatic vertebrates such as fishes and amphibians inhabit rice fields and colonize. In tropical rice paddies, fish play a crucial role in aquatic ecosystems (Fernando 1996). The population ecology of fish species inhabiting rice fields in South Asia and their contributions to agroecology have been studied by many researchers (Bambardeniya and Amerasinghe 2003). These products are economically important for the livelihood of farmers in traditional rice culture. With significant human intervention and the adoption of modern farming techniques in recent years, the effect of these practices on the fish population has severely increased the survival of the hardest species. In addition to fishes, amphibians also inhabit and colonize rice fields. The amphibians mostly comprising frogs are insectivorous, feeding on pest insects and ameliorating the problems of farmers. Species

Fig. 2.7 Rice fields fauna. (Source: PhD thesis of Dr. Antaryami Naik 2022, Sambalpur University, India)

living close to one another are natural enemies of amphibians. Several species of granivorous aquatic birds visit rice fields to feed on the crop as pests. Some species of mammals visit rice fields for food. These are small rodents, such as rats and mice that consume both rice crops and insect pests. Some common rice field fauna available in Odisha, India are presented in Fig. 2.7 (Naik 2022).

2.6 Environmental Stresses and Rice Plant Responses

The growth duration of the rice plant varies from 3 to 6 months. The duration of stress affliction depends on the culture type and environmental conditions. The plant completes its life cycle during this period, with the vegetative phase first followed by the reproductive period ending in maturity. Rice cultivation occurs across a diverse range of environmental conditions, marked by fluctuations in temperature, climate, and soil moisture levels. It is opined that plants such as rice are sturdy and have developed strategies for adapting to various environments throughout evolution (Ashikari and Ma 2015). Nonetheless, at various stages of the life cycle, rice plant survival is threatened by multiple environmental challenges; the effects become highly damaging when they coincide with critical stages of development. Indeed, adverse environmental conditions pose a significant threat to rice production, resulting in substantial yield losses across extensive regions worldwide, including even highly productive irrigated systems. Under environmental stress conditions, high crop yield is a valuable trait that is targeted for improvement via classical plant breeding and genetic engineering. The physicochemical stresses that pose a significant threat to sustainable rice production include drought, flooding, extreme temperatures (both high and low), alkalinity, limited sunlight, and mineral deficiencies, as well as common pollutants found in the air, water, and soil, among others.

Recently, researchers have investigated the mechanisms underlying how rice plants respond to various stresses at different levels, including those related to morphology, physiology, biochemistry, genetics, and molecular biology. In the realm of scientific exploration, these studies have remained dynamic, fruitful, and exceptionally inventive, yielding significant outcomes with implications for both the environment and society. This is especially crucial given the ongoing, irreversible trends of global climate change and relentless population growth (Breviario and Genga 2013). Global climate change exacerbates the incidence and frequency of stress, while an increase in population growth exacerbates the demand for food under suboptimal food production conditions.

Rice plants are regarded as an exemplary model system for studying monocotyledonous crop species. The uniqueness of rice among cereal crops lies in its parallel cultivation in two diverse and contrasting ecosystems, upland areas with minimum soil-saturated conditions and lowland areas with an abundance of water. Upland-adapted rice plants have evolved drought resistance attributes, such as early flowering and low shoot dry mass, while lowland drought-susceptible ecotypes are characterized by long flowering times, high input responsiveness, and conditions conducive to anaerobic growth. This parallel evolution has given rise to a rich diversity of cultivars. Plant breeders and molecular biologists have capitalized on the abundance of genome and genomic data accessible for a wide range of genotypes to investigate the mechanisms underlying stress tolerance. A plethora of rice lines exhibiting tolerance to environmental stresses have been generated and introduced successfully. Rice plant responses to some of the major and recurrent stresses have been discussed. Compared with those of any other crop species, the rainbow rice germplasm has shown adaptability to a wider spectrum of hydrological conditions. In this section, the effects of drought, submergence and salinity stress on rice plants, and tolerance mechanisms are discussed.

2.6.1 Drought

2.6.1.1 Occurrence of Drought

Rice habitats are hydrologically marginal for semiaquatic plant species. Rice cultivation occurs in an exceptionally diverse array of ecological niches across various regions of the world, encompassing a wide spectrum of temperatures, photoperiods, and solar radiation levels (Yoshida 1977); however, the soil moisture regime is the most crucial factor that dictates yield and growth. Compared with any other crop species, rainfed rice is known to have adapted to the most extensive variety of hydrological conditions (O'Toole 1982). Plant growth is impacted by both deficiency and excess soil water, the former being more prevalent in rice agro-systems than the latter. Selection pressure for the adaptation of plants to semiaquatic conditions is responsible for such vulnerability. An increase in the frequency of dry periods leads to the occurrence of drought and salinity in arable land systems worldwide. In addition to insufficient rainfall, drought can also be triggered by factors such as salinity, extreme temperatures (both high and low), and intense sunlight. In rice ecosystems, drought is the most damaging and widespread of all physicochemical stresses. In South and Southeast Asia, drought affects as much as 23 million hectares of rain-fed rice, posing serious concerns for the food safety of poor and marginal farmers. In these rainy, sustained habitats rice is commonly grown during the monsoon period of the year, during which the crop growth period matches the availability of seasonal rainwater. However, monsoonal rains are erratic and unpredictable. Fickle weather conditions with marked variations in the duration and intensity of rainfall impose episodes of either dry spells or heavy downpours. In the absence of rainfall for approximately 2 weeks in the rainfed system, the farmer fails to accumulate enough water in the field and manage crop growth to harness its full productive potential. Land and water temperatures, air circulation above the canopy and soil moisture retention capacity are the other causes of drought; they are influenced by both natural and man-made factors.

In general, rice plants are highly sensitive to water deficiency. The grain yield is seriously impacted. When exposed to water regimes below saturation levels, all morphological parameters, including plant dry weight, plant height, tiller number, effective panicle number, grain number, grain weight, and harvest index, decrease (Hossain et al. 2020). In various rainfed rice farming systems, seasonal drought represents a significant threat to production. The extent of grain yield loss due to drought is contingent on factors such as the growth stage of the crop, the severity of the drought stress, and the duration of the stress. Nevertheless, the sensitivity of rice plants to drought varies at different stages of growth and is not uniform throughout development. In comparison to those in the vegetative stage, rice plants in the reproductive stage are considerably more vulnerable to water stress, which is evident in the extent of yield loss. During plant development, sensitivity to water stress peaks at the reproductive stage (O'Toole 1982; Venuprasad et al. 2007) as floral fertility subsists with cell turgidity. Therefore, improving resilience to water deficit stress at the reproductive stage and during anthesis has been an agronomic target (Richards et al. 2010), and the magnitude of yield loss has been used as a selection criterion for various purposes (Kumar et al. 2014). In addition to the reproductive stage, which includes the period from panicle initiation to flowering, the mid-tillering stage is also considered sensitive to water deficit (Wepereis et al. 1996). Yang et al. (2019) reported that the imposition of water stress on rice cultivars at the flowering stage increased water use efficiency, while its impact on grain yield loss was enormous.

2.6.1.2 Symptoms of Drought on Morphological Features of Rice Plants

Rice plants are considered drought-susceptible. The factors that contribute to susceptibility are discernible in the morphology. When exposed to stress, plants close their stomata swiftly, the root system is poor at absorbing and storing water, and thin cuticular wax on the leaf surface does not prevent moisture loss. The symptoms of drought are most prominently expressed on the leaves because of the quick loss of turgor. The symptoms include rolling, drooping, yellowing, etiolation, and premature downfall of leaves. Plants not being able to access water downregulate primary production, resulting in stunted growth and delayed flowering and reproductive growth. The loss of turgor pressure in individual cells leads to a reduction in leaf size, limits stem elongation, and hampers root growth in the soil. This destabilizes water relations within the plant and diminishes water use efficiency. As the water content within plant cells decreases, the cells undergo shrinkage, and their cell walls can collapse. A decrease in cell size results in reduced cell turgor and an increase in the concentration of solutes in the protoplasm. This shrinkage of the protoplasm causes the cell membrane to become thicker. The initial biophysical impact of water stress is the loss of turgor pressure, and activities that depend on turgor, such as leaf expansion and cell elongation, are particularly sensitive to water deficit. Often, it is conceptualized that exposition to drought stress could modify the roots to grow thicker and longer for deeper penetration into the soil layer and access to water. However, such adaptations in rice are genotype-specific; many instances of root elongation occur in rice cultivars, which become susceptible to acute deficits in soil water (Kim et al. 2020).

The mathematical equation below illustrates how cell expansion is contingent on cell turgor, with a decrease in cell turgor resulting in a reduced growth rate.

$$\mathrm{GR} = m\left(\psi_{\mathrm{P}} - Y\right)$$

In the equation, GR represents the growth rate, Ψp denotes cell turgor, Y represents the yield threshold (the pressure below which the cell wall resists plastic and irreversible deformation) and m represents wall extensibility, which reflects the wall's responsiveness to pressure.

2.6.1.3 Drought Tolerance in Rice

Drought tolerance is divided into two major types according to the plant kingdom, and rice is no exception. The first type is drought tolerance, during which the plant reaches equilibrium with the suboptimal water level of the growing medium and retains the ability to perform metabolic functions while also dehydrating and performing essential metabolic functions. The second type is drought avoidance. Here, the plant does not reach equilibrium with the water status of the environment; the resistance is measured by the ability to maintain tissue hydration. This statement is synonymous with that of Kumar et al. (2016), where avoidance in the context of plants has been defined as the capacity to sustain relatively high tissue water potential even when there is a deficiency of moisture in the soil. However, the term 'avoidance' can be misleading because drought is a meteorological condition endured by all surviving plants and evaded by none (Taiz and Zeiger 2002). Plants that complete their life cycle early in the wet season before drought commences in tropical rice ecosystems can be considered true drought avoiders. Often, they are categorized as drought escape; early-maturing rice cultivars could be included in this class. Kariali et al. (2008) studied drought escape mechanisms extensively in the wild rice variety *Oryza nivara* where plants fold up their life cycle quite early before the onset of drought. An extensive account of rice plant adaptation in terms of morpho-physiological features under non-growth stage-specific and growth stage-specific conditions was described by O'Toole (1982) (Fig. 2.8).

In regions where rainfall is insufficient, the capacity of crops to endure drought is crucial for tropical rice cultivation. Drought conditions are recognized by plants when water loss surpasses the ability of roots to take up water, leading to a reduction in tissue water content that disrupts normal plant functions. Under such conditions, the shortage of plant-available water prevents the physiological need for evapotranspiration during the growing season from increasing grain yield.

The drought tolerance can be divided into two response types according to the duration of exposure. In a suboptimal environment, rice plants can experience gradual development or long-term water shortages that span days or weeks, or they may encounter short-term water deficits that fluctuate from hours to days. In both situations, plants tend to adapt appropriately by maximizing water uptake and minimizing water loss through evapotranspiration. The plants are not identically water-susceptible at different growth stages. Compared to plants in the vegetative stage, plants in the reproductive growth phase are more vulnerable to drought; in particular, events such as anthesis, fertilization, and early post-anthesis conditions are highly critical and susceptible.

Examples of short- and long-term responses are given below.

Short-term responses:
1. In the leaf: Stomatal closure, low carbon assimilation, and root signal perception.
2. In the stem: Hydraulic changes lead to a decrease in stem diameter and infringement of signal and assimilate transport.
3. In roots: Perception of drought signalling, osmotic adjustment.

Long-term responses:
1. The above-ground parts of the plant exhibited the following characteristics: stunted shoot growth, folding of the transpiration surface, spikelet degeneration, promotion of senescence, osmoregulation, photosynthetic pigment degradation, intervention with osmoprotectants, and intervention with ROS-scavenging enzymes.
2. In the below-ground parts of the plant, the following steps occurred: shrinkage of root growth, maintenance of cell turgor, and increase in water uptake capacity.

2.6.1.4 Mechanisms of Drought Tolerance

The mechanisms that contribute to drought tolerance are complex and involve multiple metabolic pathways that enable plants to detect and adapt to specific sets of conditions as needed. Drought tolerance primarily arises through processes such as osmotic adjustment, antioxidant capacity, and desiccation tolerance (Yue et al. 2006). Some of the interactions involved in this process include stomatal conductance; the denaturation of carotenoids; the accumulation of anthocyanin pigments; the intervention of osmoregulators (glycine-betaine, proline, pinitol, sucrose, etc.); the use of ROS-scavenging enzymes such as glutathione peroxidase; and the

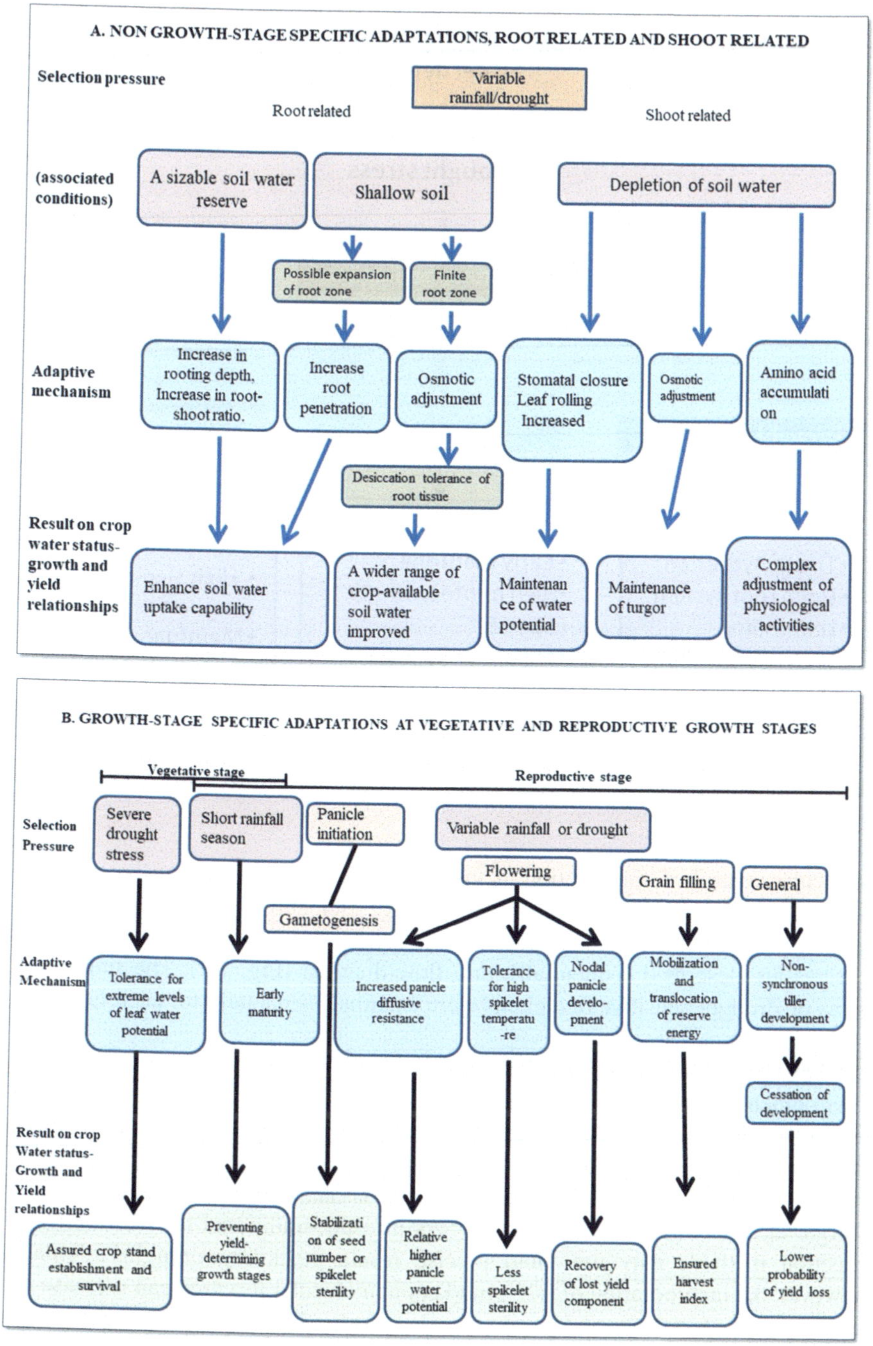

Fig. 2.8 Apparent selection pressure, adaptive mechanisms and resultant effects on crop water relation-growth and yield relationships in rainfed rice. (Adapted from O'Toole 1982)

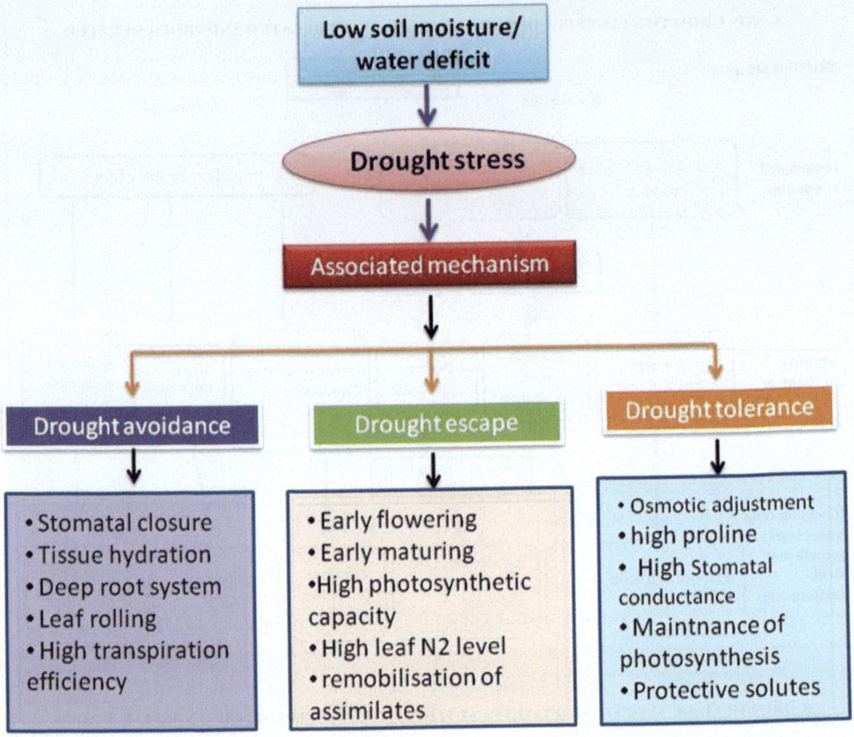

Fig. 2.9 Adaptive features of rice plants in drought-prone environments. (Adapted from Panda et al. 2021)

molecular control of transcription factors such as dehydration-responsive element-binding proteins (DREB) and ABA-responsive element-binding factor (AREB), which play pivotal roles in the plant's response to drought stress. These adaptive features are described sequentially in a flow diagram (Fig. 2.9). The impacts of stress on drought-sensitive rice plants were examined separately for each factor.

2.6.1.5 Drought Perception Signals

Environmental drought cues are detected by sensors on membranes that have not been fully elucidated. These signals are subsequently transmitted through various signal transduction pathways, ultimately leading to the expression of drought-responsive traits with corresponding gene functions and enhanced drought tolerance (Zargar et al. 2011; Panda et al. 2021). As a result, traditional hybridization and selection methods may not yield precise results regarding drought tolerance. However, the incorporation of DNA markers in molecular research can increase the accuracy of the technique and provide precise results. These molecular markers can also facilitate the screening of large quantities of drought-tolerant germplasms, which can then be utilized for crop development purposes (Panda et al. 2021). Many studies have reported quantitative trait loci (QTLs) linked to various physiological and biochemical traits (Upadhyaya and Panda 2019; Panda et al. 2021). Given the paramount significance of yield in rice cultivation, research institutions worldwide are predominantly dedicated to mapping quantitative trait loci (QTLs) associated with grain production in rice under drought-stress conditions (Panda et al. 2021). The different QTLs involved in rice drought tolerance *are qDTY1.1*, *qDTY2.1*, *qDTY2.2*, *qDTY3.1*, *qDTY6.1*, *qDLR8.1* and *qDTY12.1* (Panda et al. 2021). Many genes are differentially expressed, 5000 and 6000 of which are upregulated and downregulated, respectively (Joshi et al. 2016). Panda et al. (2021) described three categories of genes related to (a) membrane transport, (b) signalling, and (c) transcriptional control, and the authors revealed that the expression of these genes was associated with physiological, biochemical, and molecular mechanisms under drought stress in rice. Many of the genes regulated by drought exhibit dual control mechanisms involving both ABA-independent and ABA-dependent pathways. For example, the *OsJAZ1* gene negatively impacts drought tolerance in rice by influencing ABA signalling, which plays a crucial role in how plants respond to drought stress (Gupta et al. 2020; Panda et al. 2021). DNA studies employing marker-based phenotyping represent a pivotal approach for identifying genes associated with drought tolerance in rice. Despite significant advancements, only a limited number of traits have been effectively verified for their potential to confer drought resistance (Zargar et al. 2011). Hence, gaining comprehensive insight into both the structural and functional aspects of each trait related to drought tolerance in rice is crucial for the development of rice genotypes that can consistently deliver long-term benefits in drought-prone environments.

2.6.1.6 Effects of Drought on Physiological Processes

Compatible Solutes

Rice plants are considered drought-susceptible due to their characteristics, such as a limited root system, thin leaf cuticle, and rapid closure of stomata. These traits increase plant vulnerability to water scarcity and drought conditions. The physiological changes accompanying morphological aberrations include a reduction in the transpiration rate, chlorophyll content, stomatal conductance, photosynthetic rate, plant water use efficiency and leaf water potential. The early-stage closure of stomata in response to water stress indeed enhances water use efficiency. However, during severe water stress, reactive oxygen species are generated impacting both photosynthesis and respiration. These reactive molecules can be harmful because they interact with lipids, proteins, and nucleic acids, leading to structural damage and destabilization of these essential components within the plant. At the biochemical level, plants adjust their cellular metabolism in various ways to adapt to the impact of stress. This adaptation involves the synthesis of osmoregulatory substances such as proline, glycine-betaine, and a cyclic sugar known as pinitol. To achieve this goal, it is essential to increase the expression of genes that control the production of specific enzymes: Δ′1-pyrroline-5-carboxylate synthase for proline, betaine aldehyde dehydrogenase for glycine-betaine, and myo-inositol 6-O-methyltransferase for pinitol. These compounds and enzymes are critical components of a plant's strategy to respond to environmental stress and maintain the stability of its cells.

The accumulation of proline in leaves is considered to be an important adaptive feature of rice plants in drought-stress-prone environments (Xiong et al. 2012). Proline accumulates in leaves, while drought stress inhibits the growth of above- and underground plant parts of rice plants (Saha et al. 2019). Proline is an important variable amino acid that not only functions as an osmoprotectant but also functions as a nutritional source when cellular functions are destabilized. Furthermore, these mechanisms are crucial for preserving the integrity of proteins and membranes and for scavenging reactive oxygen species generated in response to stress. Like proline, glycine-betaine, which is a quaternary amine compound, serves as a highly effective compatible solute that enhances drought stress tolerance in rice. Glycine-betaine is formed from choline through a two-step process involving the enzymes betaine aldehyde and choline monooxygenase, with betaine aldehyde dehydrogenase controlling these reactions (Shirasawa et al. 2006). Reports indicate that osmolytes such as proline and glycine betaine stimulate stress tolerance mechanisms in rice by activating resistance genes, elevating the levels of both enzymatic and non-enzymatic antioxidants, preserving cell membrane integrity, maintaining the photosystem II complex, and safeguarding osmotic pressure within the cell.

In addition to the enzymes that regulate proline and glycine-betaine synthesis, the presence of myo-inositol oxygenase, a unique mono-oxygenase enzyme that maintains the myo-inositol content and is essential for the production of certain low-molecular-weight compounds, is another noteworthy aspect of drought tolerance in rice plants (Shi et al. 2017). The myo-inositol oxygenase enzyme facilitates the oxidation of myo-inositol to D-glucuronic acid. D-glucuronic acid serves as a vital sugar precursor for the formation of cell walls. D-glucuronic acid also becomes a precursor for ascorbic acid. There is evidence that overexpression of the *MIOX* gene leads to a two to threefold increase in the ascorbic acid content in *Arabidopsis* (Radzio et al. 2003), which is an antioxidant. Ascorbic acid produced by this pathway can directly reduce superoxide generated in stress-prone environments. In regions prone to drought, the roots of rice plants experience stress, and the expression of the *OsMIOX* (*myoinositol oxygenase*) gene enhances drought tolerance by actively scavenging reactive oxygen species (Duan et al. 2012) and regulating the expression of plant hormones and sugar-related genes (Shi et al. 2020).

Leaf Photosynthesis

Drought tolerance is often calibrated in terms of plant response to the leaf photosynthetic rate. In simpler terms, tolerance refers to the plant's capacity to sustain photosynthesis and produce biomass even in water-stressed soil environments. The standard pace of photosynthesis decreases in sensitive rice plants primarily because of the closure of stomata, which limits the exchange of gases (Zhu et al. 2020) and adversely affects the transpiration rate. Stomatal closure offers a good line of defence for plants when leaf expansion is complete before stress initiation. Guard cells, which are situated on the leaf epidermis, lose turgor pressure due to the direct loss of water through evapotranspiration into the atmosphere. Consequently, they shrink in size and close the opening between them. At the early stage of water stress, stomatal closure helps conserve water and thereby increases water use efficiency. When the plant's stomata close, there is a reduction in transpiration that outweighs the decrease in the intercellular CO_2 concentration. As a result, the plant can take up more CO_2 for each unit of water transpired. In contrast, water use efficiency declines under severe stress. When exposed to severe stress, mesophyll dehydration inhibits photosynthesis, impairs mesophyll metabolism, and ultimately reduces water use efficiency in susceptible rice genotypes. Loss of turgor interferes with carbon assimilation because of aberrations in the thylakoid reactions of the PSI and PSII reaction centres, leading to the inhibition of electron transport and consequential decrease in the production of high-energy compounds such as $NADPH+H^+$ and ATP; additionally, energy-rich compounds are utilized in carbon fixation reactions to facilitate the synthesis of sugars. Exposing water-stressed leaves to an enriched CO_2 supply has the potential to enhance the response of stomatal conductance and photosynthesis to water stress. The incidence of drought causing severe destabilization of the PSI and PSII systems and the resultant inhibitory effect on the transfer of absorbed light energy to drive the movement of electrons in the membrane-bound transfer chain for the generation of reducing power have been observed in several susceptible rice genotypes (Mishra and Panda 2017; Mishra et al. 2018, 2019).

Incident light overuse destabilizes the photosynthetic apparatus and degrades the pigment system of rice leaves, promoting the loss of chlorophylls. Conversely, high chlorophyll and carotene retention and high photosynthetic performance contribute significantly to the tolerance of rice to water-deficient environments, and pigment loss becomes a negative consequence of environmental stress (Syamsia et al. 2018; Nounjan et al. 2020). In sharp contrast to the drought-susceptible rice cultivar IR64, some of the drought-tolerant landraces of Koraput, India (a hub of indigenous rice), exhibit potential genetic resources for maintaining stable yields in drought-prone environments; subjected to stress, the latter do not diminish the photosynthetic rate, photochemical efficiency of PSII (Fv/Fm), membrane stability index, SPAD chlorophyll index or leaf relative water content as much as the former (Mishra et al. 2018). Some of the developmental responses of leaves of drought-tolerant cultivars to water stress for conservation of tissue hydration include deposition of waxes on leaf cuticles, possession of bristles or hairs and a decrease in leaf size. Pubescent hairs reflect incident radiation and thereby keep the leaf cooler. Wax deposition reduces cuticular transpiration, while a reduction in leaf size minimizes boundary layer resistance, permitting thermal heat dissipation of the incident light energy and keeping the leaf surface cool for low transpiration. A reduction in the number of stomata, thickening of cell walls, and less developed conducting systems are additional factors that contribute to the drought resistance of rice.

Reactive Oxygen Species

Another unavoidable outcome of drought stress in rice is the production of reactive oxygen species in various cellular compartments, including chloroplasts, mitochondria, and peroxisomes. The roots under suboptimal conditions exhibit a limitation of water deficiency. When roots experience water deficiency, they send a signal in the

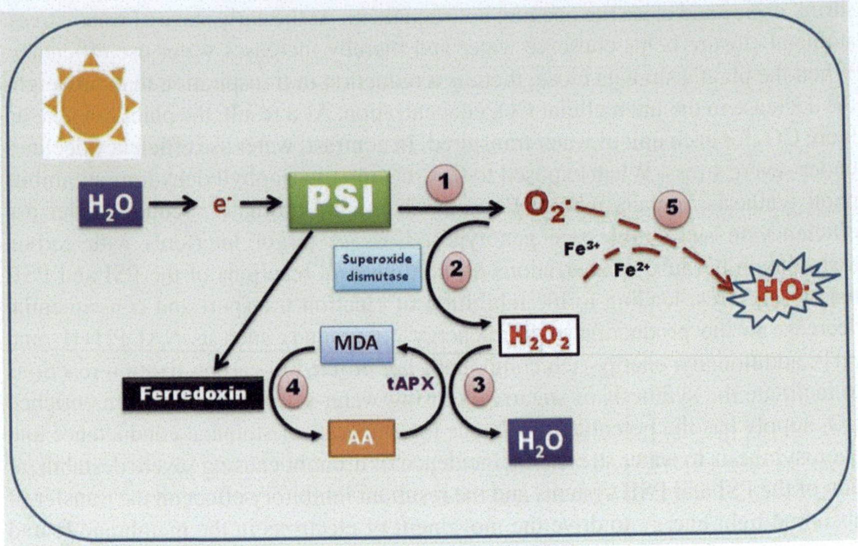

Fig. 2.10 The figure illustrating the Mehler-peroxidase reaction, also known as the water–water cycle, depicts the process in which electrons leak from the photosynthetic electron transport chain to oxygen. This leakage leads to the formation of superoxide (1), which is then dismutated by superoxide dismutase to generate H_2O_2 (2). H_2O_2 is further scavenged by a thylakoid APX (tAPX), resulting in the production of H_2O (3). This cycle results in reducing power from the formation of ascorbate from ferredoxin (4). Metal-catalyzed Haber-Weiss/Fenton reactions that result in the generation of extremely reactive hydroxyl radicals are examples of side reactions that can occur in chloroplasts (5). *PSI* photosystem I, *tAPX* thylakoid ascorbate peroxidase, *MDA* monodehydroascorbate, *AA* ascorbate. (Adapted from Cruz de Carvalho 2008)

form of abscisic acid to the leaves through the xylem sap. The leaves perceive this signal, which triggers a water-saving mode that induces stomatal closure. This closure prevents the loss of water through transpiration, conserving water within the plant. However, concomitant limitations in CO_2 entry reduce net photosynthesis and underutilize the light reaction-generated reducing power in $NADPH+H^{+}$, leading to a net deficiency of $NADP^+$ for further capture of the excited electrons in the system, thus leading to leakage of electrons to oxygen with the generation of superoxides in the water–water cycle or the Mehler peroxidase reaction (Fig. 2.10) (Cruz de Carvalho 2008). In addition, the inability of RuBP carboxylase to fix CO_2 during the dark reaction of photosynthesis triggers oxygenase activity, leading to the enhancement of the photorespiration pathway for H_2O_2 synthesis and subsequent generation of hydroxyl radicals.

Enzymes as Scavengers of Reactive Oxygen Species

Stress-tolerant plants possess an efficient and versatile scavenging system to maintain control of the generation of reactive oxygen species, which include antioxidant enzymes such as superoxide dismutase, glutathione peroxidase, catalase, and

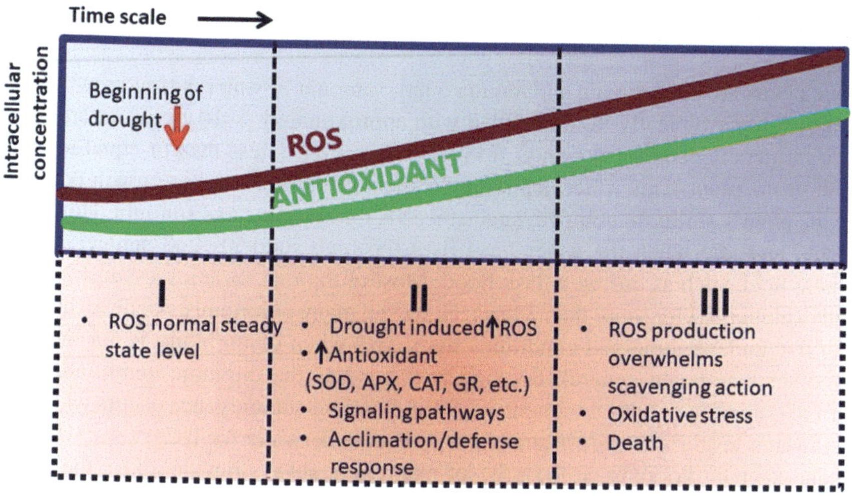

Fig. 2.11 Diagram illustrating the proposed model for the three stages of drought stress responses. Drought stress disrupts the usual steady-state levels of reactive oxygen species (ROS) (I). The increased production of reactive oxygen species (ROS) resulting from stomatal closure shifts the equilibrium upwards, thereby initiating a defence signal transduction pathway (II). Extended periods of drought stress led to heightened production of reactive oxygen species (ROS) that surpass the capacity of the antioxidant system to counterbalance stress, leading to damaging oxidative events that ultimately culminate in cell death (III). (Adapted from Cruz de Carvalho 2008)

ascorbate peroxidase. A model describing the role of drought stress perception in the generation of reactive oxygen species is provided below in Fig. 2.11.

2.6.1.7 Effects of Drought on Root Traits

Roots continue to grow in a plant until their need for assimilates (nutrients and energy) from the leaves matches the supply provided by the leaves. This balance between root demand and leaf supply regulates root growth. The inhibition of leaf expansion in drought-prone environments curtails the consumption of photosynthetic products for shoot growth and partitions them to root systems that support further root growth. This phenomenon continues until drought becomes acute, and root apices lose turgor in dry soil. Thus, mild water stress, often caused by the depletion of water in the upper soil layer, encourages the development of deeper roots in the moist soil layer below. Some drought-tolerant rice cultivars exhibit extensive and prolific root systems as an adaptive response to these conditions (Mishra et al. 2019; Kim et al. 2020). Lee et al. (2017) showed that the overexpression of *OsERF71* in rice roots changes the root structure by enlarging aerenchyma and promoting radial root growth through the co-expression of cell wall-associated and lignin biosynthesis genes for drought tolerance.

2.6.2 Flooding Stress

Rice plants are semiaquatic and require a large amount of water during growth. The crop is grown typically in fields filled with approximately 5–10 cm of water. Most rice farmers depend on flooding, if the flooding level is less than or equal to one-fifth of the plant. This water depth builds up an advantage for rice growth because of the plant's ability to compete against weeds for nutrients and sunlight. However, plants may die when the water level rises too high suddenly and submerges the whole field, such as during a flash flood. Most of the rice varieties are destroyed if fully submerged for more than 3 days. However, many agronomists believe that rice can live under submerged conditions for 8 days when comorbidity is not induced by other factors. In general, flooding is defined as the duration, depth, and pattern of complete or partial inundation. Flash flooding/submergence occurs when the field water level rises sharply and plants remain underwater for 1–2 weeks, whereas under stagnant flooding, at least 5% of plant parts always remain above the water level for a minimum period of 10 days to several months (Sarkar et al. 2014; Sarkar 2016). Flooding during the germination stage occurs due to the stagnation of water above the soil surface. If rice is cultivated under direct seeding due to heavy downpours and poor drainage systems, water may stagnate in the field and hinder crop establishment (Ray et al. 2016). The mechanism associated with tolerance to submergence at the germination stage differs from that associated with the vegetative stage and likely stagnant flooding tolerance at different depths of water (Sarkar et al. 2014; Sarkar 2016). Based on the coincidence of flooding with plant growth, flooding stress in rice production systems can be divided into four types (Sarkar et al. 2009; Singh et al. 2017) as follows:

2.6.2.1 Submergence During Germination

This type of flooding stress is common in rice fields where direct seeding is practised and where sudden heavy rainfall occurs immediately after seed sowing. Rice seeds of almost all varieties are susceptible to waterlogging for germination. When the seeds are submerged in water, they encounter hypoxia/anoxia, which results in poor or no germination or seedling death, leading to improper crop establishment. Oxygen deficiency is the main cause of submergence at the germination stage (Ray et al. 2016). The ability to germinate under oxygen deficiency, leading to proper crop establishment, is an important trait that has been found in a few rice genotypes (Ismail et al. 2009).

2.6.2.2 Vegetative Stage Submergence

In this system, plants experience flooding at the seedling or tillering stage and remain submerged for 1–2 weeks before floodwater recedes. While most rice varieties are susceptible, some varieties survive 10–14 days of complete submergence and rejuvenate growth after the withdrawal of floodwater (Xu et al. 2006; Sarkar et al. 2006).

2.6.2.3 Stagnant Flooding Up to the Booting and Reproductive Stage

Based on hydrology, stagnant flooding may differ from shallow (water depth 15–25 cm), medium-depth (water depth 26–50 cm), and deep-water (>50 to 100 cm) flooding (Sarkar 2016). In moderately deep, waterlogged rainfed lowland rice areas, water accumulates and stagnates for 1–4 months because of impeded drainage. Under stagnant flooding, plant mortality is not as conspicuous; however, yield is severely hampered by a lack of sufficient numbers of ear-bearing tillers and a high rate of sterility, where farmers obtain poor harvests varying from 0.5 to 1.5 t/ha (Kuanar et al. 2017, 2019).

2.6.2.4 Very Deep-Water/Floating Rice

Under very deep-water conditions, the shoot floats on standing water; hence, this plant is called floating rice. Water depths ranged from 1.0 to 4.0 meters and remained stagnant for a prolonged period of 1–4 months. The plant grows with increasing water levels by rapid internode elongation to keep some parts of the foliage above the water level, thereby escaping death by drowning (Kawano et al. 2008). Many traditional landraces growing in this type of ecosystem elongate as much as 25 cm per day, matching the increase in the level of floodwater (Kende et al. 1998). Very deep-water rice is eco-friendly because, unlike rice paddies, it emits the least amount of methane, a greenhouse gas released in wetland rice ecologies. The normal habitats for deep-water rice are river deltas and their flood plains in backwater swamps and natural levees highly flooded by rainwater under a tropical monsoon climate.

2.6.2.5 Flooding Injury

Flooding is common during germination in direct-seeded rice fields. Flooding also causes severe injuries in the post-seedling stage, when plants are still short and are drowned by run-off rainwater in the field in both transplanted and direct-seeded rice. In rice, submergence at the vegetative stage of growth significantly reduces grain yield, and this yield loss is attributed to a reduced survival rate, number of tillers and spikelets, plant biomass, crop growth rate, and excessive underwater elongation of the stem, which consumes reserve carbohydrates necessary otherwise for grain yield formation (Chen et al. 2021).

The injury inflicted on a plant by submergence directly varies with the characteristics of the accumulating floodwater, such as light irradiance, water depth, turbidity, silt dissolved, oxygen and carbon dioxide concentration, and temperature; these factors interact in the medium for a compound impact overall (Das et al. 2009). Submergence infringes on incoming solar radiation for photosynthesis and available oxygen levels in the water surrounding plant tissue, making the plant anaerobic. As the water level increases, injury to submerged plant parts exacerbates because of an increase in temperature and turbidity, which curtails leaf photosynthesis and increases the demand for respiratory energy. Inadequacy of light irradiance in the media affects the physiological status of plants underwater; energy deficiency inflicts severe injury, leading to accelerated plant mortality (Jackson and Ram 2003). For marginalized diffusion of gases in water, the concentration of oxygen and carbon dioxide decreases around the submerged leaves, which reduces respiration and photosynthesis, upsetting the metabolic balance of the plant. Oxygen deprivation restricts aerobic respiration. Aerobic respiration produces 36 ATP per glucose unit in the TCA cycle for growth and reproduction. In contrast, anaerobic mode under submergence harnesses only two ATP molecules per glucose unit and decreases the energy level of the cells. Dissipation in an aqueous medium is not possible, and the gaseous hormone ethylene produced in submerged organs accumulates in tissues. As the level of ethylene increases in submerged plant parts, the plant elongates, while leaf chlorophylls are degraded (Sarkar et al. 2006; Mackill et al. 2012). In addition, submergence impedes leaf photosynthesis directly through the closure of stomata, mesophyll conductance, inhibition of energy transfer, light absorption and photochemical activity in PSII, and reduced enzyme activity of ribulose-1,5-bisphosphate carboxylase oxygenase (RuBisCo) (Panda et al. 2006).

Submergence and low ambient oxygen in the medium reduce root hydraulic conductivity. The impaired hydraulic conductance and decreased uptake of water cause wilting symptoms because hydraulic conductance, i.e. the per unit driving force of water flow into roots, is directly correlated with the transpiration rate and determines the water uptake capacity. In addition to wilting symptoms, leaf chlorosis, which prompts premature senescence, is another more visible symptom of flooding stress. The uptake of other elements, such as phosphorus, potassium, magnesium, copper, zinc, and manganese, is also affected by roots because of poor absorption. A lack of driving force for mass flow in transpiration pull, poor elongation of roots limiting the surface area of acquisition and lack of chemical potential for nutrient diffusion are the factors responsible for poor mineral uptake. The impact of flooding on rice plants differs depending on the stage of development and likely the depth of water on the standing crop. Overcoming flooding stress during germination requires elongation of the coleoptile, whereas avoiding the deleterious effect of submergence at the vegetative stage can help tolerant plants grow quickly. Depending on the stagnation of water depth in rice fields, zero to greater than 20 cm of elongation per day helps the plant counterbalance stagnant flooding (Xu et al. 2006; Hattori et al. 2009; Sarkar et al. 2014; Sarkar 2016).

2.6.2.6 Strategies for Flood Tolerance

Improved rice cultivars with high yield potential are flood-sensitive and are not cultivated in flood-prone ecosystems. The susceptible varieties die within a few days of submergence (Panda et al. 2006). However, some landraces resist submergence for 2 weeks or more (Sarkar and Bhattacharya 2011). The genotypes that are tolerant to submergence at the germination stage or vegetative stage differ greatly. There is similarity in mechanism yet there are major differences as envisaged in different figures. Tolerance at submergence during germination depends on the growth of both the radicle and plumule in an anoxic state. The extension of the shoot above the water surface is compulsory for survival. However, under stagnant flooding, generally, a portion of shoots must always be placed above the water surface, and little to greater elongation is important depending on the depth of water stagnation.

The flood-tolerant deep-water landscapes of rice suitable for cultivation in marshlands have a tall spindly stature that makes them unsuitable for flood-prone lowlands because of severe lodging (Vergara et al. 2014). The plant allocates stem reserves primarily for elongation rather than for the growth of reproductive organs, resulting in a poor harvest index (Kato et al. 2014). In a flood-challenging environment, plants exhibit two contrasting strategies for resilience: either remain quiescent while submerged or exhibit an ephemeral attribute related to the escape of submergence by rapid elongation of the stem (Bailey-Serres et al. 2012); each of these two individual attributes is species-specific and specific to different landraces. One of the flood-tolerant landraces in India, FR13A, possessing a QTL for submergence *SUB1* on chromosome 9, was identified (Xu and Mackill 1996; Xu et al. 2006). This unique landrace was used for QTL mapping studies and follow-up map-based cloning of the *SUB1*gene. *SUB1* expression is induced by ethylene generated during submergence; this gene inhibits stem elongation and conserves valuable energy for plant survival after approximately 2 weeks of inundation. With marker-assisted backcrossing, this gene has been introduced into several mega rice cultivars for resiliency under flood-prone systems. However, such introgression often enhances susceptibility to stagnant flooding in semi-dwarf rice genotypes and promotes increased ethylene synthesis to the detriment of grain filling in mega rice cultivars (Kuanar et al. 2019). Similarly, in other flood-avoiding landraces, Hattori et al. (2009) identified two ethylene response factor (ERF) genes, *Snorkel1* (*SK1*) and *Snorkel 2* (*SK2*), in the QTL region of chromosome 12. The ERFs SNORKEL1 and SNORKEL2 allow rice to survive submergence under flooding by rapid elongation of the stem. QTL mapping for anaerobic seed germination in the rice cultivar Khao Hlan has been performed on chromosome 9; the gene encoding a trehalose phosphate phosphatase (*OsTPP*) has been cloned and validated functionally (Kretzschmar et al. 2015; Singh et al. 2017). Although the impact of excess water mainly involves oxygen deficiency, overproduction of ethylene and other hormonal disbalances, alcoholic fermentation leading to diminished energy production, greater consumption of carbohydrates, and reactive oxygen stress, physiological requirements differ greatly for submergence during the germination stage to submergence at the vegetative stage to stagnant flooding tolerance under different water stagnation depths.

2.6.2.7 Germination-Stage Oxygen Deficiency (GSOD) Tolerance

Basis of Tolerance

Regulation of carbohydrate metabolism

Better carbohydrate utilization through the overexpression of α–amylase, sucrose synthase, lactate dehydrogenase, aldehyde dehydrogenase, alcohol dehydrogenase, and pyruvate decarboxylase.

Regulation of Nitrogen Metabolism and Cellular pH Homeostasis

Overexpressed glutamate decarboxylase (GAD), nitrate reductase, and nitrite reductase. The GAD system maintains cellular pH homeostasis through the consumption of excess protons (H^+).

Regulation of Redox Status

Increased capacity to remove reactive oxygen species such as H_2O_2 and superoxide radicals contributes to tolerance. Overexpression of catalase, superoxide dismutase, and peroxidase genes.

Regulation of Hormone Biosynthesis

Increased ethylene release increases survival during submergence at the germination stage.

2.6.2.8 Submergence Tolerance

Basis of Tolerance

There was greater non-structural carbohydrate content, especially after submergence. Maintenance of greater structural and functional ability of chloroplasts Maintenance of greater leaf gas-film area before and under submergence. Robust seedlings counterbalance submergence better than poor seedlings.

Maintenance of greater H_2O_2 content and activities of antioxidant enzymes during submergence. H_2O_2 acts as a signalling molecule and activates different biochemical processes to induce tolerance.

Greater activities of antioxidant enzymes and antioxidant levels after withdrawal of submergence help in rapid regeneration and survival.

Suppression of ethylene production through the activity of the submergence-tolerant gene *SUB1-A*: restricts elongation, improves redox potential, and conserves non-structural carbohydrates during submergence.

2.6.2.9 Stagnant Flooding Tolerance

Basis of Tolerance

Stiff and strong culms, especially for shallow and medium-depth ecologies.

Kneeing ability in deep-water and floating ecologies. If the water depth does not exceed 75 cm, the stiff culm and plant height beyond 100 cm at the onset of flooding are important.

During the rise of water and period of stagnation, zero elongation is appropriate for shallow ecology, ±2.0 cm for medium-depth, 5–7 cm deep-water and sometimes more than 10 cm per day for floating rice.

An appropriate tiller length (tiller height must be high compared to the depth of flooding) helps in the survival of tillers and hence improves productivity.

There was a greater accumulation of non-structural carbohydrates before the onset of flooding. The expression of amylolytic agents depended on the elongation rate: control = shallow < medium depth < deep water < floating rice.

Greater O_2 Liberation to Root Zones

The area of aerenchymatous tissues: Depending on the depth of water stagnation, floating rice> deep-water>medium depth>shallow = control.

Regulation of hormone biosynthesis: Ethylene release capacity, floating rice > deep-water > medium depth > shallow = control.

A greater capacity to scavenge reactive oxygen species such as H_2O_2 and superoxide was observed. Variation between control and shallow submergence should be minimal.

2.6.3 Salt Stress

Salt stress is characterized by the accumulation of an excessive amount of salt in the soil, which ultimately hinders crop growth and can lead to plant death. According to a 2009 report by the Food and Agriculture Organization (FAO), soils affected by salt stress typically have elevated levels of soluble salts such as sodium (Na^+), magnesium (Mg^{2+}), calcium (Ca^{2+}), chloride (Cl^-), and sulphate (SO_4^{2-}) ions. Notably, more than 20% of irrigated land worldwide is affected by salt stress (Pitman and Läuchli 2002; Munns 2005). It is projected that the area affected by salt stress may increase to encompass as much as 50% of the total irrigated land area by the year 2050 (Wang et al. 2003).

2.6.3.1 Pathways of Salt Entry into Plants

Entry of various elements from the salt environment into plant cells depresses both the water and osmotic potential of the root cells. Depression of osmotic potential helps preserve cytoplasmic water, enabling essential cell metabolism. A model depicting the schematic flow of Na^+ and Cl^- ions is proposed in the figure below (Fig. 2.12). The uptake of sodium (Na^+) and chloride (Cl^-) ions in plants can occur through both apoplastic and symplastic pathways. The apoplastic pathway involves a direct continuum between the external environment and the plant's xylem tissues. Compared to symplastic uptake, apoplastic flow is insignificant but can increase severalfold when the transpirational demand is very high. It was revealed that apoplastic barrier activity is strengthened by suberization upon exposure to salinity stress, which precludes the entry of Na^+. Suberization is extensive in the salt-tolerant rice cultivar Pokkali, in which the entry of Na^+ is minimized, and the plant increases the expression of mRNAs encoding enzymes involved in suberin synthesis within 30 min of exposure to stress (Krishnamurthy et al. 2009). Cl^- also follows an apoplastic route into root cells. Studies have demonstrated that as much as 50% of chloride (Cl^-) translocation to the shoot in rice occurs via the apoplastic pathway (Shi et al. 2013).

In comparison to passive translocation in the apoplastic pathway, Na^+ and Cl^- ions use membrane transporters for entry into the root via the symplastic route. Na^+ uses a nonselective cation channel (NSCC) for plastic entry, and this route is considered to be the major route for Na^+ influx. The other membrane transporters involved in the symplastic entry of Na^+ are HKT, AKT2, PIP 2:1, NHA, LAC1, and HAK5 (Isayenkov and Maathuis 2019). However, some authors have shown that AKT, PIP2:1, and LCT1 may be involved in transporting Na^+ from the external environment into the cytosol (Keisha et al. 2018). Unlike Na^+, Cl^- is an essential element, and its uptake into the cell occurs via an H^+/Cl^- symporter or cation

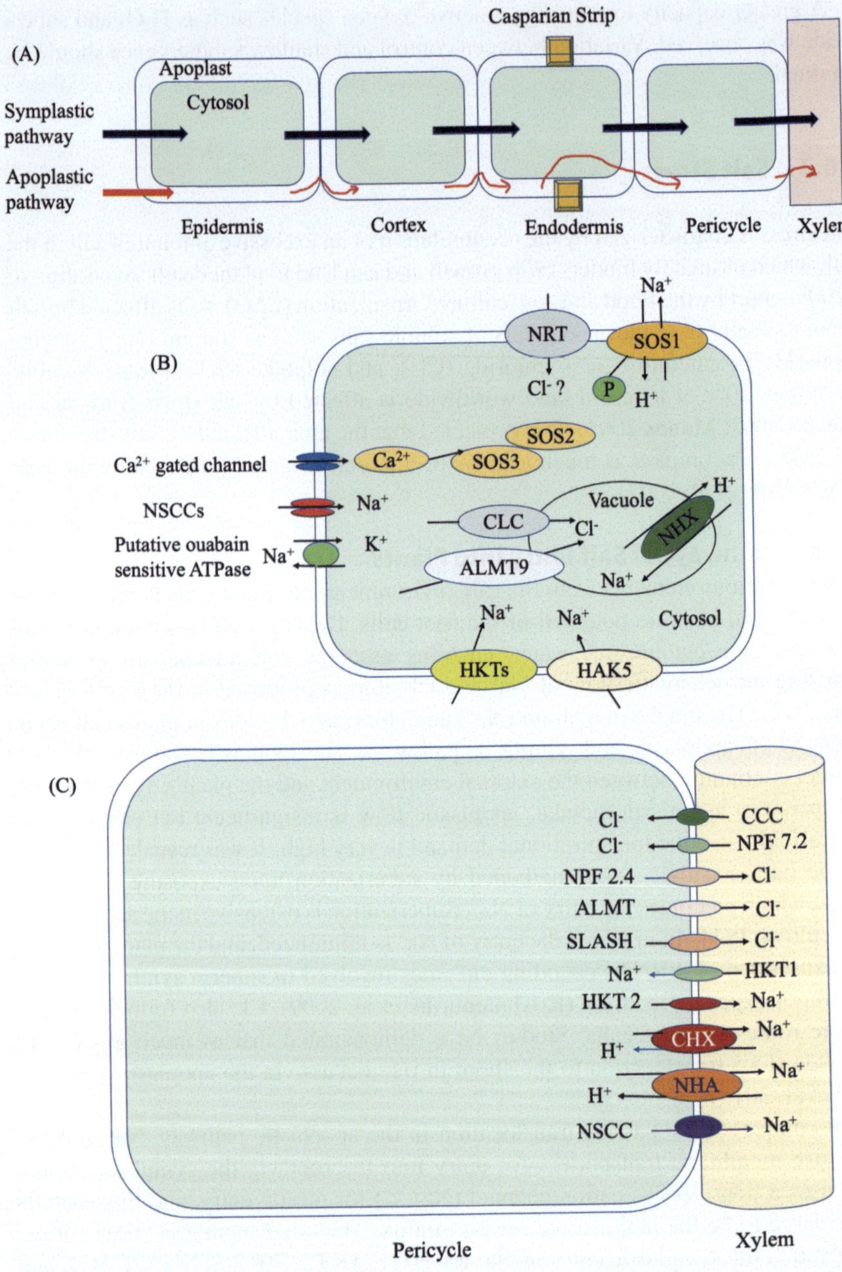

Fig. 2.12 (**A**) A possible movement of Na⁺ and Cl⁻ from the external environment to the inside of the plant system. The red and black arrows indicate apoplastic and symplastic bypass flow, respectively. (**B**) The involvement of various transporters in the movement of Na⁺ through the plasma membrane. (**C**) Movement of both Na⁺ and Cl⁻ from the pericycle to the xylem through various transporters. (Adapted from Maathuis 2014; Shabala et al. 2015; Keisha et al. 2018; Isayenkov and Maathuis 2019)

chloride cotransporter (CCC). When Cl^- is high in the external medium, the ion enters the cell passively through these anion channels because of a concentration gradient.

2.6.3.2 Salt-Sensing

The plant response to salinity stress can be divided into two phases. The first phase is due to osmotic stress, which lasts from minutes to days. The perception of stress triggers the immediate closure of stomata and inhibits cell expansion. The second phase, spanning from days to weeks, pertains to more extensive damage. Cytotoxic ions are generated in this phase, slowing cell metabolism and leading to premature senescence and eventual death. A plant cell has an extensive salt-sensing mechanism. The function of the sensor relies on the activity of specific sodium (Na^+) selective ion channels with Na^+ binding sites that modulate the gate, regulating the flow of sodium ions (Shabala et al. 2015; Chen et al. 2016; Wu 2018; Keisha et al. 2018; Isayenkov and Maathuis 2019). The proposed mechanism suggests that an increase in extracellular sodium chloride (NaCl) leads to a rapid increase in cytosolic calcium ions (Ca^{2+}). This increase in Ca^{2+} subsequently triggers a long-distance calcium signalling cascade involving calcium-dependent protein kinases (CDPKs) and calcineurin B-like proteins (CBLs) associated with CBL-interacting protein kinases (CIBLs). A schematic diagram of the early salt sensing of plant cells is given below (Fig. 2.13).

2.6.3.3 Impact of Salt Stress on Plant Growth

High salinity hampers crop growth through multiple mechanisms, including water stress, cytotoxicity due to excessive accumulation of ions such as sodium (Na^+) and chloride (Cl^-), and nutritional disorders (Chinnusamy et al. 2006; Horie et al. 2012). A schematic figure is provided here to illustrate the detrimental effects of salinity and the mechanisms employed by plants to withstand and mitigate the resulting damage (Fig. 2.14). An ionic imbalance resulting from the accumulation of sodium (Na^+) and chloride (Cl^-) ions leads to oxidative stress, which in turn promotes lipid peroxidation and increases the production of reactive oxygen species (ROS), such as superoxide radicals, hydrogen peroxide, and hydroxyl radicals. Furthermore, the accumulation of these ions in root cells reduces the uptake of water and essential nutrients from the soil. This combination of factors contributes to the overall negative impact of salinity stress on plant health. Water stress decreases osmotic potential, inducing the production of low-molecular-weight compounds known as compatible solutes. Interference in nitrogen uptake reduces plant growth, stopping reproduction. The presence of a high level of Na^+ in the intracellular space alters the Na^+:K^+ ratio, which affects the bioenergetic process of leaf photosynthesis. In direct interference, plants shut their stomata to conserve water, which resists the entry of CO_2 for photosynthesis. In addition, salinity adversely affects the activities of stomatal enzymes for CO_2 reduction in the dark during photosynthesis.

In salt-sensitive plants, salt stress reduces chlorophyll synthesis, which can also be a reason for a decrease in net photosynthesis. Light energy not used in

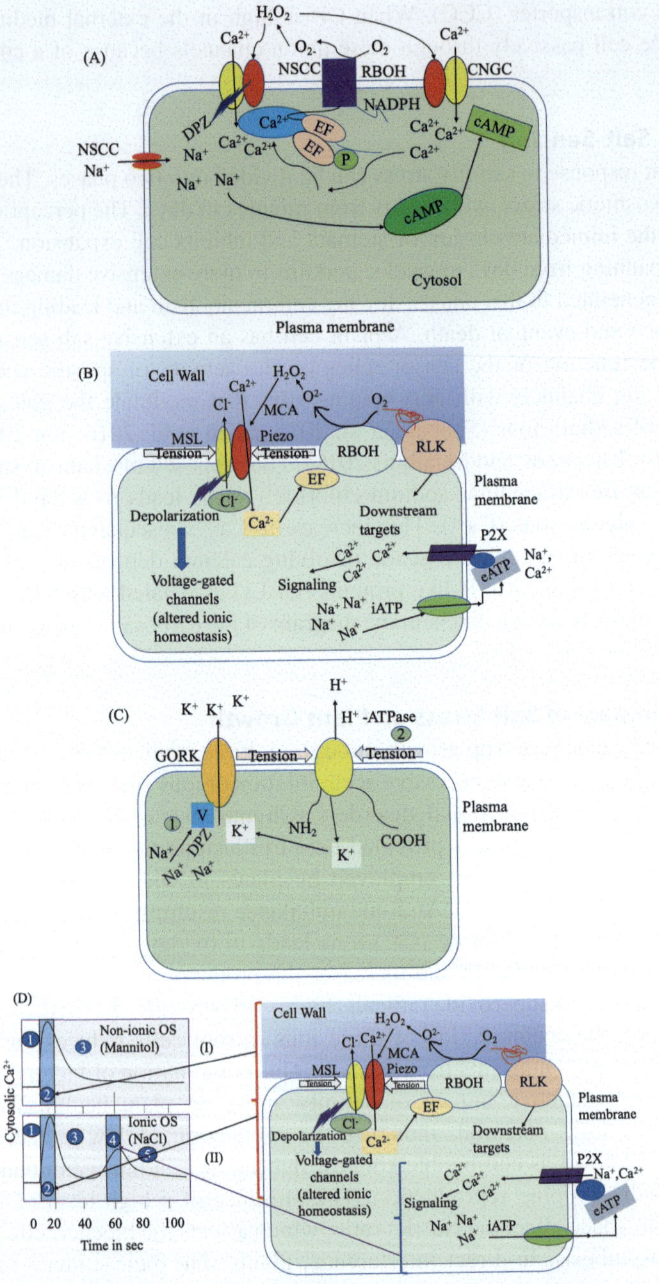

Fig. 2.13 Schematic representation of the sensing of salt stress. (**A**) Roles of NADPH oxidase, cyclic nucleotide-gated channels (CNGC) and the ROS system in the entry of Na^+ into the cell. (**B**) Salt stress sensing through mechanosensory channels (MSCs), transporters, and P2X purinoreceptors. In addition, RLK (a receptor-like protein kinase) senses the osmotic stress-induced deformation of the cell wall, activates several downstream proteins and directly affects Na^+ movement between cells. (**C**) Sensing through the H^+-ATPase/GORK tandem. (**D**) Sensing of (I) non ionic and (II) ionic osmotic stress. (Adapted from Shabala et al. 2015; Chen et al. 2016; Wu 2018; Keisha et al. 2018; Isayenkov and Maathuis 2019)

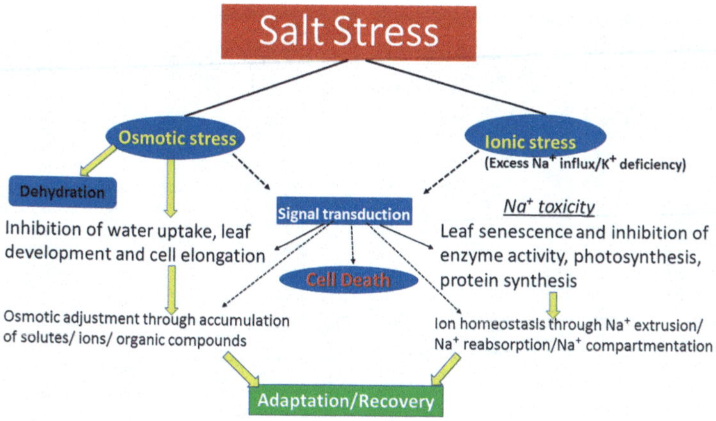

Fig. 2.14 Adaptation/recovery of plants from salt injury (Adapted from Horie et al. 2012)

photosynthesis induces oxidative stress and infringes on the growth and development of plants. Morphologically, stress has many effects, resulting in growth inhibition, a reduction in leaf and crop area establishment, inhibition of seed germination, delay in seed setting and a decrease in dry matter production. High soil salinity interferes with nitrogen uptake and ultimately reduces primary production. The most detrimental consequence of salt stress is the occurrence of ion toxicity. Excess intake of Na^+ and Cl^- causes major physiological disorders. The ions accumulate in the root zone and reduce osmotic potential, and the toxicity generated disrupts the metabolic function of cells. With the potential for folding, some ions, particularly chloride anions, become toxic at high concentrations throughout the life cycle of the plant.

2.6.3.4 Salt Stress Injury in Rice

Salt stress is a prevalent abiotic stress condition in numerous rice production regions around the world. Compared to other cereals, rice is a more saline-sensitive species (Joseph et al. 2010). The sensitivity of plants to salinity stress is very high during the seedling and reproductive stages when salinity stress severely impacts grain yield. Growing rice requires abundant fresh water. However, the entry of brackish water near seaplanes or the seasonal accumulation of salts by surface flow rainwater can have devastating effects on grain yield. On average, seasonal salinity above 1.9 dS/m can potentially reduce grain yield in rice. Osmotic stress generated by salinity because of excess accumulation of Na^+ and Cl^- in the soil decreases the osmotic potential, which decreases the uptake of water and nutrients by the root system. Subjected to stress, plants exhibit significant decrease in panicle weight and length and an increase in sterility during seed setting (Aref and Rad 2012; Chakraborty et al. 2019); additionally, stress affects primary production and inhibits the activity of starch synthase in developing grains (Abdullah et al. 2001). The various physiological traits of rice plants are sensitive to salt stress, and the irregularities caused by salt stress ultimately lead to a decrease in grain yield (Fig. 2.15).

Fig. 2.15 Impact of salt stress (8 dS/m) on panicle development. Pokkali and Chettivirippu: tolerant; Lunidhan: medium tolerant; Savita: susceptible to salt stress. (Adapted from Chakraborty et al. 2019)

The effects of high salt stress on rice physiology include either ion-specific toxicity in cell metabolism or adverse plant–water relationships. To conserve water and maintain metabolic homeostasis, plants close their stomata when subjected to salt stress. The transpiration rate decreases because of a reduction in stomatal conductance but results in a concomitant decrease in CO_2 absorption and slackening of primary production. In addition, the entry of excessive amounts of salt into plant cells eventually leads to ion toxicity in leaves, which induces premature leaf senescence and a consequential reduction in the leaf area of plants that cannot sustain growth. Furthermore, the entry of Na^+ and Cl^- ions causes severe ion imbalance because of their antagonistic competition with essential ions such as K^+, Ca^{2+}, and NO_3^-, which restricts their acquisition in the rhizosphere and causes a net deficiency of these elements in plant tissue (Hussain et al. 2017; Table 2.1). The infringement of ionic balance could be a factor responsible for the reduction in net photosynthesis. A deficiency of essential photosynthetic carbon assimilates could be a factor

Table 2.1 Imbalances of nutrients under salt stress (Adapted from Hussain et al. 2017; Sogoni et al. 2021)

Excess Na^+/Cl^-	Antagonistic effect	Synergistic effect
Na^+	–	Increase
Calcium (Ca^{2+})	Decrease	–
Potassium (K^+)	Decrease	–
Nitrogen (N)	Decrease	–
Phosphorous (P)	Decrease	–
Iron (Fe)	Decrease	–
Magnesium (Mg)	Decrease	–
Cadmium (Cd)	–	Increase
Zinc (Zn)	Decrease	–
Manganese (Mn)	Decrease	–
Boron (Bo)	Decrease	–
Silicone (Si)	Decrease	–
Na^+/K^+ ratio	–	Increase
Na^+/Ca^{2+} ratio	–	Increase
Ca^{2+}/Mg^{2+} ratio	–	Increase
Cl^-/NO_3^-	–	Increase

limiting grain filling and increasing the panicle sterility of rice plants under salt stress (Flowers and Yeo 1981). Panicle sterility could be an outcome of nutrient balance destabilization, which leads to poor grain setting and increased degeneration of spikelets (Khatun and Flowers 1995; Hasanuzzaman et al. 2009; Chakraborty et al. 2019).

2.6.3.5 Role of Phytohormones in Salt Stress Tolerance in Rice

Improving the yield of rice plants undergoing salinity stress has been one of the main goals of rice breeders. Rice plants exhibit salinity tolerance at germination, but developmental stages such as panicle initiation and anthesis are highly sensitive. Subject to salt stress, the functional roles of various phytohormones change for eventual expression in cell metabolism. In this context, the roles of various hormones, such as gibberellin, auxin, cytokinin, abscisic acid, brassinosteroids, and ethylene, have been extensively studied and discussed in the context of mitigating salt stress in plants, including rice (Ryu and Cho 2015; Hussain et al. 2017). Like in other plants, salt stress evokes two secondary stresses in rice, osmotic and ionic stresses. The signalling pathways of the two types of responses differ considerably in terms of cell metabolism and hormonal dynamics.

Ethylene

Similar to other plants, the early components of sensing and signalling mechanisms could be the generation of reactive oxygen species (ROS) and cGMP (Isayenkov and Maathuis 2019). A rapid increase in cGMP (Donaldson et al. 2004) and ROS (Isayenkov and Maathuis 2019) is detected within minutes of exposure to salinity stress. For such rapid responses as those described above, rapid production of ethylene is essential for the establishment of salt acclimation in rice plants. Ethylene signal transduction plays a crucial role in plant defence mechanisms, enabling self-adjustment and the subsequent activation of salt tolerance mechanisms (Cao et al.

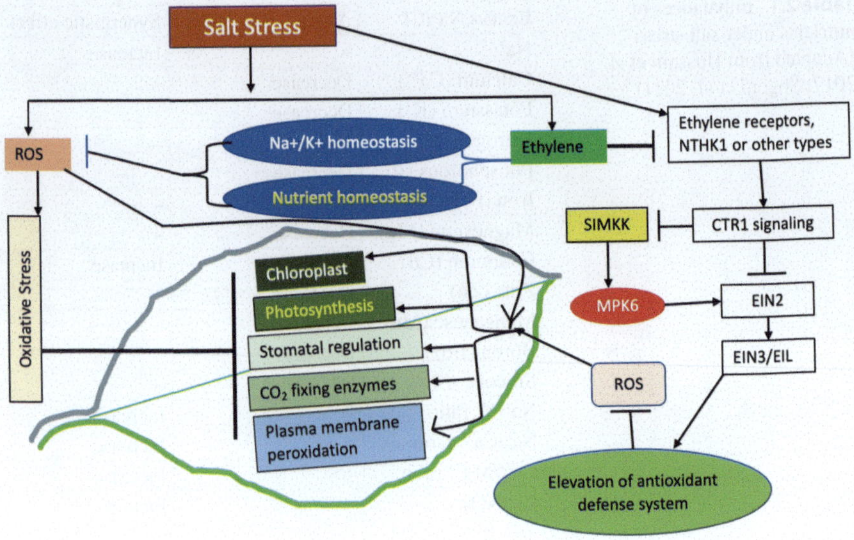

Fig. 2.16 Ethylene signalling improves salinity tolerance (Adapted from Cao et al. 2008; Tao et al. 2015; Riyazuddin et al. 2020)

2008; Tao et al. 2015; Riyazuddin et al. 2020). Salinity stress has been shown to increase ethylene production in rice, as the inhibition of ethylene action by pyridoxal 5′-phosphate (PLP) successfully improved agronomic and physiological traits such as photosynthetic activity, SPAD values, root activity, hormonal regulation, yield and yield components (Hussain et al. 2020). Observations have shown that subjecting rice plants to salt stress for 3 weeks resulted in a reduction in the number of tillers, plant height, leaf area, and overall plant biomass at the heading stage. However, the application of pyridoxal-5′-phosphate (PLP) suppressed ethylene production in the roots, leaves, and panicles of rice plants subjected to salinity stress, thereby facilitating the recovery of agronomic and physiological features affected by this stress. A model has been formulated to describe the mechanisms of salinity tolerance in rice, particularly ethylene signalling (Fig. 2.16). It is hypothesized that stimulating ethylene biosynthesis and enhancing its signal transduction pathways could bolster a plant's ability to tolerate salinity, whereas inhibiting ethylene production and signalling may render the plant more susceptible to stress (Tao et al. 2015).

Salinity-induced ethylene signalling activates the ethylene receptor signalling pathway, which is regulated by constitutive triple response 1 (CTR1), a Raf-like protein kinase. Ethylene signalling is predominantly transduced via the CTR1-EIN2-EIN3 pathway, which regulates numerous effectors linked to both plant growth and salinity tolerance. Dearth of ethylene under salt stress disrupts Na^+/K^+ homeostasis, which leads to ROS production. Overproduction of ROS negatively impacts the vital functions of plants, whereas in the presence of ethylene, the

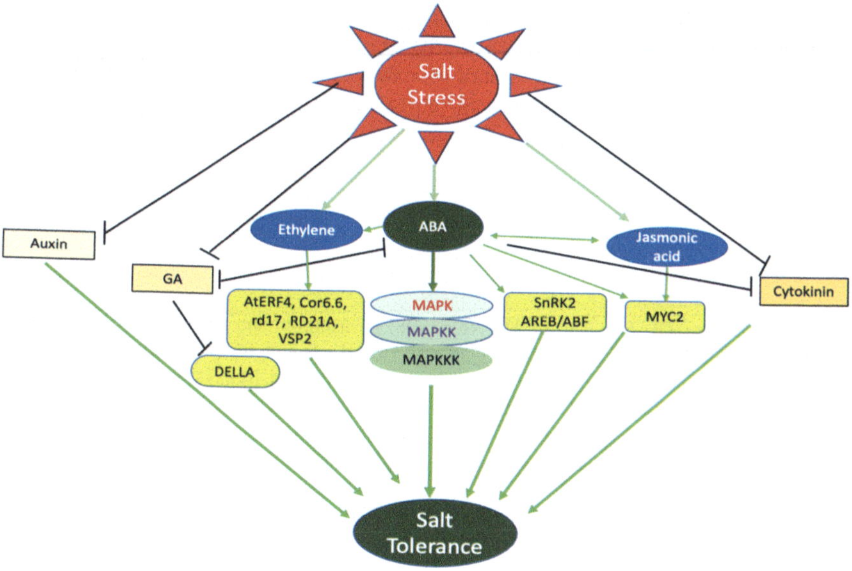

Fig. 2.17 Salt stress modulates hormone regulation to impart stress tolerance (Adapted from Kumar et al. 2013; Ryu and Cho 2015)

antioxidative system is activated, and nutrient and Na^+/K^+ homeostasis is restored, thereby diminishing the overproduction of ROS.

Abscisic Acid

In addition to ethylene, ABA is another growth-inhibiting hormone whose synthesis is stimulated in response to resilience to abiotic stresses, including salt stress. Very often, ethylene and ABA responses are intricately related. In rice plants, ABA biosynthesis increases due to the effects of osmotic stress evoked by salt accumulation (Kumar et al. 2013). Concerning ABA biosynthesis, the osmotic stress-induced signalling system has been classified into two types, and these pathways are interrelated with three other pathways, namely, the ROS pathway, Ca^{2+}/PLC pathway, and SOS pathway, which are primarily evoked by ionic stress (Fig. 2.17).

Compared to those of ethylene and ABA, the roles of other growth hormones in the salt tolerance of rice have been less explored. The role of the GA hormone is essential for the alleviation of salt-induced morbidity in rice plants (Colebrook et al. 2014). The biosynthesis of GA is rapidly reduced under salt stress in rice plants (Javid et al. 2011). When plants are subjected to stress, DELLA proteins tend to accumulate. This accumulation results in reduced growth but improved stress tolerance by decreasing the action of gibberellic acid (GA) signalling (Achard et al. 2008). Like that of GA, the IAA content declines in rice under salt stress, while growth and development are inhibited (Kazan 2013; Hussain et al. 2017). In response to salt stress, plants typically exhibit decelerated or retarded growth. It has been concluded that stress hormones such as abscisic acid, jasmonic acid, salicylic acid, and ethylene and growth-promoting hormones such as gibberellic acid, cytokinin, auxin, and brassinosteroids play critical roles in improving the adaptation to salt stress through multifaceted signalling pathways (Ryu and Cho 2015; Yu et al. 2021). Some hormones negatively control plant growth and alleviate salt tolerance, whereas others positively regulate growth processes to counteract salt injury (Fig. 2.17).

References

Abdullah M, Azura ZR, Nor Aini D, Ho NK (1998) Distribution and abundance of main insect families in the Muda areas. In: Nashriyah BN, Ho NK, Ismail BS, Ahyaudin BN, Lum KY (eds) Rice agro-ecosystem of Muda irrigation scheme, Malaysia. Malaysia Institute for Nuclear Technology Research (MINT) and Muda Agricultural Development Authority, Selangor, pp 169–170

Abdullah Z, Khan MK, Flowers TJ (2001) Causes of sterility in seed set of rice under salinity stress. J Agron Crop Sci 187:25–32

Achard P, Renou JP, Berthome R, Harberd NP, Genschik P (2008) Plant DELLAs restrain growth and promote survival of adversity by reducing the levels of reactive oxygen species. Curr Biol 18:656–660

Amersinghe FP (1993) Rice field breeding mosquitoes (Diptera: Culicidae) in a new irrigation project in Sri Lanka. Mosq Borne Dis Bull 10:1–7

Aref F, Rad HE (2012) Physiological characterization of rice under salinity stress during vegetative and reproductive stages. Indian J Sci Technol 5:2578–2586

Ashikari M, Ma JF (2015) Exploring the power of plants to overcome environmental stresses. Rice 8:10. https://doi.org/10.1186/s12284-014-0037-y

Ashraf U, Hussain S, Sher A, Abrar M, Khan I, Anjum AA (2018) Planting geometry and herbicides for weed control in rice: implications and challenges. In: Tadele Z (ed) Grasses as food and feed. https://doi.org/10.5772/intechopen.79579

Bailey-Serres J, Lee SC, Brinton E (2012) Waterproofing crops: effective flooding survival strategies. Plant Physiol 160:1698–1709

Bambardeniya CNB (2000) Ecology and biodiversity in an irrigated rice field ecosystem in Sri Lanka. PhD thesis, University of Peradeniya, Sri Lanka

Bambardeniya CNB, Amerasinghe FP (2003) Biodiversity associated with the rice field agroecosystem in Asian countries. A brief review. In: Working paper 63. International Water Management Institute, Colombo, p 24

Breviario D, Genga A (2013) Stress response in rice. J Rice Res 2:1. https://doi.org/10.4172/jrr.1000e104

Cao YR, Chen SY, Zhang JS (2008) Ethylene signalling regulates salinity stress response: an overview. Plant Signal Behav 3:761–763

Chakraborty K, Chattopadhyay K, Nayak L, Ray S, Yeasmin L, Jena P, Gupta S, Mohanty SK, Swain P, Sarkar RK (2019) Ionic selectivity and coordinated transport of Na^+ and K^+ in flag leaves render differential salt tolerance in rice at the reproductive stage. Planta 250:1637–1653

Chang TT (2003) Origin, history, and domestication. In: Smith CW, Dilday RH (eds) Rice: origin, history, technology, and production. Wiley, Hoboken, NJ, pp 3–26

Chen ZC, Yamaji N, Fujii-Kashino M, Feng Ma JF (2016) A cation-chloride cotransporter gene is required for cell elongation and osmoregulation in rice. Plant Physiol 171:494–507

Chen Y, Song J, Yan C, Hong X (2021) Effects of submergence stress at the vegetative growth stage on hybrid rice growth and grain yield in China. Chil J Agric Res 81:191–201

Chinnusamy V, Zhu J, Zhu JK (2006) Salt stress signalling and mechanisms of plant salt tolerance. Genet Eng 27:141–177

Colebrook EH, Thomas G, Phillips AL, Hedden P (2014) The role of gibberellin signalling in plant responses to abiotic stress. J Exp Biol 217:67–75

Cruz de Carvalho MH (2008) Drought stress and reactive oxygen species. Plant Signal Behav 3:156–165

Das SR (2012) Rice in Odisha. IRRI Technical Bulletin No.16, Los Banos (Phillippines), International Rice Research Institute.

Das KK, Panda D, Sarkar RK, Reddy JN, Ismail AM (2009) Submergence tolerance about variable floodwater conditions in rice. Environ Exp Bot 66(3):425–434

Donaldson L, Ludidi N, Knight MR, Gehring C, Denby K (2004) Salt and osmotic stress cause rapid increases in *Arabidopsis thaliana* cGMP levels. FEBS Lett 569:317–320. https://doi.org/10.1016/j.febslet.2004.06.016

Duan J, Zhang M, Zhang H, Xiong H, Liu P, Ali J, Li J, Li Z (2012) OsMIOX, a Myo-inositol oxygenase gene, improves drought tolerance through the scavenging of reactive oxygen species in rice (*Oryza sativa* L.). Plant Sci 196:143–151

FAO (2009) High-level expert forum—how to feed the world in 2050, economic and social development. Food and Agricultural Organization of the United Nations, Rome

Fernando CH (1995) Rice fields are aquatic, semiaquatic, terrestrial and agricultural: a complex and questionable limnology. In: Timotius KH, Goltenboth F (eds) Tropical limnology, vol 1, pp 121–148

Fernando CH (1996) Ecology of rice fields and its bearing on fisheries and fish culture. In: De Silva SS (ed) Perspective in Asian fisheries. Asian Fisheries Society, Manila, pp 217–237

Fernando CH, Furtado JI, Lim RP (1979) Aquatic fauna of world's rice fields. Wallaceana Suppl 2:1–105

Flowers TJ, Yeo AR (1981) Variability in the resistance of sodium chloride salinity within rice (*Oryza sativa* L.) varieties. New Phytol 88:363–373

Gogoi N, Baruah KK, Gogoi B, Gupta PK (2005) Methane emission characteristics and its relations with plant and soil parameters under irrigated rice ecosystem of Northeast India. Chemosphere 59:1677–1684

Gupta A, Rico-Medina A, Caño-Delgado AI (2020) The physiology of plant responses to drought. Science 368:266–269

Halwart M, Gupta MV (2004) The rice field ecosystem. Chapter 3. In: Halwart M, Gupta MV (eds) Culture of fish in rice fields. Food and Agricultural Organization and World Fish Center, Rome and Penang, pp 5–11

Hasanuzzaman M, Fujita M, Islam MN, Ahamed KU, Nahar K (2009) Performance of four irrigated rice varieties under different levels of salinity stress. Int J Integr Biol 6:85–90

Hattori Y, Nagai K, Furukawa S, Song XJ, Kawano R, Sakakibara H, Wu J, Matsumoto T, Yoshimura A, Kitano H, Matsuoka M, Mori H, Ashikari M (2009) The ethylene response factors *SNORKEL1* and *SNORKEL2* allow rice to adapt to deep water. Nature 460(7258):1026–1030. https://doi.org/10.1038/nature/08258

Heckman CW (1979) The rice field ecology in North–East Thailand. Monogr Biol 34:228

Horie T, Karahara I, Katsuhara M (2012) Salinity tolerance mechanisms in glycophytes: an overview with the central focus on rice plants. Rice 5:11. https://doi.org/10.1186/1939-8433-5-11

Hossain MZ, Sikder S, Husna A, Sultana S, Akhter S, Alim A, Joardar JC (2020) Influence of water stress on morphology, physiology and yield contributing characteristics of rice. SAARC J Agric 18:61–71

Huke RE, Huke EH (1990) Rice: then and now. International Rice Research Institute, Manila

Hussain S, Zhang J, Zhong C, Zhu L, Cao X, Yu S, James AB, Hu J, Jin Q (2017) Effects of salt stress on rice growth, development characteristics, and the regulating ways: a review. J Integr Agric 16:2357–2374

Hussain S, Huang J, Zhu C, Zhu L, Cao X, Hussain S, Ashraf M, Khaskheli MA, Kong Y, Jin Q, Li X, Zhang J (2020) Pyridoxal 5′-phosphate enhances the growth and morpho-physiological characteristics of rice cultivars by mitigating the ethylene accumulation under salinity stress. Plant Physiol Biochem 154:782–795

IRRI (1993) Rice research in a time of change. IRRI, Los Baños, Laguna

Isayenkov SV, Maathuis FJM (2019) Plant salinity stress: many unanswered questions remain. Front Plant Sci 10:80. https://doi.org/10.3389/fpls.2019.00080
Ismail AM, Ella ES, Vergara GV, Mackill DJ (2009) Mechanisms associated with tolerance to flooding during germination and early seedling growth in rice (*Oryza sativa* L.). Ann Bot 103(2):197–209
Jackson MB, Ram PC (2003) Physiological and molecular basis of susceptibility and tolerance of rice plants to complete submergence. Ann Bot 91(2):227–241
Javid MG, Ali S, Foad M, Seyed AMMS, Iraj A (2011) The role of phytohormones in alleviating salt stress in crop plants. Aust J Crop Sci 5:726–734
Joseph B, Jini D, Sujatha S (2010) Biological and physiological perspectives of specificity in abiotic salt stress response from various rice plants. Asian J Agric Sci 2:99–105
Joshi R, Wani SH, Singh B, Bohra A, Dar ZA, Lone AA, Pareek A, Singla-Pareek SL (2016) Transcription factors and plants response to drought stress: current understanding and future directions. Front Plant Sci 7:1029
Kariali E, Kuanar S, Mohapatra PK (2008) Individual tiller dynamics of two wild *Oryza* species in contrasting habitats. Plant Prod Sci 11(3):355–360
Kato Y, Collard BCY, Septiningsih EM, Ismail AM (2014) Shoot growth and other traits associated with tolerance of stagnant flooding in rice. AoB Plant 6:plu058
Kawano R, Doi K, Yasui H, Mochizuki T, Yoshimura A (2008) Mapping of QTL for floating ability in rice. Breed Sci 58:47–53
Kazan K (2013) Auxin and the integration of environmental signals into plant root development. Ann Bot 112:1655–1665
Keisha M, Mukherjee S, Bhatla SC (2018) Mechanisms of sodium transport in plants—progresses and challenges. Int J Mol Sci 19:647. https://doi.org/10.3390/ijms19030647
Kende H, Vander Knaap E, Cho HT (1998) Deepwater rice: a model plant to study stem elongation. Plant Physiol 18:1105–1110
Khatun S, Flowers TJ (1995) Effects of seed set in rice. Plant Cell Environ 18:61–67
Kim Y, Chung YS, Lee E, Tripathi P, Heo S, Kim KH (2020) Root response to drought stress in rice (*Oryza sativa* L.). Int J Mol Sci 21(4):1513
Kretzschmar T, Pelayo MAF, Trijatmiko KR, Gabunada LFM, Alam R, Jimenez R, Mendioro MS, Slamet-Loedin IH, Sreenivasulu N, Bailey-Serres J, Ismail AM, Mackill DJ, Septiningsih EM (2015) A trehalose-6-phosphate phosphatase enhances anaerobic germination tolerance in rice. Nat Plants 1:15124
Krishnamurthy P, Ranathunge K, Franke R, Prakash H, Schreiber L, Mathew M (2009) The role of root apoplastic transport barriers in salt tolerance of rice (*Oryza sativa* L.). Planta 230:119–134. https://doi.org/10.1007/s00425-009-0930-6
Kuanar SR, Ray A, Sethi SK, Chattopadhyay K, Sarkar RK (2017) Physiological basis of stagnant flooding tolerance in rice. Ric Sci 24(2):73–84
Kuanar SR, Molla KA, Chattopadhyay K, Sarkar RK, Mohapatra PK (2019) Introgression of *Sub1 (SUB1) QTL* in mega rice cultivars increases ethylene production to the detriment of grain-filling under stagnant flooding. Sci Rep 9:18567. https://doi.org/10.1038/s41598-019-54908-2
Kumar K, Kumar M, Kim SR, Ryu H, Cho YG (2013) Insights into the genomics of salt stress response in rice. Rice 6:27. https://doi.org/10.1186/1939-8433-6-27
Kumar A, Dixit S, Ram T, Yadav RB, Mishra KK, Mandal NP (2014) Breeding high-yielding drought-tolerant rice: genetic variations and conventional and molecular approaches. J Exp Bot 65:6265–6278
Kumar A, Basu S, Ramegowda V, Pereira A (2016) Mechanisms of drought tolerance in rice. In: Sasaki T (ed) Achieving sustainable cultivation of rice, vol 11. Burleigh Dodds Science, Cambridge. https://doi.org/10.19103/AS.2106.0003.08
Lacey LA, Lacey CM (1990) The medical importance of Riceland mosquitoes and their control using alternatives to chemical insecticides. J Am Mosq Control Assoc 6(Supplement 2):1–93
Lee DK, Yoon S, Kim YS, Kim JK (2017) Rice *OsERF71*-mediated root modification affects shoot drought tolerance. Plant Signal Behav 12:e1268311

Li H, Liang X, Lian Y, Xu L, Chen Y (2009) Reduction of ammonia volatilization from urea by floating duckweed in flooded rice fields. Soil Sci Soc Am J 73:1890. https://doi.org/10.2136/sssaj2008.0230

Lim RP, Abdullah FM, Frenando CH (1984) Ecological studies of Cladocera in the rice fields of Tanjung, Karang, Malaysia subjected to pesticide treatments. Hydrobiologia 113:99–103

Maathuis FJM (2014) Sodium in plants: perception, signalling, and regulation of sodium fluxes. J Exp Bot 65:849–858

Mackill DJ, Ismail AM, Singh US, Labios RV, Paris TR (2012) Development and rapid adoption of submergence-tolerant (Sub1) rice varieties. Adv Agron 115:299–352

Mishra SS, Panda D (2017) Leaf traits and antioxidant defence for drought tolerance during the early growth stage in some popular traditional rice landraces from Koraput, India. Ric Sci 24(4):207–217

Mishra SS, Behera PK, Kumar V, Lenka SK, Panda D (2018) Physiological characterization and allelic diversity of selected drought-tolerant traditional rice (*Oryza sativa* L.) landraces of Koraput, India. Physiol Mol Biol Plants 24(6):1035–1046

Mishra SS, Behera PK, Panda D (2019) Genotypic variability for drought tolerance-related morpho-physiological traits among indigenous rice landraces of Jeypore tract of Odisha, India. J Crop Imp 33:254–278

Munns R (2005) Genes and salt tolerance: bringing them together. New Phytol 167:645–663

Naik A (2022) Biodiversity of beneficial and harmful fauna in rice fields of western Odisha. PhD thesis, Sambalpur University, India

Naylor R (1996) Invasion in agriculture: assessing the cost of golden apple snail in Asia. Ambio 25:443–448

Nounjan N, Mahakham W, Siangliw JL, Toojinda T, Theerakulpisut P (2020) Chlorophyll retention and high photosynthetic performance contribute to salinity tolerance in rice carrying drought tolerance quantitative trait loci (QTLs). Agri 10(12):620. https://doi.org/10.3390/agriculture10120620

O'Toole JC (1982) Adaptation of rice to drought-prone environments. In: Drought resistance in crops with emphasis on rice. IRRI, Los Baños, pp 195–213

Panda D, Rao DN, Sharma SG, Strasser RJ, Sarkar RK (2006) Submergence effects on rice genotypes during seedling stage: probing of submergence driven changes of photosystem 2 by chlorophyll *a* fluorescence induction O-J-I-P transients. Photosynthetica 44:69–75. https://doi.org/10.1007/s11099-005-0200-1

Panda D, Mishra SS, Behera PK (2021) Drought tolerance in rice: focus on recent mechanisms and approaches. Ric Sci 28(2):119–132. https://doi.org/10.1016/J.RSCI.2021.01.002

Paudel YP, Pradhan S, Pant B, Prasad BN (2012) Role of blue-green algae in rice productivity. Agric Biol J N Am 3:332–335

Pitman MG, Läuchli A (2002) Global impact of salinity and agricultural ecosystems. In: Salinity: environment-plants-molecules. Springer, Dordrecht, pp 3–20

Prot JC, Rahman ML (1994) Nematode ecology, economic importance and management in rice ecosystems in South and South–East Asia. In: Teng PS, Heong KL, Moody K (eds) Rice pest science and management. International Rice Research Institute, Manila, pp 129–140

Radzio JA, Lorence A, Chevone BI, Nessler CL (2003) L-gulono-1,4-lactone oxidase expression rescues vitamin C-deficient *Arabidopsis* (*vtc*) mutants. Plant Mol Biol 53:847–844

Ray S, Vijayan J, Sarkar RK (2016) Germination stage oxygen deficiency (GSOD): an emerging stress in the era of changing trends in climate and Rice cultivation practice. Front Plant Sci 7:671. https://doi.org/10.3389/fpls.2016.00671

Richards RA, Rebetzke GJ, Watt M, Condon AG, Spielmeyer W, Dolferus R (2010) Breeding for improved water productivity in temperate cereals: phenotyping, quantitative trait loci, markers and the selection environment. Funct Plant Biol 37:85–97

Riyazuddin R, Verma R, Singh K, Nisha N, Keisham M, Bhati KK, Kim ST, Ravi Gupta R (2020) Ethylene: a master regulator of salinity stress tolerance in plants. Biomol Ther 10:959. https://doi.org/10.3390/biom10060959

Rozilah I, Ali AB (1998) Aquatic insect population in the Muda rice agro-ecosystem. In: Nashriyah BN, Ho NK, Ismail BS, Ahyaudin BA, Lum KY (eds) Rice agro-ecosystem of the Muda irrigation scheme Malaysia. Malaysia Institute of Nuclear Technology Research (MINT) and Muda Agricultural Development Authority (MADA), Selangor

Ryu H, Cho YG (2015) Plant hormones in salt stress tolerance. J Plant Biol 58:147–155

Saha S, Begum HH, Nasrin S (2019) Effects of drought stress on growth and accumulation of proline in five rice varieties (*Oryza sativa* L.). J Asiat Soc Bangladesh Sci 45(2):241–247

Sarkar RK (2016) Stagnant flooding tolerance in rice: endeavours and achievements. NRRI Research Bulletin no. 11. ICAR-NRRI, Cuttack

Sarkar RK, Bhattacharya B (2011) Rice genotypes with SUB1 QTL differ in submergence tolerance, elongation ability during submergence and re-submergence and regeneration growth at re-emergence. Rice 5:7. https://doi.org/10.1007/s12284-011-9065-z

Sarkar RK, Reddy JN, Sharma SG, Ismail AM (2006) Physiological basis of submergence tolerance in rice and implications for crop improvement. Curr Sci 91:899–906

Sarkar RK, Panda D, Reddy JN, Patnaik SSC, Mackill DJ, Ismail AM (2009) Performance of submergence tolerant rice (*Oryza sativa*) genotypes carrying the Sub1 quantitative trait locus under stressed and non-stressed natural field conditions. Indian J Agric Sci 79:876–883

Sarkar RK, Das KK, Panda D, Reddy JN, Patnaik SSC, Patra BC, Singh DP (2014) Submergence tolerance in rice: biophysical constraints, physiological basis and identification of donors. CRRI Research Bulletin no. 7. ICAR-CRRI, Cuttack

Shabala S, Wu H, Bose J (2015) Salt stress sensing and early signalling events in plant roots: current knowledge and hypothesis. Plant Sci 241:09–119

Shi Y, Wang Y, Flowers TJ, Gong H (2013) Silicon decreases chloride transport in rice (*Oryza sativa* L.) in saline conditions. J Plant Physiol 170:847–853. https://doi.org/10.1016/j.jplph.2013.01.018

Shi F, Dong Y, Zhang Y, Yang X, Qiu D (2017) Overexpression of the PeaT1 elicitor gene from *Alternaria tenuissima* improves drought tolerance in Rice plants via interaction with a Myo-inositol oxygenase. Front Plant Sci 8:970. https://doi.org/10.3389/fpls.2017.00970

Shi F, Dong Y, Wang M, Qiu D (2020) Transcriptomics analyses reveal that OsMIOX improves rice drought tolerance by regulating the expression of plant hormones and sugar-related genes. Plant Biotechnol Rep 14:339–349

Shirasawa K, Takabe T, Takabe T, Sachie Kishitani S (2006) Accumulation of glycine-betaine in rice plants that overexpress choline monooxygenase from spinach and evaluation of their tolerance to abiotic stress. Ann Bot 98(3):565–571

Simpson RA, Roger PA, Oficial R, Grant IF (1994) Effect of nitrogen fertilizer and pesticide management on flood water ecology in a wetland rice field. II. Dynamics of microcrustacean and dipteran larvae. Biol Fertil Soils 17:138–146

Singh A, Septiningsih EM, Balyan HS, Singh NK, Rai V (2017) Genetics, physiological mechanisms and breeding of flood-tolerant rice (*Oryza sativa* L.). Plant Cell Physiol 58(2):185–197

Smith RJ (1983) Weeds of major economic importance in rice and rice yield losses due to weed competition. In: Proceedings of the Conference on Weed Control in Rice. International Rice Research Institute, Philippines and International Weed Science Society, Manila

Sogoni A, Jimoh MO, Kambizi L, Laubscher CP (2021) The impact of salt stress on plant growth, mineral composition, and antioxidant activity in *Tetragoniadecumbens* mill.: an underutilized edible halophyte in South Africa. Horticulturae 7:140. https://doi.org/10.3390/horticulturae7060140

Syamsia IA, Noerfitryani Nadir M, Reta Kadir M (2018) Paddy chlorophyll concentrations under drought stress conditions and after endophytic fungus application. In: ICFSS-2017 IOP Conference series: earth and environmental science. IOP, Bristol, p 156. https://doi.org/10.1088/1755-1315/156/1/012040

Taiz L, Zeiger E (2002) Plant Physiology, 3rd edn. Sinauer Associates, Sunderland, MA

Takagi M, Sugiyama A, Maruyama K (1995) Effect of rice culturing practices on the seasonal occurrence of *Culex tritaenieorhynchus* (Diptera: Culicidae) immature in three different types of rice growing areas in Central Japan. J Med Entomol 32:112–118

Takagi M, Sugiyama A, Maruyama K (1996) Effect of rice plant covering on the density of mosquito larvae and other insects in rice fields. Appl Entomol Zool 31:75–90

Taniguchi M, Toyota K, Kimuni M (1997) Seasonal variations in microcrustaceans and microbial flora on their surface in the overlying water of a Japanese paddy field. Soil Sci Plant Nutr 43:651–664

Tao J-J, Chen H-W, Ma B, Zhang W-K, Chen S-Y, Zhang J-S (2015) The role of ethylene in plants under salinity stress. Front Plant Sci 6:1059. https://doi.org/10.3389/fpls.2015.01059

Upadhyaya H, Panda SK (2019) Drought stress responses and its management in rice. In: Hasanuzzaman M, Fujita M, Nahar K, Biswas JK (eds) Advances in rice research for abiotic stress tolerance. Elsevier, Amsterdam, pp 177–200

Venuprasad R, Lafitte HR, Atlin GN (2007) Response to direct selection for grain yield under drought stress in rice. Crop Sci 47:285–293

Vergara GV, Mackill DJ, Ismail AM (2014) Variation in tolerance of partial stagnant flooding in rice. AoB Plants 6:plu055

Wang W, Vinocur B, Altman A (2003) Plant responses to drought, salinity and extreme temperatures: towards genetic engineering for stress tolerance. Planta 218:1–14

Wepereis M, Kropff M, Maligaya A, Toung T (1996) Drought stress responses of two lowland rice cultivars to soil water status. Field Crop Res 46:21–39

Whitton BA (2007) Soils and rice fields. Chapter 8. In: The ecology of cyanobacteria, pp 233–255

Wu H (2018) Plant salt tolerance and Na^+ sensing and transport. Crop J 6:215–225

Xiong J, Zhang L, Fu G, Yang Y, Zhu C, Tao L (2012) Drought-induced proline accumulation is uninvolved with increased nitric oxide, which alleviates drought stress by decreasing transpiration in rice. J Plant Res 125(1):155–164. Epub 2011 Mar 12. https://doi.org/10.1007/s10265-011-0417-y

Xu K, Mackill DJ (1996) A major locus for submergence tolerance mapped on rice chromosome 9. Mol Breed 2:219–224. https://doi.org/10.1007/BF00564199

Xu K, Xu X, Fukao T, Canlas P, Maghirang-Rodriguez R, Heuer S, Ismail AM, Bailey-Serres J, Ronald PC, Mackill DJ (2006) *Sub1A* is an ethylene-response-factor-like gene that confers submergence tolerance to rice. Nature 442:705–708. https://doi.org/10.1038/nature04920

Yang X, Wang B, Chen L, Li P, Cao C (2019) The different influences of drought stress at the flowering stage on rice physiological traits, grain yield, and quality. Sci Rep 9:3742. https://doi.org/10.1038/s41598-019-40161-0

Yoshida S (1977) Rice. In: Alvim PT, Kozlowski TT (eds) Ecophysiology of tropical crops. Academic Press, New York, pp 57–87

Yu J, Mao C, Zhong Q, Yao X, Li P, Liu C, Ming F (2021) OsNAC2 is involved in multiple hormonal pathways to mediate the germination of rice seeds and establishment of seedlings. Front Plant Sci 12:699303. https://doi.org/10.3389/fpls.2021.699303

Yue B, Xue W, Xiong L, Yu X, Luo L, Cui K, Jin D, Xing Y, Zhang Q (2006) Genetic basis of drought resistance at reproductive stage in rice. Genetics 172:1213–1228

Zargar A, Sadiq R, Naser B, Khan F (2011) A review of drought indices. Environ Rev 19:333. https://doi.org/10.1139/a11-013

Zhu R, Wu FY, Zhou S, Hu T, Huang J, Gao Y (2020) Cumulative effects of drought-flood abrupt alternation on the photosynthetic characteristics of rice. Environ Exp Bot 169:103901

Evolutionary Dynamics of Rice

3

Abstract

In the categorization of monocotyledonous flowering plants, the *Poaceae* family stands out as the most significant global provider of food. Furthermore, they are positioned within the top five families of flowering plants with respect to both species diversity and abundance. They grow on all continents of the world except Antarctica in widely contrasting ecosystems and account for 24% of the Earth's vegetation. *Oryza* is a plant genus within the *Poaceae* family, encompassing important food crops such as *Oryza sativa* (Asian rice) and *Oryza glaberrima* (African rice). The plants are both annual and perennial and grow as tall wetland grasses. Throughout the evolutionary history of rice, perspectives related to eco-genetic settings of cultivated rice and its wild ancestor provide crucial information underplaying differences in various morphological features that transited with the progress of time. The progenitor wild rice harboured traits that empowered its survival in natural conditions, but they had little importance for human use and were not suitable for cultivation. There is significant phenotypic variation among rice cultivars empowered by domestication within the wide spectrum of ecosystems worldwide. Domestication of rice increased the size of the organs that are consumed and carry economic importance. The phenotypic plasticity is typical to the plant, amenable to modification in response to variation of habitat. The plant generates reproductive structures like tillers and spikelets in abundance when conditions allow and sacrifices many of them first for lack of resources and second for survival under exigency of exposure to unfavourable conditions. The morphological resemblances and molecular research findings drawn from various cultivars of *Oryza sativa* show distinctive traits indicative of whole-genome duplication in all members. This gene duplication event appears to have taken place shortly before the emergence of the common ancestor. The perspectives related to eco-genetic settings of cultivated rice and its wild ancestor provide

© The Author(s), under exclusive license to Springer Nature Singapore Pte Ltd. 2025

P. K. Mohapatra et al., *Tillering Behavior of Rice Plant*,
https://doi.org/10.1007/978-981-97-5235-5_3

crucial information underplaying differences in various morphological features that transited with the progress of time.

Keywords

Rice taxonomy · Habitats · Low land · Deep water · Upland · Domestication syndrome · Gene duplication

Scientific Classification

Rice antiquity extends to more than 10,000 years old. Since ancient times the plant has been cultivated on earth providing food, nutrition, and employment to mankind. The scientific name of rice is *Oryza sativa*, where the meaning of rice is derived from the classical Latin language and *sativa* means cultivated. Scientifically rice plants have been classified as follows (Fig. 3.1A):

Oryza genus comprises three cultivated species, namely *sativa*, *glaberrima*, and *javanica*, native to Asia, Africa, and tropical Asia, respectively. It is accepted that tropical *javanica* is a sub-species of *O. sativa*.

3.1 Evolutionary Dynamics of *Oryza sativa*

3.1.1 *Poaceae* Family

In the categorization of monocotyledonous flowering plants, the *Poaceae* family stands out as the most significant global provider of food. Furthermore, they are

Kingdom:	*Plantae*
Clade:	*Tracheophytes*
Clade:	*Angiosperms*
Clade:	*Monocots*
Clade:	*Commelinids*
Order:	*Poales*
Family:	*Poaceae*
Subfamily:	*Oryzoideae*
Tribe:	*Oryzeae*
Subtribe:	*Oryzinae*
Genus:	*Oryza*

(A)

(B)

Fig. 3.1 Scientific classification of rice (**A**) and photograph showing complete rice plant (**B**) (Photo gallery of E. Kariali, Sambalpur University)

positioned within the top five families of flowering plants regarding both species diversity and abundance. They grow on all continents of the world except Antarctica in widely contrasting ecosystems and account for 24% of the Earth's vegetation. The *Poaceae* consists of annual or perennial, hermaphrodite, monoecious or dioecious herbs or trees in the bamboos. The roots can be fibrous, adventitious, branched or stilt. The family comprises about 12,000 species in roughly 800 genera, cultivated for food that includes cereals, the primary source of starchy food the world over. The identifying attributes are based mostly on leaf characteristics. The leaves are either evergreen or deciduous and are arranged in basal or alternate phyllotaxy. In size, the leaf is much longer than wide and embellish parallel venation. It consists of a cylindrical shaped sheath at the base that wraps around the subtending internode with a lamina projecting above it. The stem could be underground rhizome or cylindrical with conspicuous nodes and internodes; the leaf base is attached to the nodes of the stem (Fig. 3.1B). In addition to sharing morphological resemblances, molecular research findings drawn from various cereal species show distinctive traits indicative of whole-genome duplication in all members of the family. This gene duplication event appears to have taken place shortly before the emergence of the common ancestor of the grass family, as described by Kellog (2009). The duplicated genes have undergone divergence in their patterns of expression and functions, leading to the evolution of specific characteristics and contributing to the diversity observed in grass inflorescences (Malcomber et al. 2006; Preston and Kellogg 2007).

3.1.2 Genus *Oryza*

Oryza is a plant genus within the *Poaceae* family, encompassing important food crops such as *O. sativa* (Asian rice) and *O. glaberrima* (African rice). The plants are both annual and perennial and grow as tall wetland grasses. Leaf anatomical characteristics distinctive of the sub-family are the presence of fusoid and arm cells common to all members. Morphologically, *Oryza* is distinguished by its single-flower spikelets in which the glumes are nearly suppressed completely. Each spikelet contains two sterile lemmas, which simulate glumes. Species of the genus *Oryza* have been divided into three genomic types.

The history of domestication events of the three sub-species cohabiting in sharply contrasting ecological niches is not identical in the evolutionary history. Non-domesticated rice *O. nivara* (Fig. 3.2C, D), which is considered a modified annual version of *O. rufipogon*, is the immediate precursor for the *indica* sub-species, while the perennial stock of *O. rufipogon* (Fig. 3.2A, B) was the immediate precursor of *japonica* (temperate *japonica*) and *javanica* (tropical *japonica*) sub-species. The *indica* sub-species became self-pollinated in tropical conditions shortening the time of exposure for pollens to high temperatures during fertilization, but the *japonica* sub-species under cooler environments were out-crossing types in the absence of any challenge for morbidity. It was proposed that the two major sub-species *indica* and *japonica* diverged around 04 million years ago (Molina et al. 2011). The

Fig. 3.2 The variances in morphological characteristics between wild and cultivated rice. Plants in the field (**A**), panicle (**B**) and grain with awn and endosperm (**E**) of wild rice *Oryza rufipogon*, Plants in the field (**C**), panicle (**D**) and grain and endosperm (**F**) of *Oryza nivara* and plants in the field (**G**), panicle (**H**) grain and endosperm (**I**) of cultivated rice *Oryza sativa*. (Source: Photo-gallery of P. K. Senapati and E. Kariali)

phylogenetic pattern of domestication events within the various sub-species leading to speciation has remained mysterious, but a model has been proposed by Kellog (2009) enumerating the events prominently into three distinct types.

3.1.3 Genome Evolution in Asian Rice *Oryza sativa*

The domestication of rice stands as one of the most significant and ancient achievements of mankind, profoundly influencing the advancement of human civilization.

Studying the genetic mechanisms behind the rice domestication process has the potential not only to reveal its ancient origins but also facilitate the development of new genetic varieties through breeding. In rice evolution, domestication leading to species diversification and speciation are important issues. In the evolutionary hierarchy of plant taxonomy, *O. sativa* (Fig. 3.2G, H, I) holds a prominent position among angiosperms. Among the monocots, rice is attractive as a model species, which is second to the model dicot plant *Arabidopsis* among the angiosperms. Two salient features contribute to the choice of rice for model species, such as it represents taxonomically distinct monocots and is a crop species of high economic demand. The status of model organisms among the monocots has gained preference in the selection of *O. sativa* for the study of genome sequencing and genome evolution. It has the smallest genome of all the cereals with 430 million nucleotides. While having a common ancestor 50 million years ago, maize and wheat have genome sizes 6 to 38 times larger than rice.

Hence, for monocot evolution rice provides crucial information on the central comparative genomics core and merits attention for studies on gene evolution. The rice genome has a haploid chromosome number of 12 ($2n = 24$), which contains 430 million DNA base pairs. To put it differently, the rice genome comprises a total length of 430 Megabases (Mb), which translates to approximately 1500 centimorgans (cM). As per the Os-Nipponbare-Reference-IRGSP-1.0 data, the genome comprises 37,869 loci, encompassing 35,679 protein-coding and 2190 non-protein-coding loci. In comparison, model dicot plant *Arabidopsis* has a much smaller genome of 125 million base pairs. So long rice plants have been the subject of rigorous genetic studies on yield, genetic resistance to diseases, hybrid vigour and adaptive responses by rice scientists. There is the existence of a multitude of varieties adapted to a very wide range of ecological niches varying in land topology, moisture regime, average temperature and radiation potential that embody rice habitats from dry soil in temperate climates to deep-water cultures in tropical regions. *O. sativa* boasts an impressive biodiversity, with a staggering variety count exceeding 140,000. This uniqueness provides an exceptional opportunity for studying evolutionary biology and gaining insights into how genetic diversity influences the differences in traits between various species and cultivars.

3.1.4 Gene Duplication: A Factor for Speciation and Domestication

In plant genomes, gene duplication is a widespread phenomenon. It plays an essential part in the rice plant's adaptation to the different environmental conditions throughout its evolutionary history. Studies on genome evolution through biotechnological interventions have attracted rice because the crop provides 21% of food calories and 15% of proteins for humans. Scientists always look for potential sources out of the past and existing rice gene pools for harnessing genotypes amenable to modification and gain in yield. It is suggested that conducting a comparative analysis of genome organization, including genes, intergenic regions, and transposable elements within the *Oryza* genus, could provide valuable insights into genome evolution, speciation, and domestication as proposed by Sakai et al. (2014). Evolution by gene duplication is an event where a gene or a part of it replicates into two identical copies that cannot be distinguished from each other. The replicated genes provide raw genetic materials for mutation, drift, and natural selection in various environments. Bereft of gene duplication, the plasticity of a species or genome in adapting to a new environment would be severely constrained or limited. Therefore, gene duplication is an important process of rice evolution. It was a front runner in ecotype diversification enabling environmental adaptation and domestication, and finally, the development of a new variety stable to the specific environment in the past and the process continues unabated as of now.

Jacquemin et al. (2011) opined that gene duplication and subsequent evolution of gene pairs have shaped the present-day genome of most of the individuals of the plant kingdom and *O. sativa* is not an exception. In their examination of the rice genome (*O. sativa* sp. *japonica*), scientists identified a complete genome duplication event approximately 50–70 million years ago, shared among all cereal crops, as well as a more recent segmental duplication occurring between chromosomes 11 and 12, dating back to five million years ago. In rice, ancient polyploidy produced duplicated gene pairs and maintained persistent similarity in genetic sequences over an extended duration. It was concluded that high sequence gene conversion owing to updated duplication could be responsible for the evolution. Gene conversion fosters species diversity by expediting mutation through recombination, thereby establishing the basis for natural selection, as indicated by Guo et al. (2013). The authors have given a diagrammatic format for the phylogenetic relationship and divergence times among three *Oryza* species and *Brachypodium distachyon*. They belong to the BEP clade (*Bambusoideae-Ehrhartoideae-Pooideae*) monocots. Similar observations have been made by other workers on the timings of rice evolution.

According to Vaughan et al. (2003), *Oryza* is an ancient genus related to bamboo and its ancestors were present 65 million years ago (Prasad et al. 2005). The *Oryza* genus diversified in the period between 8 and 14 million years ago and diversification of the AA genome occurred more recently over the past two million years ago. In a recent study, Wei et al. (2021) employed comparative genomics and phylogenetic analyses to revise the assessment of gene conversion rates among duplicated genes originating from a polyploidization event that occurred 100 million years ago in both the *indica* and *japonica* sub-species of Asian cultivated rice, *O. sativa*, to investigate gene conversion events and their influence on genome evolution. It was concluded that conversion between 100-million-year-old duplicated genes contributes to rice sub-species divergence. The authors investigated gene duplication and conversion patterns in three rice sub-species genomes, one *japonica and* two *indica*. The key findings of the study reveal that a significant portion, approximately 5.19–5.77%, of genes underwent duplication across the three varieties due to whole-genome changes after the separation of *indica* and *japonica*, which occurred approximately 0.4 million years ago. Furthermore, within this group, 7.77–9.53% of genes displayed partial conversion. The frequency of gene conversion was notably elevated at the ends of chromosomes, particularly in chromosomes 11 and 12.

This ongoing gene conversion resembled the patterns observed in ancient duplicates and created additional opportunities for further gene conversion, ultimately hastening the divergence of rice species. Following polyploidization, chromosome realignments are linked to continuous gene conversion events. These realignments prevent recombination and impede the conversion of duplicated genes in homologous regions. Converted genes exhibit more similar expression patterns compared to non-converted duplicates. Of utmost significance, gene conversion has a notable influence on various biological functions associated with multiple genes, particularly affecting metabolic catalytic activities.

3.1.5 Change of Plant Architecture in Domestication

Throughout the evolutionary history of rice, perspectives related to eco-genetic settings of cultivated rice and its wild ancestor provide crucial information underplaying differences in various morphological features that transited with the progress of time. The progenitor wild rice harboured traits that empowered its survival in natural conditions, but they had little importance for human use and were not suitable for cultivation. Domestication of rice increased the size of the organs that are consumed and carry economic importance. Table 3.1, based on Vaughan et al. (2008), provides a comparison of the physical characteristics between wild and cultivated rice.

As previously stated, Asian rice and African rice, each experienced distinct domestication processes originating from their wild ancestors, *O. rufipogon* and *O. barthii*, respectively. In the course of domestication, several traits are selected, altogether known as domestication syndromes, which makes the domesticated species exhibit distinct differences in appearance compared to their wild ancestors. The syndrome contributed to the transformation of grass-like features of wild rice into a refined crop. The process of selection of essential traits in the domestication process has had significant consequences for crops, leading to a moderate reduction in genetic diversity across the genome compared to their wild ancestors. In a recent

Table 3.1 Shape and size of grain and panicle of wild and cultivated rice (Source: Vaughan et al. (2008)

		Category				
		Wild			Cultivated	
		Species name			Species name	
Characters	Size	*O. barthii*	*O. nivara*	*O. rufipogon*	*O. sativa*	*O. glaberrima*
Panicle length	Longest (cm)	40	35	40	43	32
	Shortest (cm)	9	9	12	10	14
Grain length	Longest (cm)	12.3	10.5	11	14.4	10.5
	Shortest (cm)	7.7	5.2	4.8	3.5	6.1
Grain width	Widest (mm)	3.5	3.4	3.2	4.9	3.6
	Narrowest (mm)	2.2	1.5	1.1	1.4	2.2

study conducted by Swetha et al. (2022), numerous miRNAs with differential targeting were identified between wild and domesticated rice lines. The study illustrated how domestication led to alterations in sRNA-mediated cascade silencing as part of the evolutionary process in *indica* rice. The domesticate grows in well-managed fields under reduced competition relative to the wild state and hence the expression of many genes necessary for survival under stress-prone environments is attenuated. There have been some instances where domestication events included the loss of natural means of seed dispersal, delayed seed germination and devices for protection and anchorage and gain in the size of edible parts of the plant favoured by humans. These syndromes include morphological traits like change over from prostrate to erect form, loss of seed dispersal, improvement of grain size, more determinate growth, and increased apical dominance, while physiological traits are reduction of seed dormancy, loss of seed shattering, amelioration of bitter substances in edible seed, attenuation of photoperiodic sensitivity, and synchronization of flowering (Civan and Brown 2017). These attributes have become appropriate for human consumption due to the occurrence of loss-of-function mutations in the genes responsible for these traits. The mutations of the genes mostly by artificial selection not only ensured rice evolution but also contributed immensely to the diffusion and spread of rice cultivation into various geographic locations of the world.

3.1.5.1 Intermediate Forms

There are two parallel series of rice domestication. The initial sequence involves the domestication of *O. sativa*, originating from *O. rufipogon*, while the subsequent sequence pertains to the emergence of *O. glaberrima* from its precursor, *O. breviligulata* (Oka 1988). Because of its immense contribution to the food security of mankind, the former has received greater attention from rice scientists for the exploration of evolutionary biology. In *rufipogon-sativa* and *breviligulata-glaberrima* series, intermediate wild-cultivated forms have been noted in areas where there is considerable accumulation of genetic diversity. One of the most prominently studied areas is the Jeypore tract in Odisha (India). Close sympatric association of the crop with wild progenitor and a primitive cultural condition maintained over a long period could have stabilized the ecosystem increasing the diversity of crop species and genesis of many intermediate forms. Thus, the intermediate forms are exclusive products of introgression between wild and cultivated rice. The panicle of an intergrade (C) derived from the natural cross-pollination of *O. rufipogon* (A) and traditional rice genotype *Meghadambaru* (B) has been depicted in the figure below (Fig. 3.3).

3.1.5.2 *Oryza rufipogon*

The shared perennial wild rice and the precursor to contemporary cultivated rice is *O. rufipogon*, previously referred to as *O. perennis* by Oka (1988). This taxon also includes *O. nivara*, i.e. very often considered as an annual form of *O. rufipogon* (Chang 1976; Matsuo and Hoshikawa 1993). *O. rufipogon* is a tufted grass that grows in shallow water habitats. It has a creeping and branched rhizome and the length of the rhizome varies according to the nature of the moisture regime of the

(A) (B) (C)

Fig. 3.3 Panicles of *Oryza rufipon* (**A**), traditional rice Meghadambaru (**B**) and Intergrade formed naturally through cross-pollination (**C**). (Source: Photo gallery of P. K. Senapati and E. Kariali)

habitat. Rhizome bears many nodal tillers far away from the main stalk. The plant propagates vegetatively when individual tillers gain independence from the main stalk. The ligule elongates to a certain length before splitting into two distinct parts at its sharp tip. The panicles exhibit an open and loosely arranged structure, bearing slender spikelets that drop off. Anthers are elongated and display a yellowish-white hue. Initially white, the stigma gradually changes to a purple colour as it matures. The grains are slender and possess long, barbed awns. At maturity, it takes on a dark purple colour and tends to shatter easily, falling to the ground. Similar to traditional rice, the plant is strongly photoperiodic (short photoperiod type). Essential differences from cultivated rice are the tendency for the mode of cross-pollination for fertilization, mature seed to shatter from panicle to ground and their prolonged seed dormancy, which inhibits germination in saturated soil. These are features preferred in natural selection for easy and wide dispersion maintaining wide genetic diversity (Vaughan et al. 2003). The presence of growth inhibitors in the hull is responsible for seed dormancy (Barik et al. 2017). The seeds can retain dormancy and viability for up to 3 years under field conditions. Seeds fall near the parent plant anchoring to moist soil with the help of the awn and remain dormant until the next seasonal cycle for germination and reproduction. Rice seeds can disperse over longer distances through various means, including contamination with other seeds and due to human activities, as well as through water and soil movement. The plant is highly tolerant to flooding and also to acid soils.

3.1.5.3 *Oryza nivara*

The adaptive changes that occurred during the evolutionary transition from the perennial *O. rufipogon* living in persistently marshy habitat to the annual form

O. nivara, growing at the edge of wetlands or seasonally dry habitat (Vaughan and Morishima 2003), and then to the cultivated annual rice *O. sativa* was reflected in the changeover of various morphological features. As surmised, both the wild species evolved in areas characterized by different soil moisture regimes and *O. nivara* evolved from *O. rufipogon* after a habitat shift (Sang and Ge 2007). Despite their ecological distinctions, both species share comparable morphological traits and display a relatively limited level of genetic divergence. Frequently, *O. nivara* is considered a sub-species or ecotype of *O. rufipogon*, a classification noted by Vaughan and Morishima (2003) and Bautista et al. (2006). Being an intermediary, *Oryza nivara* shares some characteristics unique to its ancestor and descendant. Unlike its ancestor *O. rufipogon*, *O. nivara* is an annual, short to intermediate-height grass-bearing compact panicles with large size spikelets; the traits are almost similar in cultivated rice. However, grain shattering and possession of barbed awn are features retained suggesting descendants from *O. rufipogon*. Both the wild species belong to the AA genome group and are cross compatible and they grow in close sympatric association with *O. sativa*. The association could have allowed gene flow towards the latter species with the introgression of key domestication alleles (Kovach et al. 2007). According to Sweeny and McCouch (2007), the genetic loci cohabiting with the new alleles selected from the wild species appear fixed within the domesticated rice, but they vary in wild species. Many of the domesticated traits are polygenic. Domesticated phenotypes are not favoured by natural selection. Although stable in wild species, the traits do not survive long without artificial selection in domesticated genotypes. The authors defined domestication traits as those chosen by humans; they occur at frequencies higher in cultivated than wild species and negatively affect plants' ability to complete the life cycle bereft of human assistance. Numerous phenotypic differences exist between wild and cultivated rice species, and the mechanism by which these traits transition from the former to the latter is elaborated upon below (Fig. 3.4).

3.1.6 Types of Domestication Syndromes

Liu et al. (2018) in a review categorized domestication syndrome genes into two types based on their distribution and expression patterns in cultivated versus wild rice. In the first step, genes of type 1 regulate prostrate growth, seed shattering, and seed dormancy and are the results of a sudden loss of function mutation, which permits better growth and higher yield. On the other hand, during the second step, type II syndrome genes possess potent functional alleles in cultivated rice, inherited through introgression from their wild ancestors, while weaker or non-functional alleles emerge within cultivars. This applies to genes related to traits like grain size and flowering time. Sudden loss of function mutations in prostrate-erect transition, attenuation of dormancy or loss of seed shattering makes the genotype amenable for domestication, while in the case of adaptive traits that govern the geographic expansion of rice, like photoperiodic sensitivity, the evolutionary process involves a

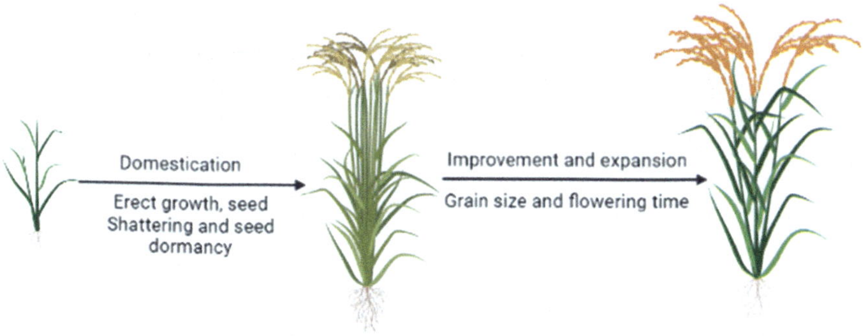

Fig. 3.4 Two-step transition of wild to modern rice

gradual transition from strong alleles to weak alleles and eventually to non-functional alleles due to the accumulation of loss-of-function mutations (Fig. 3.4).

Gross and Zhang (2014) with support of archaeological and genetic evidence suggested the chronology of rice domestication occurred 10,000 years BP (Before Present); the possible places of geographic origin were the Gangetic River basin and southern slopes of the Himalayas, India harbouring rich biodiversity of *Oryza* species and Yangtze River basin of south China. In the Yangtze River basin rice, preferably *O. sativa,* and *japonica* was cultivated as early as 8000 BP (Before Present), but the loss of grain-shattering trait was not fixed until 1000 years later. Gross and Zhang (2014) proposed that domestication genes moved from *japonica* to *indica* sub-species through hybridization in a model given below (Fig. 3.5).

3.1.6.1 Transition from Prostrate to Erect Architecture

The shift from a prostrate (creeping) growth form to an erect (upright) form (Fig. 3.6) is a fundamental prerequisite for domestication and represents a crucial occurrence in the evolution of rice. This significant step played a crucial role in enhancing the plant's architecture, making it well-suited for cultivation and unlocking yield benefits. Typically, the nativity of wild rice *O. rufipogon* and to a lesser extent that of *O. nivara* are marshy habitats and a prostrate stolon growth during the vegetative period suits the purpose for easy propagation, which becomes irrelevant and unsuitable for crop cultivation. Erect tillers coming out of stolon nodes increase the population size, enlarge area coverage, repress the growth of weeds, and develop resistance to trampling and grazing by wild animals in their native environment. By contrast, *O. sativa* maintains an erect growth habit persistently in the entire growth period from seedling to maturity (Fig. 3.6).

Some workers (Jin et al. 2008; Tan et al. 2008; Wang and Li 2008) have identified a semi-dominant gene *PROG1* (*PROSTRATE GROWTH1*) responsible for the prostrate growth of wild rice and the genetic locus resides on chromosome 7 and encodes a solitary Cys2-His2 zinc-finger transcription factor protein. Loss of function mutation identified in the *prog1* variant of *O. sativa* inactivates the gene expression leading to erect growth and an increase in grain number and yield. Tan et al. (2008)

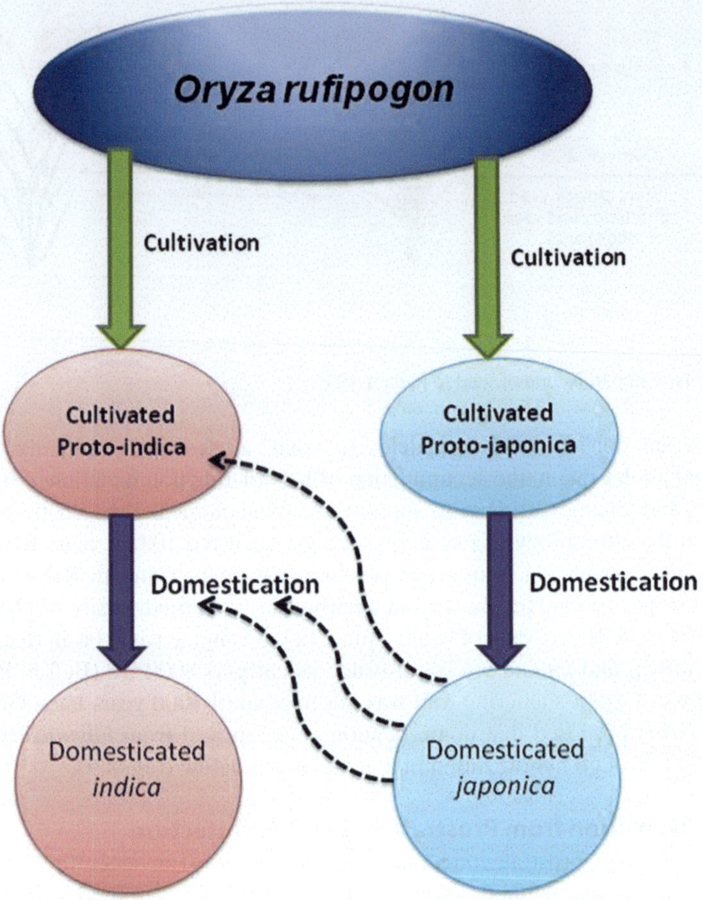

Fig. 3.5 This figure illustrates possible outcomes for the development of *japonica* and *indica* subspecies from a genetically diverse *Oryza rufipogon* population, with different colours representing genetic differentiation. The dotted lines depict the plausible timeline for the transmission of domestication alleles from *japonica* to *indica* via hybridization. Within the ultimate domesticated *indica* circle, the colours (light red) represent inputs from the original *O. rufipogon* populations (dark blue) and the incorporation from *japonica* (light blue). (Adapted from Gross and Zhang 2014)

found an identical mutation in the coding region of *prog1* in a wide range of *indica* and *japonica* rice varieties collected from 17 countries to conclude its fixation in the course of domestication. In wild rice, *PROG1* produces a wide tiller angle favouring prostrate stolon growth, whereas the mutated allele *prog1* of cultivated rice results in a small tiller angle facilitating erect stem growth. Wu et al. (2018) discovered a 110-kilobase (kb) deletion situated on the short arm of chromosome 7 in *O. sativa*. This deletion was near the *PROG1* gene, which includes a repeated cluster of seven zinc-finger genes. The deletions played a pivotal role in facilitating the crucial transition from the low-yield and prostrate growth architecture of *O. rufipogon*, leading

Fig. 3.6 The transition from a prostrate growth habit in *Oryza rufipogon* (**A**) to a semi-erect form in *O. nivara* (**B**) and finally to a fully erect form in *O. sativa* (**C**) was observed under simulated conditions in cemented pots at the Botanical Garden of the School of Life Sciences, Sambalpur University. (Source: Photo gallery of E. Kariali)

to the erect growth and high-yield traits of *O. sativa* (Fig. 3.7). The particular site on the rice chromosome linked to the domestication of rice plant architecture is designated as '*RICE PLANT ARCHITECTURE DOMESTICATION*', abbreviated as *RPAD*. A similar deletion of 113-kb was also noted in African cultivated rice. It is suggested that the removal of these tandem repeats may have played a role in the parallel evolution of both African and Asian domesticated rice, where *PROG7 (PROSTRATE GROWTH 7)* retained at the *RPAD* locus, instead of the *PROG1* gene of Asian rice, plays a critical role for prostrate growth (Hu et al. 2018).

In a recent article, the molecular biology of prostrate-erect transition has been advanced further by Xu and Sun (2021). In the domestication of Asian rice, it is hypothesized that three crucial loci, namely *RPAD*, *TIG1*, and *TAC1*, have vital functions in the shift from a sprawling growth structure to an upright architectural form. The regulation of *TAC1 (TILLER ANGLE CONTROL1)*, which is a major QTL for tiller angle in rice, is also important for prostrate-erect transition (Yu et al. 2007). The *TAC1* gene encodes a protein consisting of 259 amino acids, although its specific functions are not yet understood. In its wild form, the *TAC1* gene leads to a wide tiller angle, while the mutated *tac1* variant is responsible for a small tiller angle and more compact plant architecture. Post-domestication Asian *japonica* rice has the mutated allele *tac1*; in contrast the *indica* lines and wild ancestor rice harbour the *TAC1* allele (Xu and Sun 2021). Consequently, the preference for the *tac1* allele has resulted in the compact structure characteristic of *japonica* rice, while this feature is not as conspicuous in *indica* rice. *TIG1 (TILLER INCLINED GROWTH1)* gene also contributes to the compact architecture of *O. sativa*. This gene encodes a class II transcription factor responsible for regulating tiller growth by stimulating the expression of genes involved in cell expansion, including *EXPA3*, *EXPB5*, and *SUAR39* genes. The *indica* sub-species cultivars carry mutated *tig1*alleles showing reduced expression, while the wild rice and *japonica* rice harbour *TIG1* with greater expression. *TIG1* gene acts additively with *PROG1* and *TAC1*, indicating that these

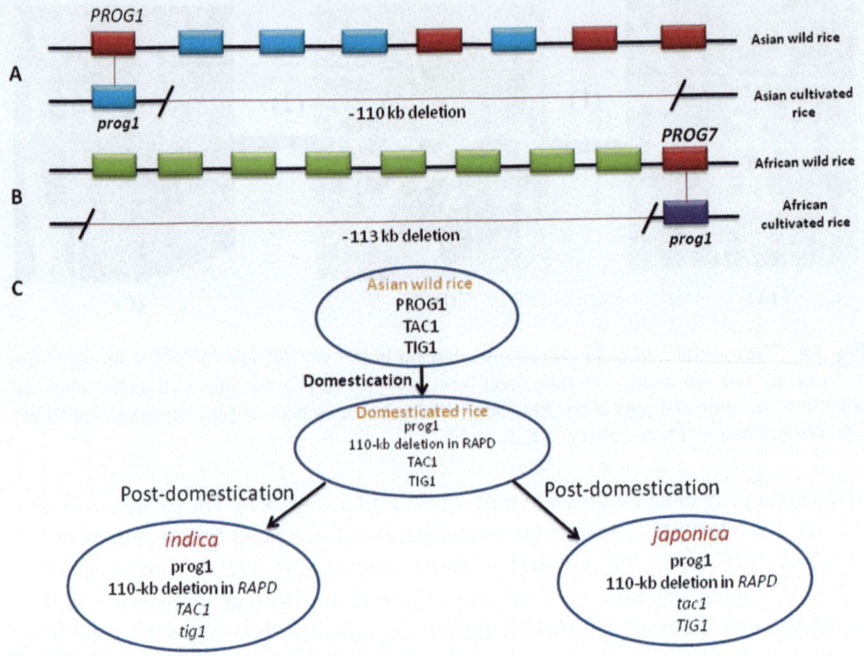

Fig. 3.7 The genetic shift from prostrate to upright growth in rice during domestication. (Fig. **A**, **B**) shows a parallel selection of the *RAPD* locus, with zinc-finger genes influencing the tiller angle. Red boxes-zinc-finger genes associated with tiller angle, purple box-weaker allele. Green boxes-non-functional zinc-finger genes/ones with unclear functions and Blue boxes represent non-functional zinc-finger genes. (**C**) The *prog1* allele, 110-kb *RAPD* locus deletion, and *tac1* allele collectively enhanced the upright growth and overall plant structure of Asian *japonica* rice. The *prog1* allele, 110-kb *RAPD* locus deletion, and *tig1* allele played a role in promoting erect growth and enhancing the overall plant architecture in Asian *indica* rice. (Adapted from Xu and Sun 2021)

three genes regulate tiller angle independently vide different mechanisms (Xu and Sun 2021) (Fig. 3.7).

3.1.6.2 Grain Shattering

Shattering of the rice grain is caused by seed abscission. This behaviour in wild rice reduces competition within the population, consequently enhancing the dispersal of offspring and overall species fitness in its natural environment. The presence of long awn in *O. rufipogon* and *O. nivara* (Fig. 3.8A, B) with large barbed panicles can fend off grazing animals, help seed dispersal and assist in planting the seeds whereas most of the cultivated rice lacks awn (Fig. 3.8C) with non-shattering attribute. Humidity changes while shattering on moist soil causes awns to flex and help anchor to the soil. Grain-shattering attribute, however, is an unfavourable trait for cultivated rice because of its detrimental impact on harvestable produce. Generally, easy-shattering wild rice forms a complete abscission layer, but the layer is formed partially or absent altogether in cultivated rice. In wild rice, an abscission layer

Fig. 3.8 Panicles with flag leaf of *Oryza rufipogon* (**A**), *O. nivara* (**B**) and high-yielding rice (**C**). (Photo gallery E. Kariali)

comprising thin-walled cells exists at the junction between the seed base and pedicel (Li et al. 2006) and it degrades leading to detachment and dropping of grains to the ground with minimal physical disturbance. It was reported that in most of the red rice ecotypes, seed shattering starts as early as 14 days after anthesis when the seed moisture content was about 25% (Noldin et al. 1999). In weedy rice, an abscission layer develops before flowering, but it degrades as flowering occurs. In contrast, in wild rice, specifically *O. rufipogon*, the degradation of this abscission layer is delayed until and after flowering, as documented by Thurber et al. (2011). Because non-shattering attribute is a major pre-requisite in the rice breeding system, genes responsible for the process have been the subject of intensive study. Several genes account for the loss of grain shattering in cultivated rice.

Genes Controlling Grain Shattering

SH4 (*Shattering4*)/*SHA1* (*Seed shattering1*) is a major QTL of large effect responsible for grain shattering in rice. It involves the encoding of a protein that possesses a DNA-binding domain of the MYB type at its N-terminus, along with a nuclear localization signal at its C-terminus. Xu and Sun (2021) reported that a mutation occurring at the 79th amino acid position within the MYB domain of *SH4/SHA1*, where lysine is substituted with asparagine, has been identified as the underlying cause of the absence of grain shattering. However, Liu et al. (2018) expressed a different type of genealogy for cultivated rice. In their study, they identified the presence of *SH4* in an F2 population that originated from a cross between an *indica* cultivar and *O. nivara*. Within this population, a specific nucleotide substitution, guanine/thymine (G/T), was observed in the first exon of *SH4*. This genetic change led to the emergence of a mutated allele, which was present in all domesticated cultivars, in contrast to the non-shattering allele found in wild rice. Regardless of the evolutionary lineage, this substitution had a detrimental impact on the gene's functionality, halting the normal creation of an abscission layer, which usually separates the grain from the pedicel naturally. This observation was made in a study conducted by Li et al. (2006). According to Liu et al. (2018), *SH4* encodes an MYB transcription factor, while *qSH1* contributes to the synthesis of a BELL-like homeodomain protein, and the presence of mutated alleles *qsh1*and *sh4* results in the loss of seed shattering because it leads to the absence of an abscission layer. Zhou et al. (2012) identified the *SH4* gene's position on chromosome 4, whereas the quantitative trait locus for *qSH1* was found on chromosome 1. They also detected a mutant variant called *seed-shattering abortion1* (*shat1*) in wild rice.

The *SHAT1* gene encodes an APETALA2-type transcription factor, which holds significant importance in regulating seed shattering by controlling the abscission zone's development at the point where the seed and pedicel meet. The expression of this gene is controlled by the tri-helix transcription factor SH4 protein. Within the signal transduction pathway, *qSH1* operates downstream of *SH4* and *SHAT1*. It plays a role in creating a positive feedback loop with these genes by ensuring the continual expression of *SH4* and *SHAT1* within the abscission zone, as described in the study by Zhou et al. (2012). Persistent and concentrated *SHAT1* and *SH4* expression in early spikelet development is essential for abscission zone identity. The mutated form of *sh4* represents a single ancestral allele, and it has become prevalent and stable in numerous cultivars of *O. sativa*, throughout the process of domestication. A parallel process of domestication was followed in African rice domestication in the *SH4/SHA1/GL4* locus, where the non-shattering allele of *GL4* was *gl4* (Xu and Sun 2021). There are some other genes identified controlling the shattering behaviour of rice. As reported by Lin et al. (2012), the *OsSH1* gene, which codes for a YABBY transcription factor, represents a minor-effect Quantitative Trait Locus (QTL) accountable for the differences in grain shattering observed between wild and cultivated rice varieties.

Three QTLs, namely *qSH1, qSH3*, and *qSH4*, have been identified as contributors to the reduction in grain shattering, as documented by Onishi et al. (2007). It's worth mentioning that the existence of a non-shattering allele at *qSH1* has a substantial impact on the shattering characteristics, causing a shift to a non-shattering phenotype, irrespective of the presence of shattering alleles at *qSH3* or other genetic loci. Fine mapping and sequence analyses revealed similarity between *qSH1* and *qSH4*. Compared with the distinct effect of *qSH1* and *qSH4* on the reduction of grain shattering, *qSH3 is* another QTL that exhibits a reduced expression in the shattering of grains in *O. sativa*. The grain-shattering behaviour of panicle in wild rice *O. rufipogon* is depicted in Fig. 3.9.

Fig. 3.9 Shattered panicle of *Oryza rufipogon*. (Source: Photo gallery of P. K. Senapati and E. Kariali)

3.1.6.3 Individual Genes for Grain Shattering

qSH1

There are some other genes identified controlling the shattering behaviour of rice. In the study conducted by Lin et al. (2012), it was found that the *OsSH1* (*O. sativa shattering1*) gene, encoding a YABBY transcription factor, a minor-effect Quantitative Trait Locus (QTL), plays a role in the variation of grain shattering between Asian wild and cultivated rice. *ObSH3*, sharing allelic similarity with *OsSH1* plays a role in the domestication of African rice by contributing to the loss of grain shattering. Three QTLs, namely *qSH1*, *qSH3*, and *qSH4*, have been identified as responsible for reducing grain shattering, as documented by Onishi et al. (2007). A non-shattering allele at *qSH1* is a major effect QTL controlling grain shattering in *O. sativa* by drastically changing shattering to non-shattering phenotype irrespective of the presence of shattering allele at *qSH3* and *qSH4* loci. Fine mapping and sequence analyses revealed similarity between *qSH1* and *qSH4*. Compared with the distinct effects of *qSH1* and *qSH4* on the reduction of grain shattering, *qSH3* exhibits reduced expression. The *qSH1* gene contains the genetic information for a BEL1-like homeobox protein. A single nucleotide change in the gene's 5′

regulatory region leads to reduced expression, resulting in the formation of an abnormal abscission layer and ultimately decreasing grain shattering. Similarly, another gene called *SH5* also codes for a BEL-1-like homeobox protein and is associated with shattering. As described by Yoon et al. (2014), it triggers seed shattering by stimulating the development of the abscission zone and inhibiting the biosynthesis of lignin. While *qSH1* primarily plays a role in cell differentiation within the abscission zone, *SH5*'s major function is to suppress lignin deposition. Yoon et al. (2017) discovered that the KNOX protein gene *OSH15* also regulates seed shattering. *OSH15* and *SH5* form a dimer that directly inhibits lignin biosynthesis genes, thereby enhancing seed shattering.

SNB

Jiang et al. (2019) discovered a shattering-resistant mutant known as '*suppression of shattering 1*' (*ssh1*) by screening wild rice introgression lines that naturally exhibited strong shattering tendencies after being subjected to mutagenesis. A single nucleotide change in *SUPERNUMERARY BRACT 1* (*SNB1*), which encodes a plant-specific transcription factor closely resembling the APETALA2-like factor encoded by the *SHAT1* gene, cause the transition from easy shattering in the wild plant to no shattering in the mutants. SNB protein binds with the promoter of *qSH1* and *SH5* for enhancement of their expression in modulating seed shattering and suppressing lignin deposition in the abscission layer. The mutated allele introgressed into modern high-yielding rice increases grain size and weight. A model summarizing genetic control of seed shattering in rice has been given below adapted from Xu and Sun (2021) (Fig. 3.10).

3.1.6.4 Loss of Awn Length

Awn is an important domestication and agronomic trait of rice plants. Awn of the rice grain is either a hair or spicule-like appendage that extends from the tip of the lemma in the floret. Awn length in rice is species-specific. One of the standout characteristics of wild rice species is the existence of elongated and bristled awns. A wild rice seed with a 7 cm long awn is highly likely to land in an upright position, with the embryo buried in muddy soil. This adaptation offers a survival advantage and facilitates self-propagation in its natural environment. Further, the awns are beneficial for wind and animal-mediated dispersal. Seeds of most of the cultivated rice cultivars are awnless or have short awns. In Asian rice, the absence of awns helps ease the process of harvesting and seed storage while also boosting grain quantity and overall yield, as noted in a study by Bessho-Uehara et al. (2021). The absence of awns also improves bulk density promoting dense packaging of seeds during storage. Awn also influences the aerodynamic properties of rice grain; the terminal velocity of the awnless rice grain is 2.5% slower than the awned grains (Masoomi et al. 2020).

The presence or absence of awns in rice is a crucial domestication trait that is closely linked to grain yield characteristics. Awn being unfavourable for harvesting and storage, was partially or eliminated by artificial selection in the transition from wild to cultivated rice (Fig. 3.11). The process of awn development is very complex

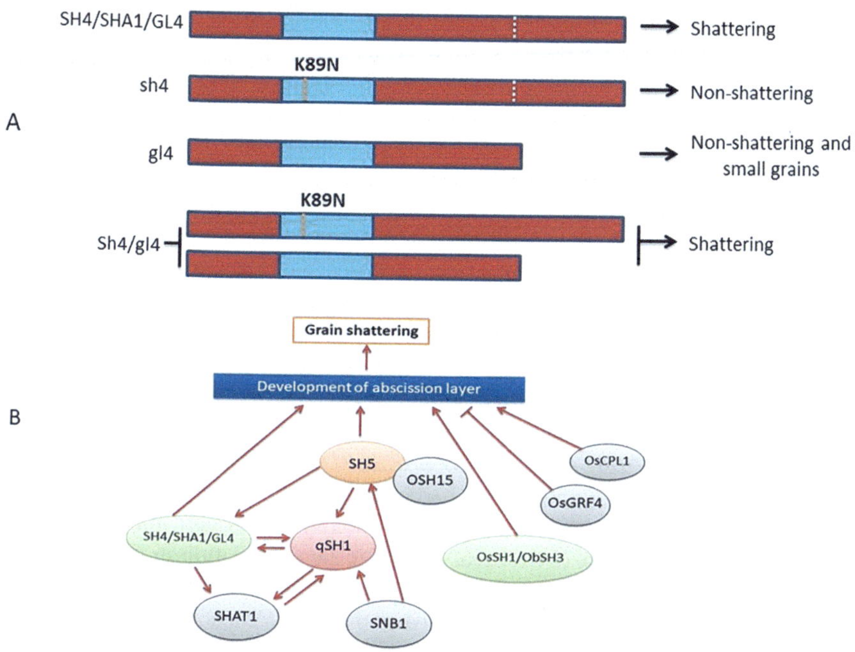

Fig. 3.10 The genetic mechanisms responsible for the reduction of seed shattering during rice domestication. (**A**) This figure depicts the parallel selection of the *SH4/SHA1/GL4*. Light blue boxes—MYB-type DNA-binding domain and white lines—C-terminal nuclear localization signal. (**B**) The genetic mechanisms of grain shattering. Light green—*OsSH1/ObSH3* and *SH4/SHA1/GL4*. The non-shattering allele of *qSH1* (in peach) was chosen in Asian *japonica* rice, and *SH5* (in orange) likely reduces grain shattering in African rice. *SH4/SHA1/GL4* has the potential to enhance *SHAT1* expression, and together with *SH4/SHA1/GL4*, it plays a role in sustaining *qSH1* expression within the abscission layer. *SH5* also contributes to the promotion of *SHAT1* and *SH4/SHA1/GL4*. *SNB* promotes *qSH1* and *SH5* expression, and *OSH15* regulates grain shattering through interaction with *SH5*. Arrowed lines represent 'promotion', while T-shaped lines indicate 'inhibition'. (Adapted from Xu and Sun 2021)

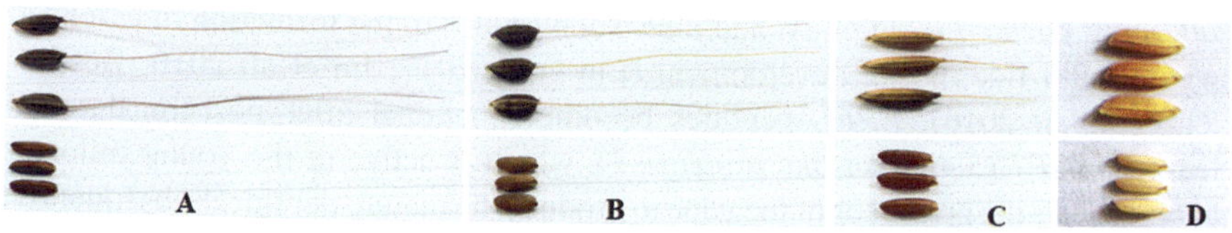

Fig. 3.11 Gradual reduction of awn length from wild rice to cultivated rice. Grains with awns and their endosperm in *Oryza rufipogon* (**A**), grains with awns and their endosperm in *O. nivara* (**B**), grains with small awns and their endosperm in *O. sativa* (**C**) and grains with no awns and their endosperm in *O. sativa* (**D**) (Source: Photo gallery of P.K. Senapati and E. Kariali, Sambalpur University)

and governed by many genes in wild rice. Numerous QTLs associated with the loss of awns during domestication have been identified in various studies, including those by Xiong et al. (1999), Jin et al. (2009), and Furuta et al. (2015), but only three of them, such as *An1*, *An2/LABA1* and *An3/GAD1/RAE2*, have been cloned and characterized as summarized by Qin et al. (2021).

An1 (*Awn1*) is a significant QTL that codes for a bHLH (basic-helix-loop-helix) transcription factor. It has a positive impact on awn size but simultaneously exerts a negative influence on grain number in rice, as demonstrated in the study by Luo et al. (2013). The gene exhibits heightened expression initially in the tip of lemma primordia and then in the subsequent development of awn primordia. This overexpression promotes continuous cell division, leading to the development of a long awn in wild rice. In cultivated Asian rice, the absence of awns is caused by either a transposon insertion in the promoter region or a 1-base pair deletion in the coding sequence of the *An1* gene. Compared with wild rice nucleotide diversity of *An1* is considerably poorer in cultivated rice indicating strong artificial selection of the locus in the course of domestication. A mutated allele of *AN1 not* only shortens awn length but also leads to an increase in grain number and yield.

Similar to *AN1*, *LABA1 (LONG AND BARBERD AWN1/AN2* (*Awn2*) is a domestication gene associated with the long-barbed awn in wild rice (Hua et al. 2015; Gu et al. 2015). *An2/LABA1*encodes a cytokinin biosynthesis activating enzyme, a homologue of *Lonely Guy* (*LOG*) that promotes awn elongation in rice but, at the same time, reduces grain production. In *O. sativa*, a single-base pair deletion within the coding sequence of *An2/LABA1* abolishes the gene's capacity to produce cytokinins, thereby hindering the elongation of awns and the formation of barbs. The researchers (Gu et al. 2015; Hua et al. 2015) showed that the *An1* gene precedes the expression of *An2/LABA1* in the course of awn development, which implies that the former is involved in an initiation while the latter promotes awn elongation. Nucleotide diversity analysis has revealed considerable artificial selection of the gene during domestication parallel to *An1*.

The selection for the mutated allele of *AN2* in cultivated rice not only reduces awn length but also increases grain yield by improving grain and tiller numbers of the plant. *An-3/GAD1/RAE2/GLA* (*Awn-3/Grain Number, Grain Length and Awn Development1/Regulator of Awn Elongation 2/Grain Length and Awn Development*) serves as the third domestication gene that controls awn development in wild rice. This gene contains a small signal peptide that falls under the *EPFL* (*Epidermal Patterning Factor-Like*) family, and it also contributes to the formation of extended awns in wild rice during development (Liu et al. 2017; Jin et al. 2016; Bessho-Uehara et al. 2016). *EPFL* peptides become functional after post-translational cleavage. *SLP1* (*Subtilisin-like protease 1*), which is active in the young inflorescence, cleaves the precursor of the gene to produce the mature peptide. Subsequently, this mature *GAD1/REF2/GLA* peptide promotes the extension of awns in wild rice. It's noteworthy that *EPFL*-family peptides possess a conserved region characterized by six cysteine residues. In the context of *GAD1/REF2/GLA* in cultivated rice, there is a frameshift deletion within a GC-rich region, which interrupts the protein's C-terminal sequence and consequently leads to the absence of two out of the six conserved cysteine residues. This alteration leads to shortened awns. Additionally, it plays a role in determining both grain size and grain number. A frameshift deletion in the coding region of *GLA* does not affect grain size. However, when a similar

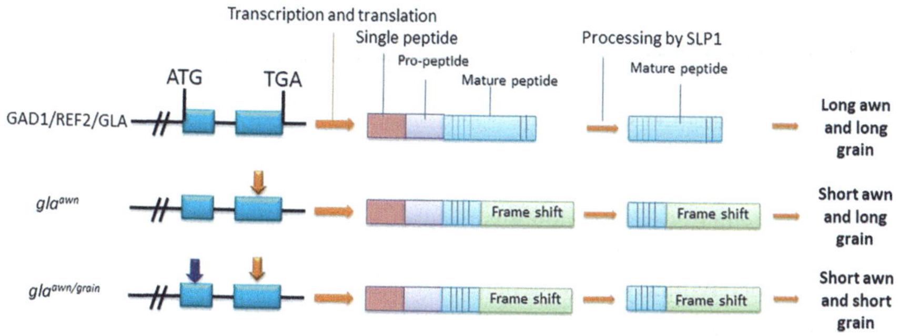

Fig. 3.12 *GAD1/REF2/GLA* controls the length of both awns and grains. *GAD1/REF2/GLA* basically encodes a short peptide of EPFL-family whose precursor is made up of a signal peptide, a pro-peptide and a mature peptide at C-terminal end. SPL1 is responsible for cleaving of the precursor of *GAD1/REF2/GLA* resulting into the mature peptide consisting of six conserved cysteine residues labelled with blue lines within the boxes. The gla^{awn} allele carries a frame-shift deletion (indicated by a yellow triangle). The $gla^{awn/grain}$ allele exhibits both a coding region frame-shift deletion and a promoter region deletion (indicated by a blue arrow), leading to both shortened awns and reduced grain size. (Adapted from Xu and Sun 2021)

deletion occurs in the promoter region, it reduces *GLA* expression and leads to shorter grains, without affecting awn development. Indeed, the mutated *GAD1/REF2/GLA* peptide with the disrupted C-terminal sequence retains its functionality in governing grain length but does not impact awn development. Conversely, the complete *GAD1/REF2/GLA* peptide is involved in the regulation of both grain length and awn development, as elucidated in the research conducted by Xu and Sun (2021) and described in Fig. 3.12.

An4

Recent research, using chromosome substitution lines from wild and *indica* rice, has unveiled *An4* as a potential QTL governing awn development in rice, as reported by Qin et al. (2021). *An4* has been precisely located within a 56-kb region on chromosome 2. This QTL plays a significant role in enhancing awn development through an additive effect, in conjunction with other awn development genes like *An-1*, *An-2/LABA1* and *An-3/GAD1/RAE2*.

3.1.6.5 Change of Grain Colour

A distinct contrast in grain hull and pericarp colour is evident between wild and cultivated rice. Grains of wild rice generally have a black hull and a red pericarp, whereas cultivated rice grains are characterized by straw-white hulls and white pericarps. Genetic mechanisms causing the shift from grain with the black hull in wild rice to straw-white in cultivated rice in the course of rice domestication have been pinpointed by researchers. Sweeny and McCouch (2007) reported cloning of the gene for pericarp colour. In the research conducted by Zhu et al. (2011), it was identified that the black hull trait in *O. rufipogon* is governed by the *Black hull 4* (*Bh4*) gene, pinpointed precisely within an 8.8 kb region on chromosome 4 of rice. Deletion of 22-base pairs in exon 3 of the *Bh4* variant led to the loss of *Bh4*'s functionality, consequently giving rise to the pale, straw-white hull colour observed in cultivated rice. Zhu et al. (2011) found nucleotide diversity 70 times higher compared to cultivated rice and attributed the reduction of nucleotide diversity of the latter to a domestication syndrome.

3.1.6.6 Reduction of Grain Dormancy

Seed dormancy in rice is agriculturally useful as it prevents the grain from sprouting in the field when weather flickering brings unseasonal rains in the pre-maturity period. Wild rice typically growing in marshy habitats does not germinate for a year after matured grains shatter to the ground because of deep dormancy. This characteristic of desynchronized germination is advantageous for survival and propagation in the natural environment. However, in cultivated rice, weak dormancy is preferred as it helps achieve consistent and uniform germination, which is essential for crop establishment in the field. Too weak seed dormancy of modern rice cultivars, subjected to erratic rainfall, has very often led to precocious seed germination impairing grain quality and yield. Loss of seed dormancy in cultivated rice has been a component of domestication syndrome. Some domesticated genes of rice have been identified (Xu and Sun 2021).

OsG overexpression increases, while its knockout reduces grain dormancy in rice. The gene is located within a selective region. In cultivated rice, there exists a single nucleotide polymorphism (SNP) within the conserved C-terminal region of *OsG*. This SNP is fixed, and the mutated allele, *OsG*, is responsible for the loss of dormancy. The selection for the *OsG* allele has played a role in reducing the dormancy of cultivated rice. Additionally, Sugimoto et al. (2010) successfully identified and cloned a QTL called *Sdr4*, which contributes to variations in seed dormancy between *indica* and *japonica* rice cultivars. The absence of functional *Sdr4* results in the total absence of seed dormancy. The activity of *Sdr4* is promoted by *OsVP1* (*O. sativa viviparous1*). *OsVP1* governs potential seed dormancy regulators while simultaneously repressing the expression of genes linked to post-germination processes. The expression of *Sdr4* significantly decreases in *OsVP1* mutants, suggesting that *Sdr4* functions as an intermediate regulator of seed dormancy within the seed maturation program. Recent findings indicate that *OsVP1* activates the expression of *Sdr4* to modulate rice seed dormancy through the abscisic acid (ABA) signalling pathway.

3.1.6.7 Transition from Open to Compact Panicle Architecture

The arrangement of flowers on a rice plant plays a crucial role in influencing its grain production, which is an important agricultural characteristic. Hence, it has been a primary focus of efforts to domesticate and enhance rice yield. The wild rice species, *O. rufipogon*, showcases characteristic spread panicle architecture. This architecture is well-suited for cross-pollination, efficient seed dispersal when mature, and provides a natural defence against pathogen infections (Grillo et al. 2009; Vaughan et al. 2003). In its natural ecological setting, these traits are advantageous for preserving a wide range of genetic variations and enabling adaptation to specific niches. Conversely, phenotypic attributes of cultivated rice are sharply different, particularly in the morphology of the panicle for making the organ more

Fig. 3.13 Change of open panicle type from *Oryza rufipon* (**A**) to increase compactness in *O. nivara* (**B**), traditional rice (**C**) and highly compact panicle in high-yielding rice (**D**). (Photo-gallery of E. Kariali)

amenable to yield benefit. The panicle of cultivated rice is compact, providing more space for grains and thereby increasing grain numbers. Mitra and Ganguli (1932) documented that the spread panicle trait in rice was governed by the interaction of two complementary dominant genes. Later, Zhu et al. (2013) identified the dominant gene *OsLG1* (*OsLIGULELESS1*) responsible for the spread panicle architecture of *O. rufipogon*. Previously, this gene was identified as a transcription factor with a significant function in forming the laminar joint at the boundary between the leaf blade and leaf sheath, consequently influencing the development of ligules and auricles (Lee et al. 2007). Through association analysis, Zhu et al. (2013) determined that a single-nucleotide polymorphism-6 (SNP-6) located 11 kb upstream of the translation initiation site in the cis-regulatory region had an impact on the expression of *OsLG1*; this led to changes in cell morphology within the panicle pulvinus regulatory region, ultimately culminating in the compact architecture of the panicle in cultivated rice. This type of mutation might have taken place in the course of domestication. The change of panicle architecture from lax to compact type during evolution has been described in Fig. 3.13.

3.1.7 Habitat for Cultivated Rice

Rice is a semi-aquatic plant. Tropical marshlands of Asia are the natural habitat for rice. Nonetheless, this crop is extensively grown across tropical, sub-tropical, and temperate regions. Coastal plains, river basins, and tidal deltas are preferred places

Fig. 3.14 Short duration upland rice cultivation. (Source: Photo gallery of E. Kariali)

for rice cultivation. In sharp contrast to other crop grasses like wheat, barley or maize, rice plant thrives under extremely moist conditions and moderate temperatures. For cultivation, the seeds are sown in prepared beds. Depending on the suitable environmental conditions, rice seedlings that are 25–30 days old are typically transplanted into paddy fields. The conventional approach to rice cultivation has involved the practice of flooding the fields either during or after the transplanting process. The paddy is enclosed by levees or bunds that remain submerged by 5–10 cm of water during most parts of the growing season. Water being the crucial factor, the success of rice cultivation depends on the inundation of fields to a depth of a few inches either by conservation of surface flowing rainwater or periodic irrigation for the greater part of the crop season. The plant also requires a long period of sunshine. Although moisture-friendly, flooding of the field is not always mandatory. All other methods of irrigation with lesser levels of soil saturation require greater efforts in weed and pest control and alternative approaches for soil fertilization. Often the plant is grown in the water-limited upland area as aerobic rice (Fig. 3.14), like other dryland cultivated cereals. Upland paddies in hilly areas are commonly grown in water-controlled terraced systems, and they have embankments for keeping fields flooded by rainwater at various elevations. In Asia, rice is cultivated in three primary types of soils, which include clays with a solid layer close to the surface, silt and clays with soft bottoms that harden upon drying, and lowland peat soils with a manageable depth.

3.1.8 Diffusion of Rice into Various Habitats

As discussed, rice growth and production are influenced by several factors, such as the environment, biotic conditions, soil properties, and cultural practices adopted by the farmers. Environmental factors depend on climate and include the amount of rainfall received, temperature, solar radiation, photoperiod, and, in many instances, coincidence of unpredictable tropical storms and cyclones. Biotic factors encompass fungi, weeds, insects, diseases, and various crop varieties. Soil factors, on the other hand, pertain to the types of soil and their respective locations or positions in the field. Depending on these factors rice cultural types have been classified into five categories.

3.1.8.1 Upland Rice

Upland rice accounts for only 5% of world rice production. Globally, approximately 20 million hectares of land are dedicated to upland rice cultivation, with roughly 60% of this acreage in Asia, 30% in Latin America, and the remaining 10% in Africa, as reported by Gupta and O'Toole (1986). Upland rice farmers are predominantly subsistence farmers who have limited alternative sources of food. Upland rice cultivars grown in Asia belong to *O. sativa* L. and some of them cultivated in Africa are derivative of *O. glaberrima* Steud. Nearly 100 million people depend on upland rice for food calories, mostly in countries of South Asia. Upland rice is cultivated in rainfed fields that are prepared and seeded when the soil is dry, similar to the cultivation practices of dryland crops such as wheat, barley or maize. The ecosystem ranges from fields that are level, steep or gently sloping and at altitudes up to 2000 m. The rainfall ranges between 1000 and 4500 mm annually. The soil in hilly terraces is weathered and infertile. The farmers use traditional landraces that do not respond to fertilizer management practices like their high-yielding counterparts. Indigenous upland cultivars were selected from nature through human selection based on the ability to compete with weeds, drought tolerance, blast resistance, and adaptation to poor soil fertility. Upland rice is characterized by its deep and robust root systems, which allow it to access water and nutrients from the relatively moist deeper layers of the soil, as indicated by Bernier et al. (2008). Additionally, it exhibits high water-use efficiency, as demonstrated by Liu et al. (2019), which contributes to the crop's stability and yield when cultivated in rainfed conditions.

The development of root traits for the improvement of water capture ability has been considered an important strategy for drought tolerance of upland rice in stress-prone environments (Fukai and Cooper 1995). Seminal and crown root development in the plant contributes effectively to the extraction of water (Luo and Zhang 2021). Studies suggest that an intricate network of regulatory pathways plays a significant role in shaping the development of larger xylem vessels, the formation of smaller aerenchyma, and an increase in non-structural carbohydrate content. These traits are essential for maintaining water conductivity and enhancing the water extraction capacity of upland rice from deeper soil layers (Yambao et al. 1992; Singh et al. 2013). The process of selection was carried out by local ethnic communities residing in proximity to the cultivation fields. As highlighted by Xiongshiyee et al. (2018), their main objectives were to boost crop yield, enhance cooking and taste qualities, and secure a reliable food source for household consumption. Besides practices of soil puddling, transplanting and continuous flooding are at risk in high water scarcity agro-ecosystems and mandate preference for indigenous rice in upland areas. Low soil fertility and poor water supply are the major constraints for upland rice production. Compared to lowland high-yielding rice, upland rice has low yield potential; the average yield comes down to 0.9 t/ha in Northern Thailand (Karladee et al. 2012). Modern semi-dwarf rice bred for irrigated land is not preferred for adoption in traditional upland areas due to constraints in crop management. However, lowland cultivars grown in upland systems with appropriate irrigation and soil fertilization can furnish better yield than indigenous upland rice (Liu et al. 2019).

Biodiversity of Upland Rice

Although disadvantaged in yield, upland rice has a rich biodiversity and traits of interest for human selection are many. Various traits related to functional properties have been pinpointed, including traits associated with more anthocyanin content (Sutharut and Sudarat 2012), elevated γ-Aminobutyric acid (GABA) levels (Karladee and Suriyonga 2012), phenolic compounds (Vichapong et al. 2010), and high ferulic acid content (Tian et al. 2004; Wanyo et al. 2014).

3.1.8.2 Lowland Rainfed Rice

Drought- and flood-prone favours medium-depth soil submergence in a lowland rainfed rice ecosystem. The primary and distinctive characteristics of rainfed lowland rice cultivation are as follows: (1) The crop is not irrigated, relying solely on natural rainfall. (2) The soil surface of the crop is intentionally flooded for a brief period at certain stages of the crop's growth cycle, differentiating it from upland rice cultivation. (3) The depth of flooding does not surpass 50 centimetres, signifying that it does not entail deep-water conditions (as per IRRI 1984).

According to Mackill et al. (1996), rainfed lowland rice refers to rice that is transplanted and cultivated in fields that have been levelled and bunded. These fields are intentionally shallowly flooded with rainwater during the rice cultivation process. Lowland rainfed rice cultivation is the major agricultural activity in the rice-growing belts from India to Cambodia. More than 50% of the land area in countries like Bhutan, Thailand, Nepal, Bangladesh, and Myanmar and almost 40% of rice lands in India, Cambodia, Laos, and the Philippines follow lowland rainfed rice cultivation practices (Mackill et al. 1996). In 1991, the region allocated to this specific type of rice ecosystem in Asia encompassed roughly 36.4 million hectares, with an average grain yield of 2.30 metric tons per hectare, as reported by IRRI (1997). This constituted about 25% of the total rice cultivation area. In contrast, irrigated rice areas held the largest share globally, accounting for 54% of the cultivated rice area. Rainfall is the major environmental factor determining grain yield in rainfed lowland rice. Additionally, topography, soil fertility, and selection of appropriate rice genotypes in these ecosystems also affect the grain yield considerably.

Zeigler and Puckridge (1995) suggested that rainfed lowland rice cultivation is well-suited for slightly sloping bunded fields with non-continuous flooding, where the depth and duration of flooding can vary. In rice fields located on upper terraces, a significant portion of water is lost, particularly after heavy rainfall, as it runs off the surface towards the lower sides of the field. In addition to the field conditions,

Fig. 3.15 Photograph showing cultivation of lowland rainfed rice in natural conditions from Jamunadhipa village of Sundargarh district in Odisha, India. (Source: Photo gallery of E. Kariali)

changes in rainfall patterns, and the use of specific rice genotypes are also important yield determinants on a year-to-year basis. Thus, the selection of a suitable rice cultivar for its cultivation in this type of ecosystem is a major challenge and short duration cultivar of modern high-yielding photo-insensitive type with high yield potential is the suitable option to ensure a good crop yield even under drastic fluctuating weather conditions as it has been followed in Cambodia and Laos (Wade et al. 1999). A photograph showing the cultivation of lowland rainfed rice naturally in the Jamunadhipa village of Odisha is given in Fig. 3.15.

3.1.8.3 Lowland Irrigated Rice

Rainfed lowland rice is cultivated in both dry and wet seasons, and this choice depends on the nature of the available irrigation water. Rice crops cultivated on the fields that are either flooded or irrigated, are generally included under lowland irrigated rice. Though the lowland rainfed rice fields are subjected to drying of the habitats due to less rainfall, lowland irrigated rice fields are usually located in the low-lying areas where plants are naturally or artificially irrigated and plants are often flood-prone. The crop is cultivated in bunded fields and artificial water supply is the only source of water in the dry time whereas natural rainfall is the additional source of water during the wet season. In the Indian context, the primary rice cropping systems are rice-rice, rice-rice-rice, and rice-wheat. These systems are prevalent, and the average rice yield typically falls within the range of about 4–5 metric tons per hectare (t/ha^{-1}). Generally, the plants in this type of ecosystem belong to late maturation groups and have good cooking quality. If the plants are transplanted in straight lines or rows (Fig. 3.16), weeding can easily be done by a rotary weeder and susceptibility towards lodging in this type of rice culture is minimal. Yield instability, habitat degradation due to imbalanced use of nutrients and poor water management practices are the major problems in this type of rice cropping system causing a reduction in grain yield.

Fig. 3.16 Lowland irrigated rice at Regional Research and Technology Transfer Station, Chipilima. (Source: Photo gallery of P. K. Senapati and E. Kariali)

3.1.8.4 Deep-Water or Floating Rice

Rice cultivars grown in ecological habitats where the water depth is more than 50 cm at least for 1 month is called deep-water rice. Deepwater rice is broadly categorized into two types such as (1) traditional tall cultivars and (2) floating rice. Traditional tall cultivars are generally grown when water depth is in the range of 50–100 cm whereas floating rice has the remarkable ability for quicker stem and leaf sheath elongation and prefers to grow in very deep-water habitats when the water level is more than 100 cm. The deep-water rice cultivars grow in a submerged condition where production of growth retarding hormone ethylene is accelerated. The hormone in turn triggers the expression of the *Sub1A* gene followed by the expression of downstream molecules encoded by *SLR1* and *SLRL1* genes leading to inhibition of stem elongation and the plants sustain the stress effects due to the utilization of storage carbohydrates in the vegetative structures. This strategy followed by the rice varieties is called the 'quiescent strategy'. On the other hand, floating rice cultivars can elongate their stem very fast and keep some of their leaves above the water level to carry out normal photosynthesis for efficient growth. The central part of the stem also contains a hollow space through which oxygen is transported to the submerged parts to perform aerobic respiration. This mechanism of submergence tolerance is called the 'Escape strategy'.

In this type of rice agroecosystem, plant hormones like ABA and GA play an important role. Submergence-induced ethylene production degrades endogenous ABA and also inhibits new ABA biosynthesis whereas it promotes GA biosynthesis in deep-water rice and regulation of balance between the two hormones is critical for stress tolerance. GA is responsible for cell division and cell elongation, especially in the intercalary meristem and the elongation zone of the internode and

Fig. 3.17 Deep-water floating rice grown in Hirakud dam reservoir and visit to plant site by boat. (Source: Photo gallery of P. K. Suna and E. Kariali)

causes stem elongation. In addition to stem elongation, ethylene also plays an important role in the adventitious root formation in the nodal portions of the stem. The hormone induces the killing of epidermal cells at the site of root emergence, helps in the penetration of the roots from the epidermis and promotes the growth of roots at the nodes. These roots help in the absorption of water and mineral nutrients from the surrounding environments, and due to this ability, often the plants can grow effectively as floating rice even if it is detached from the main root. In the present context, thousands of hectares of land are covered with water creating deep-water habitats, which make them agro-ecologically unproductive. However, with the advent of present modern knowledge in biotechnology and genetic engineering, several submergence tolerant deep-water rice cultivars have been developed and it is possible now to make these unutilized lands agro-ecologically more productive zones. Several traditional rice landraces are also grown in the deep-water habitat of the world-famous Hirakud dam reservoir, India by the local farmers (Fig. 3.17).

3.1.8.5 Coastal Wetland Rice

Coastal wetland ecosystems are characterized by high concentrations of salts, and cultivation of high-yielding rice cultivars is almost impossible in such a situation. This is because of the high concentration of salts in the crop environment, which in turn is the most important detrimental abiotic stress factor responsible for the reduction of both agricultural productivity and quality of produce especially in high-yielding rice cultivars. Though most of the high-yielding rice cultivars are highly sensitive to salt stress in the coastal wetland ecosystems, few cultivars exhibit salt tolerance, which is attributable to the accumulation of osmolytes, antioxidant defence and ROS detoxification, ion homeostasis and compartmentalization, salt exclusion and extrusion, and expression of many novel salt tolerant genes. According to Wang et al. (2012), several genes responsible for the transport of sodium from the roots to the shoots, such as *OsHKT1;1*, *OsHAK10*, and *OsHAK16*, show increased activity in older leaves of rice when compared to younger leaves. Conversely, genes responsible for excluding sodium ions (Na^{+}) from the root xylem vessels, like *OsHKT1,5*, and *OsSOS1*, exhibit reduced activity. This downregulation results in the accumulation of more sodium ions in the older leaves than in the younger ones, promoting improved plant growth. This adaptation can make rice suitable for cultivation in coastal wetland habitats. The achievements obtained by scientists relating to salt-tolerant rice cultivars at present are promising, but not sufficient quantitatively and both traditional and modern techniques are to be employed for the development of more novel salt-tolerant rice genotypes, which may be suitable for cultivation in the coastal wetland agro-ecosystems across the globe. This may be of immense help for the maintenance of livelihood for a considerable fraction of the human population and also for the improvement of the economic standard of the people worldwide.

3.1.9 Habit of Rice Plant

Cultivated rice, which belongs to the monocot plant group, is typically grown as an annual grass. The plant is very flexible in adaptation and grows well under both flooded and rainfed conditions. In tropical lowland fields maintaining soil saturation around the year, the plant survives as perennial and can produce ratoon crops for up to several years. The plant has vegetative organs: roots, stem, tillers, and leaves, and reproductive organs comprising a terminal determinate panicle made up of spikelets (Fig. 3.19).

3.1.9.1 Vegetative Organs

Roots

The root system of rice is relatively shallow, especially under soil submergence. Approximately, 95% of the roots of the rice plant are typically located in the uppermost 0.2 m of soil. The plant develops a fibrous root system that includes seminal roots, lateral roots, and nodal roots. Root characteristics are given below in bulleted format.

(a) After germination, the radicle pierces through the coleorhizae to form the seminal root.
(b) Lateral roots originate from the seminal root of the rice plant.
(c) As seedlings grow and stem nodes are formed, adventitious roots arise from the basal nodes. Most of the roots are damaged in transplantation, but new roots are formed sometime after in the new environment.
(d) Root growth is optimum in clayey soils. Organic and phosphatic manures promote root growth.

(A) (B)

Fig. 3.18 Fibrous root system in irrigated rice (**A**) and nodal roots in deep-water rice (**B**). (Source: Photo gallery of P. K. Suna and E. Kariali)

(e) In deep-water rice, nodal roots develop from higher nodes of the stem and they absorb nutrients from the surrounding water.
(f) Root growth is promoted in the surface soil layers under submergence, while roots penetrate deeper layers during moisture deficiency.

The fibrous root system in irrigated rice (A) and nodal roots in deep-water rice (B) are depicted in Fig. 3.18.

Stems

Rice plants typically reach a height of approximately 4 ft, and the main stem consists of multiple nodes and internodes, primarily concentrated within the upper 5–6 nodes. The internodes are smooth and hollow. Internode length increases in a hierarchical pattern from the base upwards; the last internode subtending the reproductive structure (peduncle) is the longest of all (Pattanaik and Mohapatra 1988). The shorter the internode, the plant becomes more lodging resistant. Each node possesses a single leaf and a bud that has the potential to grow into a tiller. Salient features of rice stem are described below and depicted in Fig. 3.19.

(a) The stem, also known as the culm, is composed of a series of nodes and internodes in rice plants.

Fig. 3.19 Rice plant showing root system, stem internodes, tillers, and panicle. (Source: Photo gallery of P. K. Senapati and E. Kariali)

(b) The internodes are hollow, cylindrical, and erect, and the nodes are solid.
(c) The number and length of nodes and internodes are variety- and environment-specific. The number varies from 10 to 20.
(d) The internodes are very long in deep-water or floating rice.

Leaves

The leaves are elongated and flat, emerging from the nodes of a hollow stem in an alternate phyllotaxy, with one leaf per node. The last leaf enclosing the panicle is named the flag leaf. Specific to the cultivar, leaf architecture may be erect, oblique or drooping. The sheath is the part that connects the leaf to the stem node. The junction between the leaf and sheath has a collar, where there are two elements, ligule and auricle. The auricle is a 2–5 mm appendix, crescent-shaped and hairy. The ligule is a membranous structure and its length and size are varietal-specific. Among plants of the family *Poaceae*, rice is unique in having both ligule and auricle in the leaf, which distinguishes it from weeds at the seedling stage. The salient features of rice leaves are:

(a) Leaves arise one per each node on the culm in alternate phyllotaxy. Leaf number is variety-specific.
(b) The topmost leaf located just below the panicle in a rice plant is known as the flag leaf. Primarily it contributes to grain growth in the panicle while leaves below the flag leaf normally supply photosynthetic assimilates for stem and root growth.
(c) Lamina of the leaf is subtended by leaf sheath and they are joined by a circular collar.
(d) Leaf sheath wraps around the stem above the node.
(e) Lamina is a flat organ. Rice varieties can vary in several characteristics, including their length, breadth, width, colour, thickness, shape, angle, and the presence of pubescence.
(f) The upper surface of a rice leaf typically displays parallel venation, while the underside is smooth and features a prominent ridge known as the midrib.
(g) Rice plants have paired ear-like appendages known as auricles, which are located on either side of the base of the leaf lamina.
(h) The juncture or junction between the leaf sheath and the lamina is often referred to as the ligule in rice plants.

A photograph of rice leaf (A) and a model diagram of rice leaf (B) are depicted in Fig. 3.20.

Tillers

Each leaf on the main stem has one bud in its axil arising from the stem node that can develop into a branch called tiller. Usually, the bud in a rice plant is situated near the upper portion of the node and is enveloped or enclosed by the leaf sheath. Tillers arise from the main culm in alternate order. Each tiller is an independent plant unit. Tiller growth and development are asynchronous; early tillers dominate over the new tillers. Primary tillers come from the lower culm nodes and can produce secondary tillers, which, in turn, may give rise to tertiary tillers. All the tillers after a definite growth period become photosynthetically independent, but some tillers especially the tillers formed late fail to produce panicles and thus are unable to contribute towards the overall grain yield of the plant.

Fig. 3.20 Photograph of a rice leaf (**A**) and a model diagram of a rice leaf (**B**). (Source: Photo gallery of P. K. Senapati and E. Kariali)

3.1.10 Reproductive Organs

3.1.10.1 Panicle

The flowering organ at the top part of the rice plant is a complex racemose inflorescence bearing determinate type panicle sub-tended by the last stem internode called a peduncle. There are varietal differences in length, shape, and angle of the panicle to the stem axis. Panicles are composed of a central axis, i.e. rachis with primary ramifications named primary branches. On the lower section of the primary branches, one can typically find 2–3 secondary branches. Pedicellate spikelets are located at the nodes and tips of both the primary and secondary branches. At the top of a rice tiller is the panicle, an inflorescence that grows on the uppermost culm internode. This panicle holds rice spikelets, which mature into grains. The base of the panicle is typically narrow and ring-like, referred to as the neck, and is used to measure culm and panicle length. The panicle axis is usually uninterrupted and hollow, except at the nodes where branches emerge. The bulges or enlargements on the panicle axis where the branches are connected are referred to as panicle pulvinus. At each node along the main panicle axis, primary branches originate, and these

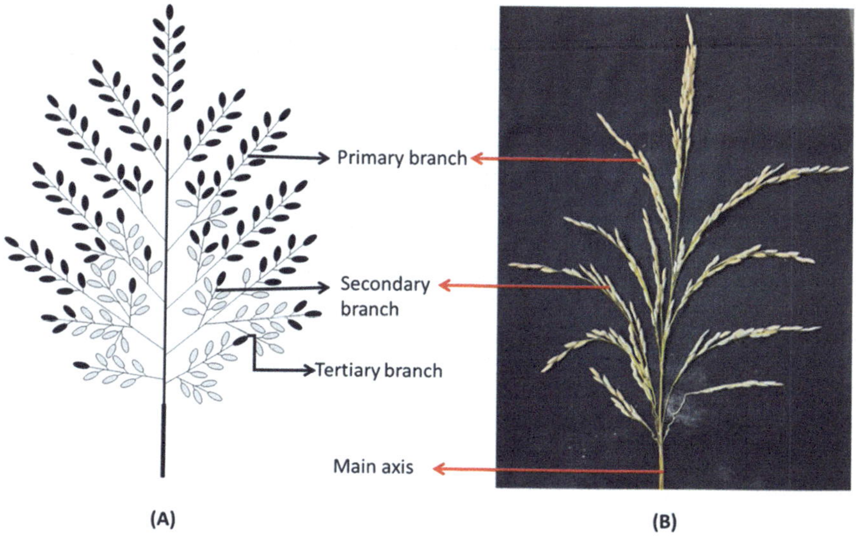

Fig. 3.21 Diagrammatic representation of a panicle (**A**) and photograph of a panicle (**B**). (Source: Photo gallery of Dr. E. Kariali)

primary branches, in turn, carry secondary branches. Primary branches can be present either as solitary branches or in pairs. The panicles bear spikelets, most of which develop into grains, and these spikelets are found on both primary and secondary branches. The spikelet serves as the basic structural unit of the inflorescence or panicle, consisting of the pedicel and the floret. The floret is connected to the pedicel, and the rudimentary glumes create the cup-like tip of the pedicel. These rudimentary glumes constitute the lowermost portion of the spikelet and are typically detached from the rest of the spikelet during threshing. Sterile lemmas are small, bract-like extensions connected to the floret. The rachillae, which is a small axis, carries a single floret and is situated between the sterile lemmas and the floret. A model of panicle architecture showing the branching pattern of the rachis (A) and a photograph of the panicle (B) is shown in Fig. 3.21.

3.1.10.2 Spikelet

Spikelets are structural inflorescence units of rice panicles, each one is genetically competent to bear a grain at maturity. It has three florets, two sterile ones at the base and one terminal fertile bisexual floret interlocked by lemma and palea in the hull consisting of six stamens, two fleshy lodicules and one pistil. The lower two sterile florets basal to the fertile one are rudimentary glumes and empty glumes positioned on the axis successively. Spikelets are located on small rachillae at the tips of the branches of the panicle. In the case of rice, each spikelet contains a solitary flower that is enclosed by the lemma and palea. Inside the flower, the six stamens are arranged in two sets of three each. The filaments of these stamens are very slender and delicate, and they undergo rapid growth during anthesis (the flowering period) (Fig. 3.22A). The anthers in rice are linear in shape. The pistil is composed of a

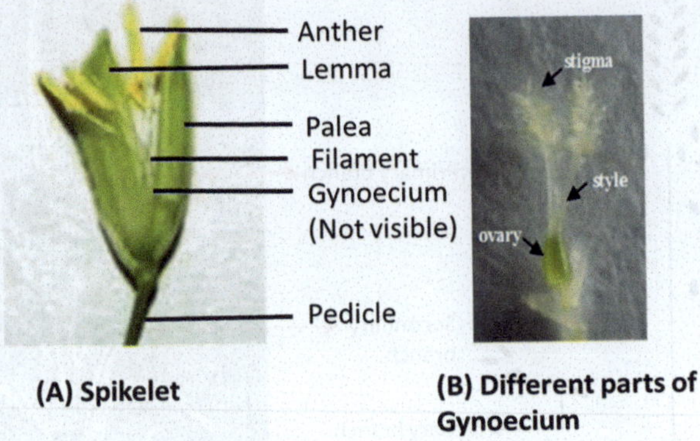

Fig. 3.22 Spikelet of rice showing different floral structures (**A**) and different parts of Gynoecium (**B**). (Source: Ph.D. thesis of Kaushik Das, 2018, Sambalpur University, India)

solitary ovary and two styles, each with a feathery and outwardly extending stigma (Fig. 3.22B). The ovary in rice is a tri-carpellary, meaning it has three carpels, and it is single-celled with a single ovule. The ovule is positioned at the base of the ovary, a condition known as basal placentation. In numerous rice varieties, the feathery stigmas are frequently coated with pollens as the anthers release pollens at the time of flower opening. While self-pollination is the predominant mode of fertilization, there is also some degree of natural cross-pollination, which can range from 0.1 to 4.0%.

3.1.10.3 Metabolic Diversification Deciding Organ Differentiation

Metabolic diversification deciding the fate of cultivar-specific organ differentiation is believed to be a crucial factor in a plant's ability to adapt to a wide range of environmental conditions over its evolutionary history. Gene duplication is a frequent event in plant genomes, and it supplies the genetic material that environmental selection can act upon. This process plays a pivotal role in the diversification of plants, ultimately facilitating their ability to adapt to a wide range of environmental challenges. After gene duplication events, the fate of gene duplicates depends on whether they can acquire beneficial functions at either the protein structure or gene transcriptional level. The retention of species-specific gene copy numbers is frequently suggested as a mechanism that helps different plants address their unique environmental challenges. This process enables plants to adapt to specific ecological niches and thrive in diverse environments.

References

Barik JK, Mohapatra PK, Pradhan SK, Kariali E (2017) De-hulling facilitates seed germination and ethylene production in dormant and non-dormant rice genotypes. J Agric Sci 155:1508–1519

Bautista NS, Vaughan DA, Jayasuriya AHM, Liyanage ASU, Kaga A, Tomooka N (2006) Genetic diversity in AA and CC genome *Oryza* species in southern South Asia. Genet Resour Crop Evol 53:631–640

Bernier J, Atlin GN, Serraj R, Kumar A, Spaner D (2008) Breeding upland rice for drought resistance. J Sci Food Agric 88:927–939
Bessho-Uehara K, Wang DR, Furutaa T, Minamia A, Nagaia K, Gamuyaoa R (2016) Loss of function at RAE2, a previously unidentified EPFL, is required for awnlessness in cultivated Asian rice. Proc Natl Acad Sci U S A 113:8969–8974
Bessho-Uehara K, Yamagata Y, Takashi T, Makino T, Yasui H, Yoshimura A, Ashikari M (2021) Exploring the loci responsible for awn development in rice through comparative analysis of all AA genome species. Plants (Basel) 10(4):725. https://doi.org/10.3390/plants10040725
Chang TT (1976) The origin, evolution, cultivation, dissemination, and diversification of Asian and African rice. Euphytica 25:425–441
Civan P, Brown TA (2017) Origin of rice (*Oryza sativa* L.) domestication genes. Genet Resour Crop Evol 64:1125–1132
Fukai S, Cooper M (1995) Development of drought-resistant cultivars using physio-morphological traits in rice. Field Crop Res 40:67–86
Furuta T, Komeda N, Asano K, Uehara K, Gamuyao R, Angeles-Shim RB, Nagai K, Doi K, Wang DR, Yasui H, Yoshimura A, Wu J, McCouch SR, Ashikari M (2015) Convergent loss of awn in two cultivated rice species *Oryza sativa* and *Oryza glaberrima* is caused by mutations in different loci. G3 5:2267–2274. https://doi.org/10.1534/g3.115.020834
Grillo MA, Li C, Fowlkes AM, Briggeman TM, Zhou A, Schemske DW, Sang T (2009) Genetic architecture for the adaptive origin of annual wild rice, *Oryza nivara*. Evolution 63:870–883
Gross BL, Zhang Z (2014) Archaeological and genetic insights into the origin of domesticated rice. PNAS 111:6190–6197
Gu B, Zhou T, Luo J, Liu H, Wang Y, Shangguan Y, Zhu J, Li Y, Sang T, Wang Z, Han B (2015) An-2 encodes a cytokinin synthesis enzyme that regulates awn length and grain production in rice. Mol Plant 8:1635–1650
Guo H, Lee TH, Wang X, Paterson AH (2013) Function relaxation followed by diversifying selection after whole-genome duplication in flowering plants. Plant Physiol 162(2):769–778
Gupta PC, O'Toole TC (1986) Upland rice. A global perspective. International Rice Research Institute, Manila, Philippines
Hu M, Lv S, Wu W, Fu Y, Liu F, Wang B, Li W, Gu P, Cai H, Sun C, Zhu Z (2018) The domestication of plant architecture in African rice. Plant J 94:661–669
Hua L, Wang DR, Tan L, Fu Y, Liu F, Xiao L, Zhu Z, Fu Q, Sun X, Gu P, Cai H, McCouch SR, Sun C (2015) *LABA1*, a domestication gene associated with long, barbed awns in wild rice. Plant Cell 27:1875–1888
IRRI (1997) Rice Almanac, 2nd edn. International Rice Research Institute, Manila, p 181
IRRI—International Rice Research Institute (1984) Annual report for 1983
Jacquemin J, Chaprro C, Laudie M, Berger A, Gavory F, Goicoechea JL, Wing RA, Cooke R (2011) Long-range and targeted ectopic recombination between the two homologous chromosomes 11 and 12 in *Oryza* species. Mol Biol Evol 28(11):3139–3150
Jiang L, Ma X, Zhao S, Tang Y, Liu F, Gu P, Fu Y, Zhu Z, Cai H, Sun C, Tan L (2019) The APETALA2-like transcription factor SUPERNUMERARY BRACT controls rice seed shattering and seed size. Plant Cell 31:17–36
Jin J, Huang W, Gao JP, Yang Y, Shi M, Zhu MZ, Luo D, Lin HX (2008) Genetic control of rice plant architecture under domestication. Nat Genet 40:1365–1369
Jin FX, Kim DM, Ju HG, Ahn SN (2009) Mapping quantitative trait loci for awnness and yield component traits in isogenic lines derived from an *Oryza sativa*/*O. Rufipogon* cross. J Crop Sci Biotechnol 12:9–15. https://doi.org/10.1007/s00122-003-1270-8
Jin J, Hua L, Zhu ZF, Tan LB, Zhao XH, Zhang WF, Liu F, Fu Y, Cai H, Sun X, Gu P, Xie D, Sun C (2016) *GAD1* encodes a secreted peptide that regulates grain number, grain length and awn development in rice domestication. Plant Cell 28:2453–2463
Karladee D, Suriyonga S (2012) γ-Aminobutyric acid (GABA) content in different varieties of brown rice during germination. Sci Asia 38:13–17
Karladee D, Boonsit P, Suriyong S, Jamjod S (2012) Population heterogeneity of upland rice in northern Thailand. Thai J Agric Sci 45:99–104

Kellog EA (2009) The evolutionary history of Ehrhartoideae, Oryzeae, and *Oryza*. Rice 2:1–14
Kovach MJ, Sweeney MT, McCouch SR (2007) New insights into the history of rice domestication. Trends Genet 23:578–587
Lee J, Park JJ, Kim SL, Yim J, An G (2007) Mutations in the rice ligule-less gene result in a complete loss of the auricle, ligule, and laminar joint. Plant Mol Biol 65:487–499
Li C, Zhou A, Sang T (2006) Rice domestication by reducing shattering. Science 311:1936–1939
Lin Z, Li X, Shannon LM, Yeh CT, Wang ML, Bai C, Peng Z, Li J, Trick HN, Clemente TE, Doebly J, Schnable PS, Tuinstra MR, Tesso TT, White F, Yu J (2012) Parallel domestication of the *Shattering1* genes in cereals. Nat Genet 44:720–724
Liu LL, Li YZ, Wei MY, Wu ZS, Liu F, Qiu YF, Luo J, Li R, Qin B (2017) The causal deletions in the second exon of An-3 closely associated with awn development and rice yield. Genes Genom 39:1205–1213
Liu H, Li Q, Xing Y (2018) Genes contributing to the domestication of rice seed traits and its global expansion. Genes 9:489. https://doi.org/10.3390/genes9100489
Liu H, Zhan J, Hussain S, Nie L (2019) Grain yield and resource use efficiencies of upland and lowland rice cultivars under aerobic cultivation. Agronomy 9:591. https://doi.org/10.3390/agronomy9100591
Luo LJ, Zhang QF (2021) The status and strategies on studying drought resistance of rice. Chin J Rice Sci 15:209–214. (Chinese)
Luo J, Liu H, Zhou T, Gu B, Huang X, Shangguan Y, Zhu J, Li Y, Zhao Y, Wang Y, Zhao Q, Wang A, Wang Z, Sang T, Wang Z, Han B (2013) An-1 encodes a basic helix-loop-helix protein that regulates awn development, grain size, and grain number in rice. Plant Cell 25:3360–3376
Mackill DJ, Coffman WR, Garrity DP (1996) Rainfed lowland rice improvement. International Rice Research Institute, Manila, Philippines
Malcomber ST, Preston JC, Reinheimer R, Kossuth J, Kellogg EA (2006) Developmental gene evolution and the origin of grass inflorescence diversity. Adv Bot Res 44:425–481
Masoomi M, Mahmoudi A, Alizadeh MR, Abdollahpour S (2020) Effect of awn on aerodynamic properties of paddy grains (*Oryza Sativa* L.). Agric Eng Int J 22:208–216
Matsuo T, Hoshikawa K (1993) Science of the rice plant, vol 1. Morphology, Food and Agriculture Policy Research Center, Tokyo
Mitra S, Ganguli PM (1932) Some observations on the characters of wild rice hybrids. Indian J Agric Sci 2:271–279
Molina J, Sikora M, Garud N, Flowers JM, Rubinstein S, Reynolds A, Huang P, Jackson S, Schaal BA, Bustamante CD, Boyko AR, Purugganan MD (2011) Molecular evidence for a single evolutionary origin of domesticated rice. Proc Natl Acad Sci U S A 108(20):8351–8356
Noldin JA, Chandler JM, McCauley GN (1999) Red rice (*Oryza sativa*) biology. I. Characterization of red rice ecotypes. Weed Technol 13:12–18
Oka HI (1988) Origin of cultivated rice. Chapter 2. Japan Scientific Society Press/Elsevier, Tokyo/Amsterdam, Oxford, Tokyo, pp 15–22. ISBN 4-7622-1544-9
Onishi K, Takagi K, Kontani M, Tanaka T, Sano Y (2007) Different patterns of genealogical relationships found in the two major QTLs causing reduction of seed shattering during rice domestication. Genome 50:757–766
Pattanaik PK, Mohapatra PK (1988) Role of assimilates and phosphates in the control of internode elongation in tall and dwarf Indica Rice varieties. J Exp Bot 39:1367–1380
Prasad V, Stromberg CAE, Alimohammadian H, Sahni A (2005) Dinosaur coprolites and the early evolution of grasses and grazers. Science 310:1177–1180
Preston JC, Kellogg EA (2007) Conservation and divergence of *APETALA1/FRUITFULL*-like gene function in grasses: evidence from gene expression analyses. Plant J 52:69–81
Qin B, Lu T, Xu Y, Shen W, Liu F, Xie X, Li Y, Wang K, Li R (2021) Identification and characterization of An-4, a potential quantitative trait locus for awn development in rice. BMC Plant Biol 21:298
Sakai H, Tanaka T, Antonio BA, Itoh T, Sasaki T (2014) The first monocot genome sequence: *Oryza sativa* (rice). Adv Bot Res 69:119–135

Sang T, Ge S (2007) Genetics and phylogenetics of rice domestication. Curr Opin Genet Dev 17:533–538

Singh A, Shamin M, Singh KN (2013) Genotypic variation in root anatomy, starch accumulation, and protein induction in upland rice (*Oryza sativa*) varieties under water stress. Agric Res 2:24–30

Sugimoto K, Takeuchi Y, Ebana K, Miyao A, Hirochika H, Hara N, Ishiyama K, Kobayashi M, Ban Y, Hattori T, Yano M (2010) Molecular cloning of SDR4, a regulator involved in seed dormancy and domestication of rice. Proc Natl Acad Sci U S A 107:5792–5797

Sutharut J, Sudarat J (2012) Total anthocyanin content and antioxidant activity of germinated coloured rice. Int Food Res J 19:215–221

Sweeny M, McCouch S (2007) The complex history of the domestication of rice. Ann Bot 100:951–957

Swetha C, Narjala A, Pandit A, Tirumalai V, Shivaprasad PV (2022) Degradome comparison between wild and cultivated rice identifies differential targeting by miRNAs. BMC Genomics 23:53

Tan L, Li X, Liu F, Sun X, Li C, Zhu Z, Fu Y, Cai H, Wang X, Xie D, Sun C (2008) Control of a key transition from prostrate to erect growth in rice domestication. Nat Genet 40:1360. https://doi.org/10.1038/ng.197

Thurber CS, Hepler PK, Caicedo AL (2011) Timing is everything: early degradation of abscission layer is associated with increased seed shattering in US weedy rice. BMC Plant Biol 11:14

Tian S, Nakamura K, Kayahara H (2004) Analysis of phenolic compounds in white rice, brown rice and germinated brown rice. J Agric Food Chem 52:4808–4813

Vaughan DA, Morishima H (2003) Biosystematics of the genus *Oryza*. In: Wayne Smith C, Dilday RH (eds) Rice: origin, history, technology and production. Wiley, Hoboken, NJ, pp 27–65

Vaughan DA, Morishima H, Kadowaki K (2003) Diversity in the *Oryza genus*. Curr Opin Plant Biol 6:139–146

Vaughan DA, Lu BR, Tomooka N (2008) The evolving story of rice evolution. Plant Sci 174(4):394–408

Vichapong J, Sookserm M, Srijesdaruk V, Swatsitang P, Srijaranai S (2010) High performance liquid chromatographic analysis of phenolic compounds and their antioxidant activities in rice varieties. LWT- Food Sci Technol 43:1325–1330

Wade LJ, Fukai S, Samson BK, Ali A, Mazid MA (1999) Rainfed lowland rice: physical environment and cultivar requirements. Field Crop Res 64(1–2):3–12

Wang Y, Li J (2008) Rice rising. Nat Genet 40:1273–1275

Wang H, Zhang M, Guo R, Shi D, Liu B, Lin X, Yang C (2012) Effects of salt stress on ion balance and nitrogen metabolism of old and young leaves in rice (*Oryza sativa* L.). BMC Plant Biol 12:194

Wanyo P, Meeso N, Siriamornpun S (2014) Effects of different treatments on the antioxidant properties and phenolic compounds of rice bran and rice husk. Food Chem 157:457–463

Wei C, Wang Z, Wang J, Ten J, Shen S, Xiao Q, Bao S, Feng Y, Zhang Y, Li Y, Sun S, Yu C, Wang Y, Zhou T, Xu W, Yu J, Li W, Wang J (2021) Conversion between 100-million-year-old duplicated genes contributes to rice sub-species divergence. BMC Genomics 22:460

Wu Z, Zhao S, Li X, Zhang B, Jiang L, Tang Y, Tang Y, Zhao J, Ma X, Cai H, Sun C, Tan L (2018) Deletions linked to *PROG1* gene participate in plant architecture domestication in Asian and African rice. Nat Commun 9:4157. https://doi.org/10.1038/s41467-018-06509-2

Xiong LZ, Liu KD, Dai XK, Xu CG, Zhang QF (1999) Identification of genetic factors controlling domestication-related traits of rice using an F2 population of a cross between *Oryza sativa* and *O. rufipogon*. Theor Appl Genet 98:243–251. https://doi.org/10.1007/s001220051064

Xiongshiyee V, Rerkasem B, Veeradittakit J, Saenchai C, Lordkaew S, Thebault-u-that C (2018) Variation in grain quality of upland rice from Luang Prabang Province, Lao PDR. Ric Sci 25:94–102

Xu R, Sun C (2021) What happened during the domestication of wild to cultivated rice? Crop J 9:564–576

Yambao EB, Ingram KT, Real JG (1992) Root xylem influence on the water relations and drought resistance of rice. J Exp Bot 43:925–932

Yoon J, Cho LH, Kim SL, Choi H, Koh HJ, An G (2014) The BEL1-type homeobox gene *SH5* induces seed shattering by enhancing abscission-zone development and inhibiting lignin biosynthesis. Plant J 79:717–728

Yoon J, Cho LH, Antt HW, Koh HJ, An G (2017) KNOX protein OSH15 induces grain shattering by repressing lignin biosynthesis genes. Plant Physiol 174:312–325

Yu B, Lin Z, Li H, Li X, Li J, Li Y, Wang X, Zhang Z, Zhu Z, Zhai W, Wang X, Xie D, Sun C (2007) TAC1, a major quantitative trait locus controlling tiller angle in rice. Plant J 52:891–898

Zeigler RS, Puckridge DW (1995) Improving sustainable productivity in rice-based rainfed lowland systems of south and Southeast Asia. Geo J 35:307–324

Zhou Y, Lu D, Li C, Luo J, Zhu B-F, Zhu J, Shangquan Y, Wang Z, Sang T, Zhou B, Han B (2012) Genetic control of seed shattering in rice by the APETALA2 transcription factor shattering abortion. Plant Cell 24:1034–1048

Zhu BF, Si L, Wang Z, Zhou Y, Zhu J, Shangguan Y, Lu D, Fan D, Li C, Lin H, Qian Q, Sang T, Zhou B, Minobe Y, Han B (2011) Genetic control of a transition from black to straw-white seed hull in rice domestication. Plant Physiol 155:1301–1311

Zhu Z, Tan L, Fu Y, Liu F, Cai H, Xie D, Wu F, Wu J, Matsumoto T, Sun C (2013) Genetic control of inflorescence architecture during rice domestication. Nat Commun 4:2200. https://doi.org/10.1038/ncomms3200

Staging of Rice Plant Growth and Development

4

Abstract

The growth of rice plant can be divided into three stages: vegetative, reproductive, and ripening. The full cycle of growth varies from 70 to 170 days depending on the cultivar. The duration of the vegetative and reproductive phases is identical, but the duration of the vegetative phase differs between cultivars. The vegetative phase extends from germination to panicle initiation, during which the plant increases in size by the increase in root formation, leaf number, and tiller number. The reproductive phase covers plant growth from panicle initiation to anthesis. It is characterized by the elongation of the main culm, degeneration of unproductive tillers, the emergence of flag leaves, booting, heading, and the emergence of anthers. In the ripening phase, assimilates are stored in developing grains as reserve food; over time, the grains lose water and harden physically to reach maturation. An assortment of MADS-box genes controls the development of roots, leaves, and floral structures. The MADS-box homeotic genes defining the floral organ identity of rice spikelets are divided into five classes encapsulated in a scientific model adopted for angiosperms called the ABCDE flower model. The identity of the genes depends on which organ they affect. The patterns of MADS-box gene expression in the meristem that is responsible for floral organ identity are regulated by upstream promoter regions and transcription factors. A rice panicle is a complex determinate type of racemose branching inflorescence borne on the uppermost internode of the culm named the peduncle. The main axis bears 8–10 primary branches and 2–3 secondary branches on each primary branch. Pedicellate spikelets are borne on the top and nodes of branches and bear a grain each upon maturity. The rice spikelet is hermaphrodite. The pollens released from the anthers travel in the air within the hull for self-fertilization. Like in other angiosperms, double fertilization in rice results in the development of endosperm and embryo, and interactions between them affect seed setting

© The Author(s), under exclusive license to Springer Nature Singapore Pte Ltd. 2025

P. K. Mohapatra et al., *Tillering Behavior of Rice Plant*,
https://doi.org/10.1007/978-981-97-5235-5_4

and development. The role of the *PTB1* (*POLLEN TUBE BLOCKED 1*) gene is crucial for the regulation of panicle set and yield.

Keywords

Panicle · Phenology · Spikelet · MADS-box · ABCDE model · Meristems

4.1 Introduction

Rice plants are annual grasses that have a growth period ranging from 3 to 6 months depending on the environmental factors and genotype. Rice is a short-day plant, and its main cultivation is winter rice. In tropical environments, winter rice is produced between June and July under long-day conditions. The plant perceives light below the critical photoperiod in October to induce flowering and reach maturity in November–December. In contrast, day-neutral rice can be grown throughout the year. Rice cultivated in the Rabi season is also known as summer rice. It is sown in the field from November to February and harvested during March and June. Understanding the rice growth cycle is crucial for harnessing yield benefits in crop management. The most common practice of rice growing begins by flooding fields with rain or irrigation water a few weeks before sowing. There are two planting methods used: transplanting and broadcasting. During transplanting, the plants are grown in a nursery bed and subsequently transplanted to the fields to provide uniform spacing between individual entities. Conversely, seeds are thrown into flooded fields, resulting in spatial morphological heterogeneity in broadcast cultivation.

4.2 Crop Stages

The three crop stages of rice were significantly different during cultivation. They are vegetative, reproductive, and ripening. The vegetative period extends from germination to panicle initiation, with a lag phase of approximately 60 days. The reproductive growth phase extends from panicle initiation to anthesis and is characterized by main culm elongation. The other characteristics include a decrease in tiller number; the emergence of flag leaves; and the booting, heading, and anthesis of the spikelets. This phase lasts for 25–35 days and involves the storage of assimilates in the stem and underground parts in the form of starch and protein reserves. The maturity phase extends from anthesis to grain hardening. The duration is approximately 25–35 days, during which grain filling occurs with the accumulation of starch and other organic materials in the developing endosperm. Depending on the cultivar and agricultural practice, the full cycle lasts between 70 and 170 days, when the durations of the reproductive and maturity periods are nearly identical but the duration of the vegetative period varies significantly (https://icar-nrri.in/released-varieties/). Sattari, an early maturing variety, matures in just 70 days, whereas a long-duration traditional variety takes as long as 170 days. According to the literature, the rice

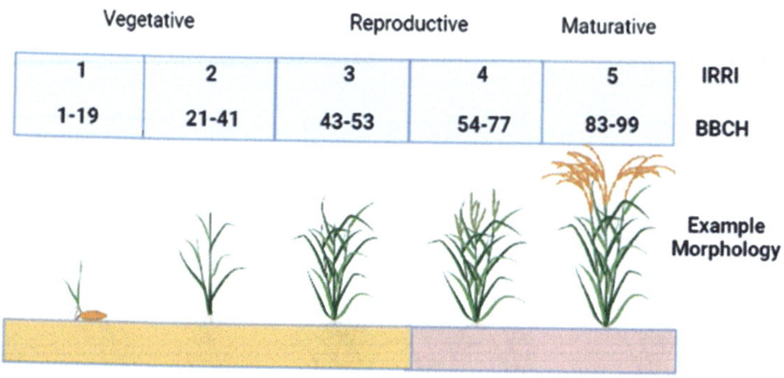

Fig. 4.1 Picture showing the growth cycle of rice plants according to the IRRI and Biologische Bunndesanstalt, Bundessortenamt und CHemische Industries (BBCH) scale and sample examples

growth cycle is calibrated on two distinct scales: (1) International Rice Research Institute (IRRI) (GRiSP 2013) and (2) Biologische Bundesanstalt, Bundessortenamt und CHemischeIndustrie (BBCH) (Lancashire et al. 1991). The IRRI scale divides the growth cycle into five phases; in contrast, the BBCH scale uses 100 stages between 0 and 99 described in Fig. 4.1.

Yujugullu et al. (2017) used an all-weather effective remote sensing technique, the synthetic aperture radar (SAR) sensor, for detecting temporal phenological changes in rice plants. For growth stage estimation, a meta-model-based inversion algorithm was proposed by these workers, where the approach is designed to provide estimates of crop morphology and the corresponding growth stage from a continuous growth scale. The SAR technique is superior to optical remote sensing techniques, which are susceptible to atmospheric factors. SAR systems have all-weather day and night imaging capabilities and play a significant role in environmental monitoring because of their sensitivity to physical alterations in monitored plants. The phenological changes occurring in the plant during the three growth phases of the rice life cycle imparted their effect on the SAR system responses.

4.2.1 Vegetative Period

4.2.1.1 Seedling Stage

The vegetative period begins with germination and continues through the tillering stage. In the seedling phase, germination begins with the appearance of the white tip of the coleoptiles. Subsequent growth of the coleoptile occurs during continuation. Seedling emergence takes place when the tip of a seedling emerges from the soil or water surface. At the germination stage, the coleoptile reaches approximately 1 cm, after which the first leaf emerges. The first leaf is small and whitish; it lacks a blade and collar and consists of a leaf sheath only. It is also known as a prophyll leaf and grows approximately 2 cm long. Before the prophyll elongates, the second leaf emerges in succession. This leaf has a leaf blade and is competent for photosynthesis. The third and other succeeding leaves emerge after each preceding leaf has

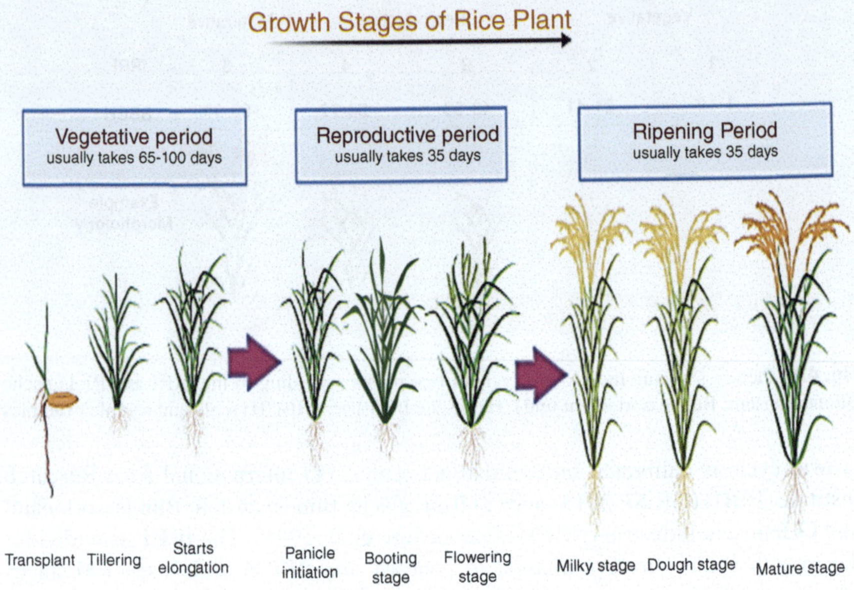

Fig. 4.2 Growth stages and different phenotypic changes in rice plants on a temporal scale

completed elongation. The recommended ages for transplanting are 18–22 days for short duration, 25–30 days for medium duration, and 25–40 days for long-duration cultivars. Transplanting young plants ensures that a uniform plant stand is helpful for crop management and that the advantage of head starts over emerging weeds.

4.2.1.2 Tillering Stage

A rice tiller is a distinct grain-bearing branch that develops from the leaf axil of the non-elongated main shoot or other tillers during the second part of the vegetative period. Later, the tiller becomes independent from the mother stem (culm) through its adventitious roots. In the vegetative period, tiller initiation starts at the end of the seedling growth stage. Tiller formation begins with the appearance of the first tiller from the axil of the second leaf of the main culm. The coleoptiles and first leaf nodes of the culm do not bear tillers. At this stage, the nodes of the stem are compressed close to the ground, and the internode length is less than 0.04 inches. The axillary bud emerges from the leaf sheath and establishes independent photosynthetic activity. Theoretically, each culm node has the potential to produce a leaf, a tiller, or roots, but the tiller and roots emerge later than the leaf. The roots are composed of nodal roots, with each node bearing approximately 5–25 roots. A tiller and its roots initiate growth when the leaf from the third node above emerges. For example, when the fifth leaf emerges, the second node's tiller and root start growing. This hierarchy is maintained for the generation of other tillers. The tillering stage after germination and seedling growth followed by initiation of reproductive phase and grain maturity is presented in a model in Fig. 4.2.

Fig. 4.3 Tillers in stubbles of rice plants grown in the School of Life Sciences, Sambalpur University, India (Source: Photo gallery of P.K. Senapati and E. Kariali)

4.2.1.3 Tillering Capacity

The tillering capacity of rice is cultivar-specific and varies widely In the same environment rice varieties differ greatly in tillering ability (Yoshida and Hayakawa (1970). Tiller number and growth are also influenced significantly by crop management, where biotic and abiotic factors such as plant spacing, soil fertility, weed competition, and damage from pests determine tiller number and growth. In addition, the tillering period was significantly longer for late-stage varieties than for early-stage varieties. Plants grown in dense stands have a short tillering duration, which is favourable for the synchronization of maturity among tillers and ease of harvesting. Under low plant density, the tillering duration increases, resulting in the formation of a greater number of tillers. Very often, late tillers become ineffective and senesce early without giving any yield advantage. Thakur et al. (2010) reported that the number of tillers produced before anthesis varied between 13 and 36 per hill in the System of Rice Intensification Practice (SRI), whereas in recommended management practices, the number of tillers produced ranged from 6 to 16 per hill. Stubbles of rice plants showing profuse tillering behaviour are presented in Fig. 4.3.

4.2.1.4 Root Growth

The root system is vital for addressing the complexity of the environment in which plants survive. The root system provides anchorage of the aerial shoot to its growth substrate, and the identity of the shoot in the community is specified. The roots enable the absorption of water and nutrients for plant growth and monitor the environmental conditions of the soil medium for water content, nutrient levels, and toxic materials, thereby ensuring an optimal response in the development of the shoot. Throughout the growth period, the root system is continuously redesigned by the initiation and elongation of new roots. Unlike dicot plants, which exhibit a tap root system, rice plants have a fibrous root system consisting of seminal, nodal, and lateral roots. The seminal root (radicle root) is formed from the radicle when it protrudes through the coleorhizae, the enveloping lemma and the palea of the germinating seeds. The seminal root does not extend much but instead gives rise to many lateral roots with root hairs. As seedlings grow, many adventitious roots develop from the basal nodes, and a high density of nodal (adventitious) roots establishes a fibrous system. As the seminal roots die with time, the adventitious roots become the main organ for various functions of the root system. A rice plant showing the root system and tillers is presented in Fig. 4.4. The nodal roots of higher nodes help plants acquire nutrients from the surrounding water (Fig. 3.18).

Fig. 4.4 Roots and tillers of rice plants. (Source: Photo-gallery of P. K. Senapati and E. Kariali, Sambalpur University, India)

4.2.1.5 Genomics of Root Development

The genomics of root development in rice has been of paramount importance to breeders for the improvement of root systems' pursuit of designer rice for efficient nutrient absorption and thereby high yield gain. The genetic control of root development has been elucidated in rice through the use of genetic mutants and QTL analyses (Meng et al. 2019). A plethora of genes are involved in the development of roots in rice plants. Many of the genes are phytohormone related and are most prominently controlled by the hormone IAA and to a lesser extent, by ethylene and cytokinin. The regulatory effects of the genes that are primarily influenced by IAA on root elongation and the initiation and emergence of adventitious and lateral roots in rice plants were summarized by Meng et al. (2019) and presented in Fig. 4.5.

4.2.1.6 MADS-Box Genes in Root Development

Several MADS-box genes regulate rice plant development. They control a variety of important morphological and physiological functions, inclusive mainly of floral organ identity but also of features such as root development, flowering time, and seed development. Shah et al. (2022) reported that *OsMADS1* is involved in flowering time and seed development and that *OsMADS26* is expressed in roots, panicles, seeds, shoots, and leaves. The overexpression of *OsMADS26* results in stress-related responses evident in the reduction of root growth and development (Lee et al. 2008a). MADS-box genes influence the formation of primary and lateral roots, root density, and root elongation. In particular, *OsMADS57* regulates long-distance nitrate translocation from roots to shoots and the elongation of seminal and adventitious roots (Huang et al. 2019). Overexpression of this gene through the corresponding transcription factor positively controls high salinity tolerance; under salt stress, gene overexpression increases the rate of seed germination and root elongation (Wu et al. 2021). Similarly, another gene, *OsMADS25,* is involved in the regulation of root system development through auxin signalling (Zhang et al. 2018). This gene is a key positive regulator of primary and lateral root formation in rice in the presence of nitrate accumulation (Yu et al. 2015). In comparison, overexpression of *OsMADS27* leads to the production of more lateral roots and shorter primary roots depending on the availability of nitrates; the gene improves salt tolerance by changing ABA signalling (Chen et al. 2018).

4.2.1.7 Leaf Growth and Development

Leaves develop on the rice stem on alternate phyllotaxy plates with one leaf per node. The last leaf enclosing the panicle is the flag leaf (also the panicle leaf). Leaves are the most important structures responsible for photosynthesis, and their arrangement on the stem is intended for optimum and efficient absorption of incident solar energy. Thus, leaf morphology is an important trait of rice plant architectural design. Usually, the leaves of high-yielding rice plants exhibit flat and erect postures, but traditional rice plants exhibit drooping leaves. The leaf structure has an unstratified epidermis, which has a thick cuticle with prominent wax deposition. Several bulliform cells are present on the adaxial epidermis. In addition, the leaf is amphistomatic, possessing equal numbers of stomata on either side. The mesophyll

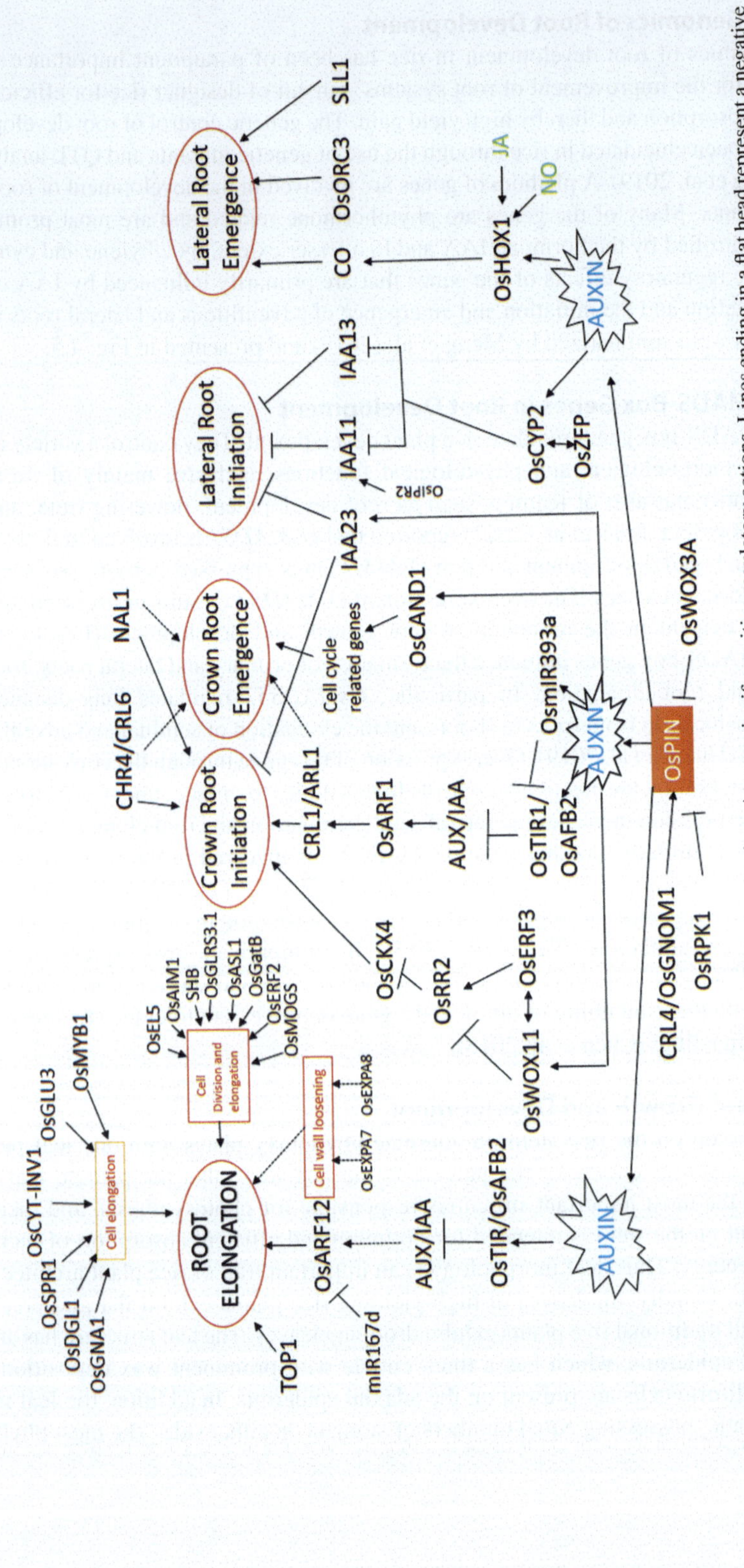

Fig. 4.5 Root development and its molecular regulatory mechanism. Arrows signify positive regulatory actions. Lines ending in a flat head represent a negative regulatory action. The dashed lines represent interactions that have not been experimentally confirmed. Double-headed arrows indicate that two proteins interact. (Adapted from Meng et al. (2019)

distribution in the leaf anatomy is homogeneous, and vascular bundles are surrounded by a bundle sheath of parenchyma cells. The mesophyll tissue is composed of lobed chlorenchymatous cells, the length of which is two times greater than the width; the long axis is oriented perpendicular to the vascular tissues. The leaf anatomy of rice plants is intermediate between that of C3 and C4 grasses (Sage and Sage 2009), where 66% of the protoplast volume is occupied by chloroplasts and these cells cover approximately 95% of the cell periphery. Chloroplasts, mitochondria, and peroxisomes are intimately associated with chloroplasts/stromules. The authors demonstrated that the architecture of chlorenchyma is well-designed for enhanced diffusive conductance of CO_2 and efficient scavenging of photorespiratory CO_2. An extensive chloroplast/stromule sheath directs photorespiratory CO_2 to exit cells via the stroma, facilitating CO_2 absorption by Rubisco. Small-sized lobed cells in association with the chloroplast sheath provide high surface area exposure of the stroma to intercellular spaces, which in turn enhances mesophyll transfer conductance. The authors found that rice plants exhibit greater mesophyll conductance, higher stromal CO_2 content, lower CO_2 compensation points at warm temperatures and less oxygen sensitivity of photosynthesis than do cooler temperature grasses. Simultaneously, the vein length per leaf, intercellular space volume, and mesophyll thickness were found to be intermediate between those of C3 and C4 grasses, suggesting the distinct possibility of introducing major changes in *Kranz* anatomy into rice leaf.

The leaf contains two prominent parts, the lamina and sheath, as well as special parts such as the ligule and auricle. The sheath is the basal part that subtends the lamina. The sheath originates from the stem node and surrounds it. It protects the underdeveloped tiller bud between the leaf and the stem. The swelling at the base of the leaf sheath is the sheath pulvinus. The blade is the upper expanded part of the leaf; it is a chlorophyllous, long, and ribbon-like structure bearing several parallel veins that are typical of the grass family. The dorsal side of the leaf bears a ridge in the middle midrib. On either side of the base of the leaf lamina, there exists a pair of ear-like small appendages called auricles. At the inside juncture between the sheath and blade, there is one papery membrane structure ligule. The length, colour, and shape of the leaf and ligule are cultivar-specific in rice.

4.2.1.8 MADS-Box Genes in Leaf Development

The MADS-box genes identified during rice leaf development are mostly related to brassinosteroid homeostasis. *OsMADS22* and *OsMADS55* encode transcription factors that negatively regulate brassinosteroid responses (Lee et al. 2008b). The expression of *OsMADS55* reaches its peak level in mature leaves, while another gene, *OsMADS47,* exhibits profound expression in seedling leaves. Among the loss-of-function triple mutants, *osmads22, osmads47,* and *osmads55* are two genes downstream of brassinosteroid synthesis, namely, *OsXTR1* and *OsBLE3*, which upregulate activity, implying that the dominant alleles of these genes interfere with the signal transduction of the hormone, decreasing leaf development.

4.2.2 Reproductive Period

4.2.2.1 Panicle Initiation

Panicle initiation is the first stage in the reproductive growth phase of rice plants. The initiation of panicle primordia at the summit of the stem denotes the beginning of panicle formation. The reproductive growth phase continues until anthesis. The stem apex alternatively produces single ridges on its flanks during the vegetative period, which in turn establishes a leaf initial on each ridge. The transition to reproductive growth marks the end of single ridge initiation, during which double ridges are produced. The transition is incumbent on the achievement of certain physiological and environmental conditions (Yoshida 1981). The panicle primordia on top of the culm can be observed under a dissecting microscope 30 days before heading. At this stage, the newly formed panicle becomes visible. The length of the young panicle at this age is approximately 2 mm (Fig. 4.6). According to Xu and Vergara (1986), the shoot apex is vegetative when the preceding leaf primordium is taller than the apical dome. The apex is reproductive when the apical dome elongates and becomes taller than the leaf primordium (flag leaf). Thereafter, the panicle grows in length and develops inside the flag leaf sheath enclosure. Subsequent growth stages are identified by the length of the panicle in the period before emergence. According to Yoshida (1981), the duration of panicle initiation to heading is slightly affected by variety and the environment, but it is considered almost constant, occurring nearly 30 days before heading. In contrast, the duration of the vegetative period differs sharply between varieties. The short-duration cultivars initiated panicle primordia much earlier than the late-maturation cultivars did.

4.2.2.2 Parts of the Rice Panicle

A rice panicle is a complex determinate type of racemose branching inflorescence consisting of the main axis and primary branches. The panicle stalk, named the peduncle, is borne on the uppermost internode of the culm. The main axis (rachis) extends from the neck node to the degenerated growing point at the base of the terminal primary branch. With time, the panicle primordium produces 8–10 primary branch primordia, one primordium on each node of the main axis, in an alternate pattern. At maturity, the lateral primary branches become loosely arranged and spaced at intervals of 2–4 cm on the central rachis. Finally, the panicle primordium degenerates and the last primordium becomes the terminal primary branch. With no additional genesis of primary branches, the rice panicle becomes a determinate type

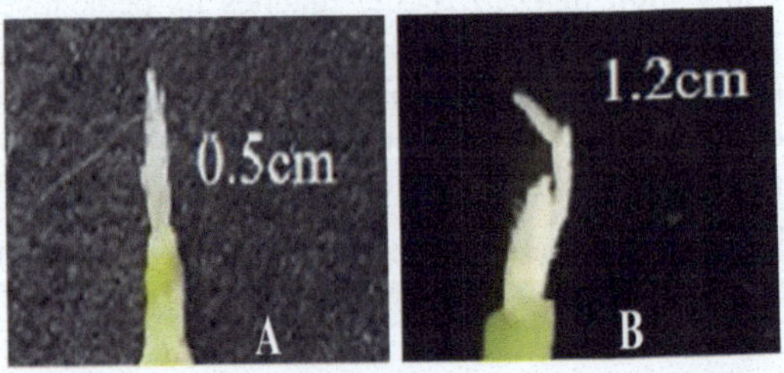

Fig. 4.6 Young rice panicle of 0.5 cm (**A**) and 1.2 cm (**B**) under a microscope. (Source: PhD thesis of Sonam Panigrahi, (2023), Sambalpur University, India)

of inflorescence. As many as three secondary branches are generated at the lower part of the primary branches. Pedicellate spikelets are borne one on each node of the secondary and residual parts of the primary branches. Both primary and secondary branches end up with a terminal spikelet. Panicle morphology is determined by the length of the rachis, the length and number of primary and secondary branches, and the number of spikelets. The length of the branches decreases in an acropetal order, which contributes to the pyramidal shape of the panicle. Panicle length is the factor most crucial for the total spikelet number, and length is positively correlated with the final spikelet number. Using high-throughput quantitative image analysis, Zhao et al. (2015) established a strong positive linkage relationship between the total length of primary branches (summation of the length of all primary branches) and the spikelet number per panicle. The spikelet number (panicle size) in turn correlates positively with grain yield (Laza et al. 2004).

4.2.2.3 Spikelet Morphology

The spikelet is a floral unit of a rice inflorescence that is genetically competent to bear a grain at maturity subject to biomass filling. Unlike those in wheat and barley, spikelets in rice are attached to nodes with stalks (pedicels). A spikelet has a single floret subtended by two pairs of sterile glumes, i.e. sterile lemmas and rudimentary glumes, in sequence (Itoh et al. 2005). The two rudimentary glumes at the base of the spikelet appear one after the other from the pedicel tip, which has a short stem axis named the rachilla. It is a small or secondary rachis branch. The rudimentary glumes of rice spikelets have no axillary buds and do not form additional florets like those of wheat or barley. In contrast, there is an additional pair of sterile glumes named empty glumes, followed by two flowering glumes, one with a lemma and one with a palea in close succession on the rachilla (Ikeda et al. 2004; Yoshida and Nagato 2011). The empty glumes are larger than the rudimentary glumes but much shorter than the flowering glumes. Flowering glumes protect reproductive parts and develop caryopses after fertilization from pathogens, insects, and abiotic stresses. The organs inside the interlocked lemma and palea consist of two lodicules, six stamens, and one pistil. The flowering glumes, rachilla, and pedicel together compose the hull, although structural features of the hull are defined by the complementary features of lemma and palea. The lemma is a five-nerved hardened bract with a filiform extension on its tip called an awn, which is not necessarily present in all cultivars. In comparison, the palea are three-nerved and closely compatible but narrower than the lemma. The male reproductive part has six stamens, each with a two-celled anther box hanging on a long slender filament. The six stamens encircle one uniovulate pistil in the next structural floral whorl. The pistil has a style and a bifid plumose stigma to facilitate the aerial mode of pollination. At the base of the ovary, two oval, small, and fleshy bodies are present. The lodicules are asymmetrically placed just outside the stamen whorl. At the time of flowering, they enlarge in shape by absorbing water, and the swelling force helps in the opening of the flower for pollination through the separation of the lemma and palea. Different parts of the spikelets including male (anthers) and female (stigma) reproductive organs and floral diagram of rice are presented below in Figs. 4.7 and 4.8.

Fig. 4.7 Rice spikelet showing different parts like lemma, palea, anther, filament, stigma, style, ovary, and pedicle (**A**) and (**B**). (Source: Ph.D. thesis of Kaushik Das, Sambalpur University, India, 2018)

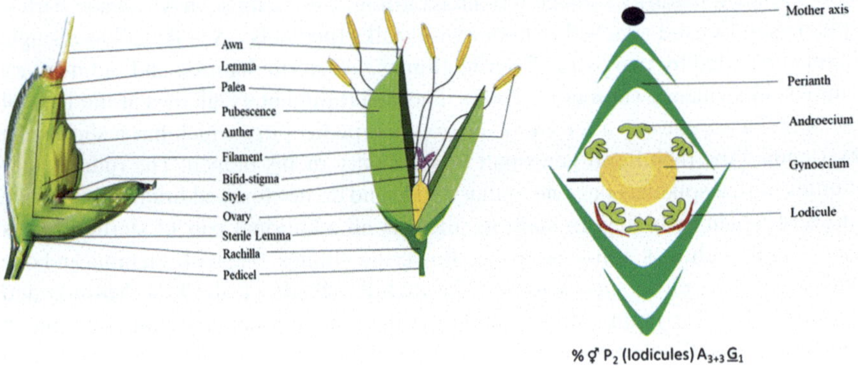

Fig. 4.8 Schematic representation showing different components of a rice spikelet and a floral diagram with floral formula

4.2.2.4 Heterogeneous Development of Spikelets

The order of development of branches and spikelets differs within rice panicles. The sequence of differentiation of primary and secondary branch primordia is acropetal, while their development on the panicle rachis follows a basipetal pattern. Within a branch, spikelet development is acropetal except for the development of the topmost spikelet (Xu and Vergara 1986). In the panicle, the sequence of flowering in the primary branches occurs in the basipetal direction from the apex to the base of the panicle. The grain-filling order followed a sequence similar to that of flowering. Branch and spikelet development in the panicle is asynchronous. The spikelets that reach anthesis early fill faster than those that reach the event later. In the panicle, the upper branches dominate the lower branches in flowering and grain filling, while within a branch, the topmost and bottom spikelets dominate metabolism during the synthesis of organic materials. Asynchrony in spikelet/branch development was enumerated in rice panicles by morphological classification by noting the time taken in days to anthesis by individual spikelets after the exertion of anthers in the spikelet at the tip of the uppermost primary branch (Mohapatra et al. 1993). The spikelets were classified into seven different groups according to the onset of anthesis. The upper branches and spikelets (groups I-III) first emerge from the flag leaf sheath enclosure and are subsequently released from physical constraints. Coming into contact with light, these spikelets fill faster and produce high-quality grains. In contrast, spikelets of the primary branches lower down the panicle axis (groups IV–VII) develop more slowly but are still retained in the panicle enclosure (Mohapatra et al. 2011) (Fig. 4.9). A longer retention time accelerates the decimation of spikelets subjected to prolonged exposure to the ethylene generated inside the boot (Kariali et al. 2022).

4.2.2.5 ABCDE Model of Flower Development

The floral architecture of rice plants is unique among the members of the grass family *Poaceae,* and the plants generate flowers distinctively different from those of common eudicots. In rice, a mature carpel is formed from three carpel primordia that develop in the central location of the spikelet. Six stamens surround the pistil. The spikelet is distinctively dissimilar from other members of the *Poaceae* family in having a pair of sterile lemmas subtending the floret comprising bract-like organs (lemma and palea). Rice flowers have no petals and possess two lodicules homologous to the petals of dicot plants. A pair of highly reduced glumes is positioned below the sterile lemmas. These organs are organized in concentric whorls and have distinct roles in sexual reproduction. The floral organ identity of these parts of the rice florets is co-ordinately defined by A-, B-, C-, D-, and E-class MADS-box genes. The identity of the genes depends on which organ they affect. Accordingly, the pattern of gene expression in the meristems responsible for flower development with the appearance of individual floral organs was encapsulated in a scientific model called the ABCDE (also ABC) model. This model was conceived by Coen and Meyerowitz (1991) for flower development in angiosperms. The genes controlling the floral organ identity of rice are mostly homologous to those of other angiosperms, especially the model plant *Arabidopsis*. Most of the genes controlling floral development in rice are likely regulated by upstream promoter regions and transcription factors (Zhang and Li 2019).

As discussed, the ABC model was based on the observation of genetic mutants with morphological defects in floral organ development. The MADS-box genes specify the identity of floral organs. The model describes how different combinations of gene products trigger the formation of individual floral components. The model proposes that class A genes specify the identity of sepals but act in combination with class B genes for petal development, while both B and C genes are responsible for stamens, and class C genes alone determine carpel identity. D class genes determine ovule identity and class E genes influence sepal, petal, stamen, and carpel specification. The original ABC model was modified to include the new ABCDE model, in which different combinations of genes control floral organ identities, such as A + E for sepals, A + B + E for petals, B + C + E for stamens, C + E for carpels

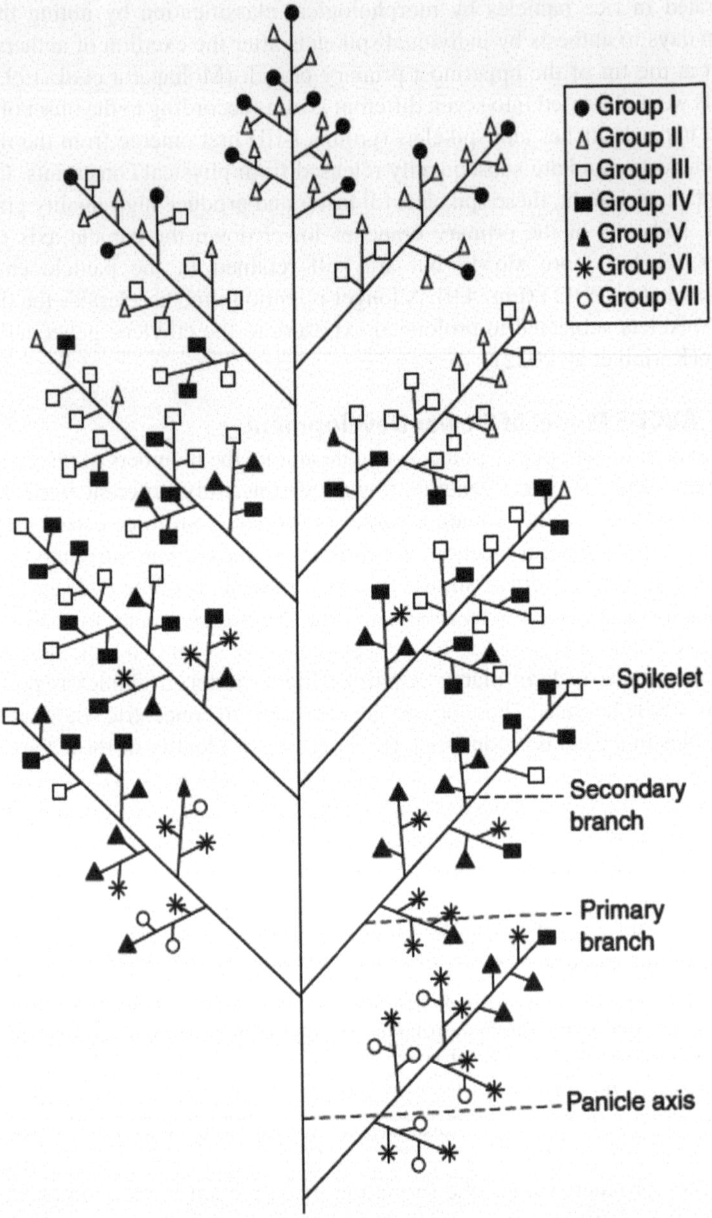

Fig. 4.9 Schematic diagram showing rice panicles with primary and secondary branches and spikelets that flowered from day 1 to 7 and classified Group I to VII (Group I on day 1 to Group VII on day 7 flowering). Groups I–III were considered superior spikelets, and Groups IV–VII were considered inferior spikelets. (Source: Mohapatra et al. 1993)

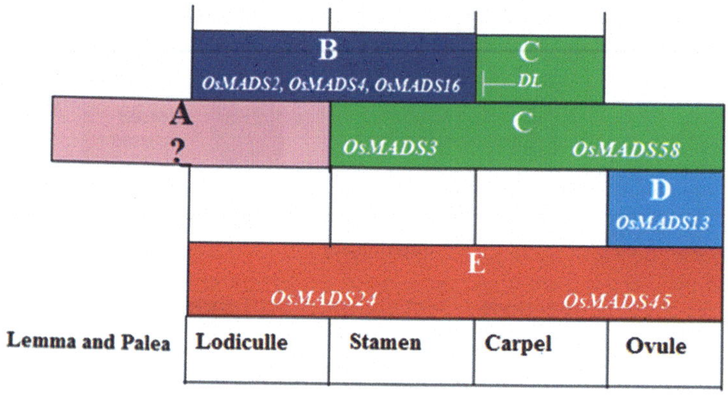

Fig. 4.10 ABCDE model for MADS-box genes in rice plants controlling flower development

and D + E for ovules (Zahn et al. 2006). The D and C class genes were new additions to the ABC group of floral homeotic genes necessary for the specification of floral organ identity in rice, where D class genes are crucial for ovule development and E class genes are necessary for the specification of all types of floral organs (Fig. 4.10). Floral induction leads to the transition of the shoot apical meristem from vegetative to flowering, allowing it to bear flowers through physiological mechanisms, probably through intrinsic hormone dynamics. At this stage, the shoot becomes committed to making a flower respond to environmental signals that trigger a change in fate. All the genes are homeotic genes translated into proteins; each protein contains a MADS-box region facilitating the binding of the protein to DNA and controlling DNA transcription and organ specification. The DNA-binding MADS domain of the proteins is approximately 60 amino acids in length.

The functions of MADS domain transcription factors (Irish 2017) are as follows: class A expression is carried out by the MADS-box gene *APETALA1* (*AP1*) together with *APETALA2* (*AP2*), which is responsible for coding a member of the ERBP family of transcription factors. Two MADS-box genes, namely, *AP3* and *PISTILATA* (*PI*), provide B-type functions, and C-class functions are regulated by the MADS-box gene *AGAMOUS* (*AG*). *AG* alone promoted carpel specification but in combination with two class *B* genes, *AP3* and *PI* promoted stamen development. The D class genes are *SEED STICK* (*STK*) or *SHATTERPROOF* (*SHP*), and the E class *SEPALLATA* subfamily genes are *SEP1*, *SEP2*, *SEP3*, and *SEP4* (Won et al. 2021). Cui et al. (2010) reported that the *SEPALLATA*-like MADS-box genes are needed for defining class E floral organ identity in sepals, petals, stamens, and carpels and for floral determinacy and for the deletion of four *SEP-like* genes, *OsMADS1*, *OsMADS5*, *OsMADS7*, and *OsMADS8*, in which knockdown results in homeotic transformation of all floral parts except the lemma into leaf-like structures. The functions of class B genes responsible for stamen and petal specification in the model plant *Arabidopsis* are conserved in rice. *SUPERWOMEN1* (*SPW1*) in rice is an *Arabidopsis AP3* orthologue. The class B gene of rice expresses only class C genes in whorls 3 and 4. Loss of function of *SPW1* in rice leads to the homeotic

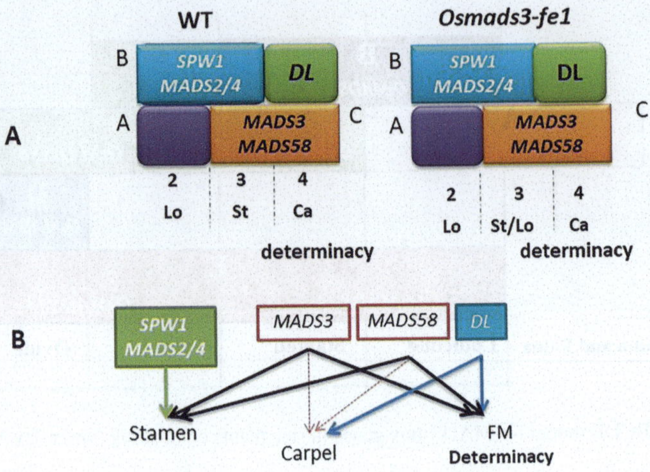

Fig. 4.11 (**A**) Uppercase-modified ABC model of rice flower development showing the location of A, B, and C class genes in whorls 2, 3, and 4. (**B**) Lowercase-Genes responsible for organ specification in rice. *Ca* carpel, *St* stamen, *Lo* lodicules, *DL* drooping leaf. (Adapted from Sugiyama et al. 2019)

transformation of stamens and lodicules into carpels and palea-like organs, respectively. Rice plants have two class C genes, *OsMADS3* and *OsMADS58,* which are in the *AG* family (Yamaguchi et al. 2006), and they contribute mainly to stamen specification and flower determinacy and partly to carpel development, although they are not key regulators of carpel development (Sugiyama et al. 2019). In the regulation of organ development, the contribution of *OsMADS3* was greater than that of *OsMADS58*. Accordingly, Sugiyama et al. (2019) proposed a revised ABC model showing genes responsible for organ specification in rice, where carpel identity is specified by the *YABBY* gene *DROOPING LEAF* (*DL*), an orthologue of *Arabidopsis* CRC because the initiation of carpel development and *DL* expression are coincidental (Yamaguchi et al. 2004). Loss-of-function mutation of *DL* prevents carpel development in rice and instead generates several ectopic stamens in whorl 4. Additionally, the prolonged expression of *OSH1* in *dl* mutants suggested that *DL* is involved in floral meristem (FM) determinacy; *OSH1* is a class 1 *KNOX* gene that is a marker of meristematic indeterminate cells (Yamaguchi et al. 2004). The expression of class C genes is not counterproductive to that of *DL,* and vice versa (Fig. 4.11).

4.3 MADS-Box Genes Specifying Panicle Architecture

In addition to the plethora of information on the MADS-box transcription factors that control various floral organ development events in rice spikelets in the ontogenic phase, there is evidence that the MADS-box is responsible for the structural development of the inflorescence, particularly the ramification of panicle branches. In *Arabidopsis*, two MADS-box genes, *APETALA1* (*AP1*) and *LEAFY* (*LFY*), suppress inflorescence branching through interaction with the *TERMINAL FLOWER1* (*TFL1*) gene. Additionally, four transcription factors encoding MADS-box genes, namely, *SEPALLATA4* (*SEP4*), *AGL24*, *SOC1*, and *SVP*, have been reported to reduce inflorescence branching by suppressing *TERMINAL FLOWER1*(*TFL1*) expression (Liu et al. 2013). The authors revealed that the *SVP*, *SOC1*, *AGL24*, and *SEP4* orthologues in rice modify inflorescence branching by regulating *TFL1*-like genes. A conserved genetic mechanism may exist in flowering plants to specify floral morphology. The LFY and AP1 transcription factors have a common set of target genes (Winter et al. 2015), and they regulate the expression of approximately 200 genes, many of which are involved in inflorescence branching and floral development. Mutant rice plants ectopically overexpressing *TFL1* exhibit high-density branching; when the *TFL1* homologues *RCN1* and *RCN2* are overexpressed, there is a significant increase in panicle branching (Nakagawa et al. 2002). *OsMADS34* interacts with the *LAX1* and *FON4* genes to regulate panicle shape, meristem activity, and branching patterns (Shah et al. 2022).

The MADS-box genes involved in spikelet initiation and structural development have also been characterized. In the inflorescence of rice plants bearing determinate spikelets, the spikelet meristem produces fixed numbers of lateral floral meristems before the terminal floral meristem closes. The *MFS1*, *SNB*, *EG1/DF1*, and *OsMADS22* genes determine spikelet determinacy or indeterminacy (Lee et al. 2007; Ren et al. 2013) because the loss of function of any of the genes induces the production of extra florets or lemmas within a single spikelet. Ren et al. (2019) showed that a mutant of *FON4 (FLORAL ORGAN NUMBER4)*, *fon4–7*, increased inner floral organ production, resulting in two florets instead of one per spikelet; the sterile lemmas gained indeterminate identity because of induced reversion and changed into fertile lemmas. In this mutant, *OsMADS1*, *OsMADS14*, and *OsMADS15* were also expressed in the lemmas and modified lemmas, and *OsMADS2* and *OsMADS6* were overexpressed in palea or palea-like organs.

In a recent review, Shah et al. (2022) reported the presence of a 75-member MADS-protein family in rice, most of which are involved in the specification of floral organs. The A-class MADS-box genes are *OsMADS14*, *OsMADS15*, *OsMADS18*, and *OsMADS20*; *OsMADS2* and *OsMADS4* belong to the B-class; *OsMADS3* is the C-class gene associated with meristem functions; and the *OsMADS13* gene is the D-class gene responsible for ovule identity. The MADS-box genes involved in the transition to flowering are *OsMADS50* and *OsMADS56*, which are counterproductive to each other; the former promotes flowering, while the latter is a suppressor. The action of *OsMADS50* is coordinated by *OsMADS14*, *OsMADS15*, and *OsMADS18*, which are all *AP1/FUL* genes that act downstream of the pathway (Lee et al. 2004). Kim et al. (2007) reported that *OsMADS51* is a flowering promoter that works upstream of *OsMADS14*, *Ehd1 (early heading date 1)*, and *Hd3a* (*heading date 3a*) in rice. Overall, these genes influence the functions of the circadian clock in rice for the induction of flowering. The core oscillator component of the circadian clock is *OsLHY* (*rice LATE ELONGATED HYPOCOTYL*), and the oscillator promotes flowering mainly by regulating the *Ehd1* and *Hd1* genes (Li et al. 2022). Prominent features of the MADS-box gene that specify organ identity

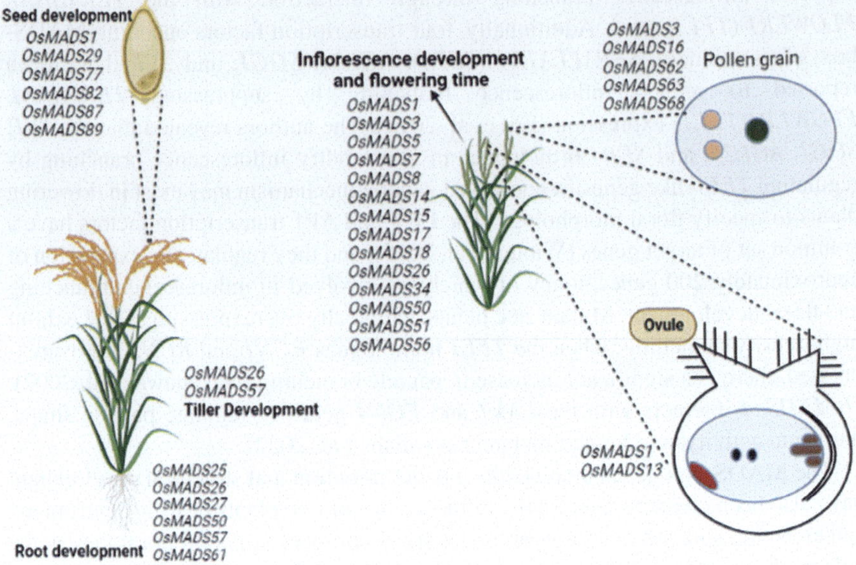

Fig. 4.12 MADS-box genes of rice plants that control the structural development of various vegetative and reproductive systems. (Adapted from Shah et al. 2022)

during the development of vegetative and reproductive parts of rice plants are described in the model (Fig. 4.12). The transcription factors of the MADS-box gene family in rice are important regulators of growth at all stages of growth, and small changes in the expression of some of these genes lead to alterations in numerous functions and the structural integrity of plant organs. The inherent potential to cause developmental shifts in a plant's life cycle provides a wide scope for a bowl of designer rice through biotechnological interventions.

4.4 Staging of Inflorescence Development

Inflorescence development in rice during the post-initiation stage of the reproductive period is determined by the activities of the rachis and branch (axillary) meristems; the activity of the former increases the primary branch number and panicle length, whereas the latter is responsible for the finalization of secondary branch and spikelet numbers. The dome of the rachis meristem generally increases in size up to the third node and thereafter decreases until it degenerates to fix the final number of primary branches. All primary branch primordia generated on the length of the rachis elongate simultaneously (Itoh et al. 2005). Overall activities of the meristems pre-ordinate the grain number and yield. Although difficult to estimate, inflorescence development during this meristem development period has been calibrated simply by measuring the increase in panicle length (Yoshida 1981). The length of the inflorescence at the stage of establishment of the rachis meristem identity varied between 0.05 and 0.01 mm. It elongates temporally as the primary branch primordia form in the sequence. Primary branch primordia elongate and generate higher-order branches, while the rachis continues to elongate and reaches a length of 0.6–0.9 mm. Spikelet meristems are established on the branches, which differentiate into two rudimentary glumes, two empty glumes and a lemma and palea. When the panicle length is 1.5–40 mm, two lodicules, six stamens, and a carpel are differentiated in the whorls. Thereafter, the rachis and branches elongate exponentially up to 220 mm, while anther and ovule development are completed inside the enclosure of the flag leaf sheath (boot). The panicle elongates further, leading to its protrusion from the boot. A more precise account of panicle development has been proposed by Ikeda et al. (2004) considering the results of molecular and genetic studies incorporated into morphological and anatomical investigations to identify landmark events of inflorescence and spikelet development. In this system, panicle and spikelet development were classified into nine and eight stages, respectively. Three landmark events of panicle development are discussed below.

4.4.1 Booting Stage

The booting stage starts nearly 16 days after visual panicle initiation, i.e. nearly 6 days before heading. The flag leaf sheath swells, enclosing the developing panicle inside it. The last stem internode (peduncle) undergoes rapid growth and elongates rapidly, causing the shoot up from the base of the plant bearing the panicle (Fig. 4.13A). The booting stage is also called the internode elongation stage. The event of booting continues for 6 days after swelling of the leaf sheath (Yoshida 1981). According to other authors, the entire period of panicle development inside the flag leaf sheath is booting, and the growth stage is identified by the length of the panicle. The panicle size reaches 2 inch long in the early booting stage, and the length of the panicle is 2–5 inches and greater than 5 inch in the middle boot and late boot stages, respectively. Following the booting stages, the panicle emerges from the flag leaf sheath enclosure and the spikelets undergo anthesis (Fig. 4.13B).

4.4.2 Heading Stage

The booting stage is followed by the heading stage. The heading stage occurs when a part of the panicle exserts outside the flag leaf sheath. For a population group, the heading date is marked when 50% of the panicles have exserted. The heading stage lasts until 90% of the panicle has exserted out of the leaf sheath. The heading date is an indicator of crop adaptation in a given locality because of its vital role in the determination of grain yield. Since manually estimating which plants reach the heading stage is tedious and time-consuming, Desai et al. (2019) used a fast automatic system based on computer vision and deep learning to identify field locations containing flowering panicles from ground-level RGB images of paddy rice. The authors utilized a CNN (convolutional neural network) to reduce the false positive

Fig. 4.13 Booting stage (**A**) and anthesis stage (**B**) of rice plant in a pigmented rice cultivar Lalkain. (Source: Photo gallery of P. K. Senapati and E. Kariali)

rate among the population. A new method for automatically observing the rice heading stage was proposed by Bai et al. (2018). In this new multi-classifier cascade method, rice panicle image patches were clearly distinguished from background patches of leaf, soil, and water for accurate estimation of the heading stage. The heading stages in tall traditional rice cultivar (A), wild rice *Oryza nivara* (B), and high-yielding (C) rice genotypes are presented in the figure below (Fig. 4.14).

4.4.3 Anthesis

Shortly after heading into rice, anthesis commences. Spikelets at the tip of the uppermost primary branches bloom first, and the event proceeds downwards (Fig. 4.13B). In most cultivars, anthesis occurs early in the morning (Julia and Dingkuhn 2012). Most of the spikelets reach anthesis between 10 and 12 h daytime (Zeng et al. 2017), and the flower remains open for approximately 30 min before closing. Early flower opening in the morning helps to prevent adverse effects of heat stress later in the day when the ambient temperature rises steadily. High-temperature episodes during anthesis lead to heat stress, which causes a great reduction in spikelet fertility and duration of grain filling (Shi et al. 2016). Antheses in rice are the most sensitive stage to adverse temperatures. Compared to those under normal temperature conditions, episodes of high temperature significantly decrease the percentage of opened spikelets, fertilized spikelets, and seed setting and increase the percentage of wrapped spikelets by reducing the spikelet opening angle and length of flower opening (Chen et al. 2020). This adverse impact of high temperature on

Fig. 4.14 Traditional rice cultivar Lalkain showing lateral mode of heading (**A**) and wild rice *Oryza nivara* (**B**) and high-yielding rice cultivars (**C**) exhibiting apical mode of heading from top of flag leaf sheath (Source: Photo gallery of E.Kariali)

flower opening is correlated with a concomitant decrease in soluble sugars, methyl jasmonate (MeJA), and jasmonic acid (JA) in the lodicules before and at the glume opening.

The mechanism underlying the hormonal regulation of flower opening in rice through a change in the turgor-motivation force was described by Chen et al. (2020). Spikelet opening in rice is caused by turgor-mediated lodicule expansion at the base and inside the lemma and palea. The turgor-motivation force prompting flower opening is generated by an increase in the soluble sugar content. Chen et al. (2020) reported that the soluble sugar content was greater in lodicules during spikelet opening than before spikelet opening, and the accumulation of water increased fresh weight of the lodicules by as much as 91%. The accumulation of osmotic regulatory substances, including potassium and other inorganic ions, along with soluble sugars, which result in lodicule expansions, is believed to be mediated by jasmonates. Lodicule expansion decreases due to the absence of turgor-motivation force when the accumulation of osmolytes is retarded; both retardations in water and osmolyte accumulation are caused by a deficiency of jasmonic acid (Liu et al. 2017). Chen et al. (2020) reported that the loss of fresh weight and soluble sugar, jasmonic acid, and methyl jasmonic acid contents in lodicules before and during spikelet opening shorten the length of spikelet opening and the number and percentage of opened spikelets in rice.

4.4.4 Pollination

A rice spikelet is a hermaphrodite comprising both male and female organs. It is self- or wind-pollinated, and there is no need for insect help. The preservation of both male and female parts within the hull enclosure precludes any possibility of cross-pollination for rice. Like those of other wind-pollinated crops, the floral organ of rice (spikelet) is small and inconspicuous bereft of scent or nectar that invites insect pollination. Pollination occurs under suitable conditions when pollen grains from the dehisced anther box float in the air and fall to the stigma surface by gravity (Fig. 4.15) However, the inclination of panicles in the field may change the relative position of the stigma and other pores to affect pollination. Aung et al. (2020) reported that panicle inclination significantly affects the number of total and germinated pollens on the stigma as the panicle inclination increases from 0° to 45°. The authors concluded that panicle inclination is a destabilizing factor for rice production, where uprightness is most ideal, but an inclination of ≥30° is unfavourable for grain production.

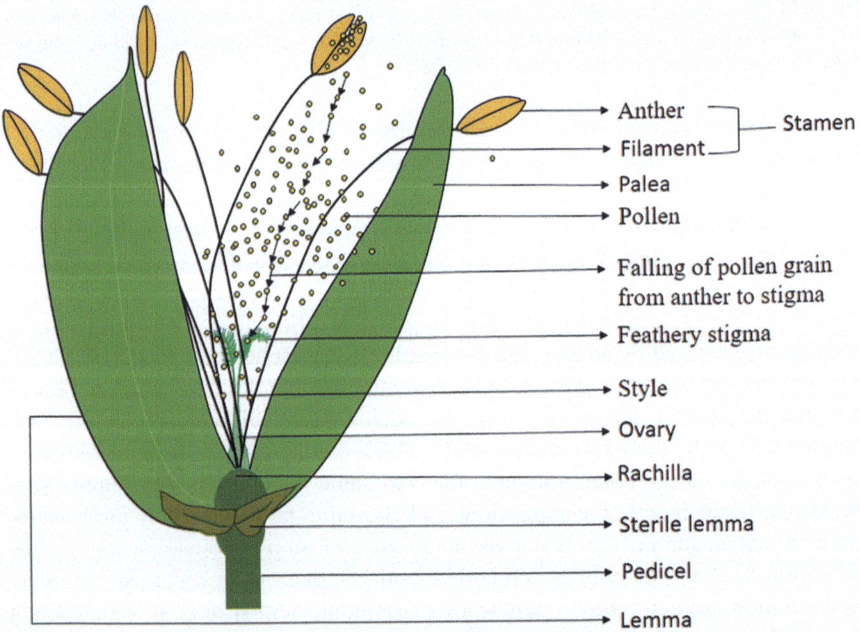

Fig. 4.15 A Model showing anther dehiscence and self-pollination in a rice spikelet

4.4.5 Release of Pollen Grains

Dehiscence of the anther box is the final feature of the anther that activates the release of pollen grains. Anther dehiscence closely matches the differentiation of pollen inside anther boxes and the opening time of the flower by the movement of lemma and palea in the hull. The driving force for anther dehiscence during the period from floret opening to pollen dispersal has been described by Matsui et al. (1999). Three anther tissues, namely, the endothecium, septum, and stomium, play crucial roles in anther dehiscence in rice. Endothecial cells undergo secondary wall thickening, which generates tensile force to break the stomium during anther wall dehydration. The septum is located between vascular bundles and two anthers locally nearby regions. The stomium is composed of a single layer of weak specialized epidermal cells and is the final breakage site for anther dehiscence. Dehiscence begins with theca splitting on the stomium in the apical part and on another wall at the base of the large locule in a turgor-driven process, whereas expansion of the split opening is a desiccatory process. The pollen grains inside the anther box swell rapidly at the time of floret opening, where they induce the theca to bulge and thereby rupture the septum. Together with pollen enlargement pressure, the inwards curvature of the locule wall adjacent to the stomium causes the splitting of the stomium in the apical part of the theca. Supported by pollen pressure, the rupture extends to the bottom of the large locule. Subsequently, the locule walls adjacent to both points of rupture dehydrate and become straight. The straitening widens the openings, permitting the exit of swollen pollen grains. A model diagram for anther dehiscence and the release of pollen is described in Fig. 4.15.

The review of Moon and Jung (2020) also highlights the role of several genes, the expression of which is critically involved in the biomechanical changes in molecular biology that occur in anther walls for successful pollen release, and they are listed below. Auxin negatively regulates endothecium lignification and jasmonic acid synthesis, and jasmonic acid controls the stomium. The genes involved in anther development are *mediating factor 17 (ARF17), MYB26, MYB108, NAC secondary wall-promoting factor 1 (NST1), NST2, FT-interacting protein 7 (OsFTIP7), OsYUCCA4*, and *homeobox 1 (OSH1)*. The host genes involved in anther dehiscence have also been elucidated; mutations in these genes result in non-dehiscent anthers. These genes included *irregular xylem1* (*IRX1*), *receptor-like protein kinase 2* (*RPK2*), *teosinte branched 1cycloidea, PCF* (*TCP24*), *Arabidopsis histidine-containing phosphor transfer factor 4* (*AHP4*), *secondary wall thickening-associated F-box 1* (*SAF1*), *cystathionine synthase domain-containing protein* (*CBSX2*), *anther dehiscence repressor* (*ADR*), and *SUMO E3 ligase1* (*SIZ1*). The genes mediating jasmonic acid-controlled stromium splitting are *MYB21* and *MYB24*.

4.5 Fertilization

4.5.1 Germination of Pollens

Seed setting in rice depends on the rapid and healthy growth of pollen tubes in the style. At pollination, the stigma receives several pollen grains. The pollen grains are recalcitrant pollens that possess high water content at the shedding stage and are sensitive to water loss. During pollination, pollens land on the stigmatic papilla surface, adhere, and hydrate. When hydration is critical, they become unviable within 5 min of exposure to air. Thus, to maintain grain setting, pollen must rapidly land on the stigma and not lose viability due to dehydration. The pollen wall is a multi-layered structure. The outer wall exine is composed of multi-layered sporopollenin; the pollen coat filling the empty cavity exine is composed of lipids, proteins, pigments, and aromatic compounds; and the innermost pollen wall intine contains cellulose, hemicellulose, pectin, and proteins. After reaching the sticky stigmatic surface of the rice gynoecium, the pollen grains germinate within 2 min. The pollen tube penetrates the stigma into the style in 5–10 min. Thirty minutes later, the end of the pollen tube reaches the bottom of the ovary. Although several pollen grains germinate and extend into tubes, only a few tubes reach the embryo sac nearly 40 min after pollination (Chen et al. 2008). A large amount of callose accumulates within 50 min after pollination. The growth rate of the pollen tube increases progressively with movement from the stigma through the style to the ovary at rates of 1500, 5000, and 5400 μm/h (Chen et al. 2008).

4.5.2 Pollen Tube Growth

Germination of compatible pollen grains occurs during the elongation of a pollen tube that emerges through the aperture of the grain. A turgor-motivated force pushes the cytoplasm to protrude through the pore and form a pollen tube. The pollen tube carrying two non-motile sperm cells descends the transmission tract of the style's retransmission tract and reaches the micropyle of the embryo sac for fertilization. It ruptures to deliver sperm cells to the embryo sac. During this process, the pollen tube communicates with the surrounding maternal tissue. Interaction between stigma, style and the pollen grain influences pollen tube elongation. The presence of high levels of calcium in the cytoplasm helps extend the pollen tube from the tip; calcium levels promote the growth of synaptic vesicles in the membrane and enlarge it from the tip. The presence of a specific cyclic nucleotide-gated channel protein, OsCNGC13, results in the identification of a novel maternal sporophytic factor needed for stylar $[Ca^{2+}]_{cyt}$ accumulation, extracellular matrix component modification, and style cell death. This factor helps pollen tubes penetrate the style for successful double fertilization and seed setting in rice plants. It is also surmised that many glycoproteins participate in the process of pollen tube growth and the success of double fertilization (You et al. 2021).

4.5.3 Embryo Sac

Rice has one ovule in the ovary. The embryo sac representing the female gametophyte develops within an ovule. It is formed from a haploid megaspore surrounded by the integument. A mature embryo sac contains one egg cell accompanied by two synergids on either side of the micropylar end and three antipodal cells at the chalazal end of the embryo sac. The central cell of the embryo sac contains two polar nuclei and a large vacuole. Hence, the embryo sac is seven-celled with eight nuclei (Fig. 4.16). The embryo sac gives rise to the endosperm and embryo after fertilization.

4.5.4 Double Fertilization

The pollen tube extends polarly in the transmitting tract of the style and reaches the micropylar end of the embryo sac in the ovule. The tube ruptures there to release the two male gametes. The male gametes enter the embryo sac through the micropyle. One sperm fuses with the egg to form a zygote, which is known as syngamy. The second gamete fuses with the two polar nuclei of the central cell to form a primary triploid endosperm nucleus. This whole event is known as double fertilization. The success of this process is crucial for seed development and grain yield. The sequence of events revealing a double fertilization mechanism in rice on a timescale has been described by Shi et al. (2018) and You et al. (2021).

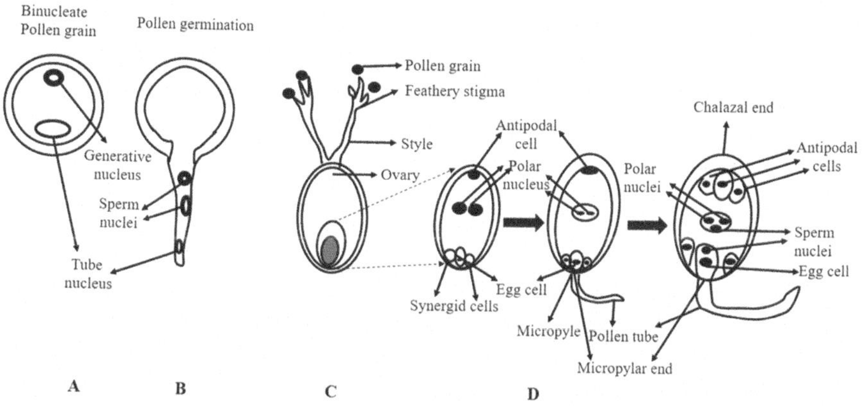

Fig. 4.16 Pollination and double fertilization of rice over time. (**A**) Binucleate pollen grain and its ability to germinate upon adhesion to the stigma. (**B**) Pollen tube growth and movement takes 1–20 min. (**C**) Events leading to fertilization and zygote formation. After 30 min, the pollen tube passes through the micropyle. (**D**) Approximately after 1 h, the pollen tube penetrates one of the synergid cells and closes to polar nuclei and egg cells. From 1.5 to 2.0 h, two polar nuclei and one sperm nucleus starts to migrate. Fusion of polar and sperm nuclei occurs at approximately 2.5 h after pollination. The free endosperm nucleus is formed from the first division of the primary endosperm nucleus. The event is accomplished at 3.0–4 h. At 5 h, the zygote is formed through the fusion of the egg cell and sperm nucleus. Within 24 h, many free endosperm nuclei are formed, and the process of seed development starts. (Adapted from You et al. 2021)

4.5.5 Seed-Setting Stage

Rice spikelets are hermaphroditic, and pollens released from anthers travel within the hull space to fertilize the female gametophyte within its enclosure, excluding chances of cross-pollination. Soon after fertilization, seed setting begins in the spikelets by initiation of grain growth. In a rice panicle, all spikelets do not mature and produce high-density good-quality grains. Premature cessation of grain growth in some developing spikelets reduces the final tally of grain numbers suitable for human consumption. These underdeveloped grains, although all fertilized, are partially filled or barren. Based on specific gravity, an index for grain quality measurement, Mohapatra et al. (1993) classified the grain quality of spikelets of rice panicles into four different categories: poor quality (floated at >1.0–1.06 sp. gr.), average quality (floated at >1.06–1.14 sp. gr.), good quality (floated at >1.14–1.20 sp. gr.) and high density grains (submerged at>1.20 sp. gr. NaCl solution). The grain setting in a rice plant can be calculated by counting the number of different types of grains via the equation given below. Hence, the seed stage is most crucial for determining the grain number, seed biomass, and yield potential. Seed-setting behaviour is influenced by biotic and abiotic factors and, above all, genetic limitations.

$$\text{Seed setting \%} = \frac{(\text{number of high density well filled grains})}{(\text{number of well filled} + \text{half filled} + \text{unfilled grains})} \times 100$$

There are reports of several genes that play important roles in the seed setting of rice via their influence on pollination. Zhou et al. (2011) identified the kinesin family-1 gene *PSS1*, which influences seed setting by controlling male gametogenesis, male meiotic chromosomal dynamics, and anther dehiscence. Li et al. (2013) reported a *POLLEN TUBE BLOCKED1* (*PTB1*) gene, encoding a RING-type E-ligase that positively regulates the panicle seed setting rate (PSSR) by promoting pollen tube growth. Xiang et al. (2019) identified a new gene, *LSSR1* (*LOW SEED SETTING RATE1*), that plays a critical role in the process of seed fertilization. The gene is predominantly expressed in rice anthers from the pre-meiotic stage to the pollen grain formation stage. A loss-of-function mutation in *lssr1* mutant lines results in low seed setting, which is caused by abnormal pollen grain germination, failed pollen tube penetration, and retarded pollen tube elongation. Several genes, such as *GS3, GS5, GW2, GW5, GW8, GL3.1, TGW6,* and *Gn1a,* for determining grain weight and size have been identified and functionally characterized. In addition, *GIF1* regulates grain filling, and *Ghd7* is a growth period-related gene. A model describing the functions of PTB1 has been presented (Li et al. 2013) in Fig. 4.17.

4.5.6 Embryo Development

Rice embryos are composed of coleoptiles, radicles, scutella, epiblasts, and shoot apical meristems (Itoh et al. 2005). Embryogenesis involves rapid cell multiplication first and cell expansion after that. Embryo development begins after the zygote in the egg nucleus divides transversely after fertilization, and daughter cells are

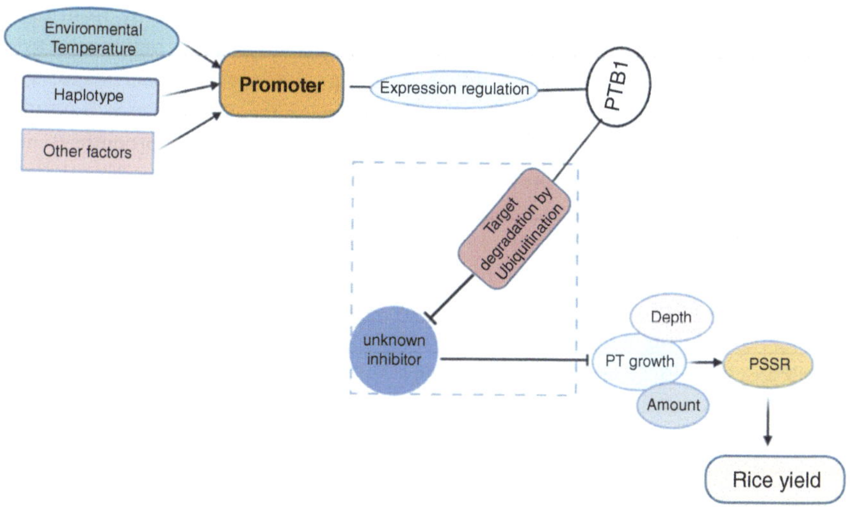

Fig. 4.17 A proposed model for the role of the *PTB1* (*POLLEN TUBE BLOCKED 1*) gene in the regulation of rice PSSR (PANICLE SEED SETTING RATE) and yield (Adapted from Li et al. 2013)

aligned at a right angle to the micropylar pole. During follow-up, there was no cell division on the antipodal side. By day two after fertilization, the embryo assumes a globular shape. On day three, there is a distinct sign of variation among the cell masses. The epidermal cells of the embryo differentiate. The coleoptile primordium and scutellum differentiate on day four. The scutellum enlarges between days 5 and 7 after fertilization. The scutellum contains two types of cells, palisade-shaped epithelial cells and oval parenchymatous cells, filled with starch granules and protein bodies. On day 5, the first leaf primordium becomes visible below the coleoptile primordium on the apical part of the embryo. The progenitor of the seminal roots, the radicle meristem, also differentiates on the basal side of the embryo at this stage. The second leaf primordium is initiated 6–7 days after fertilization. The bud scales develop at this stage to cover the coleoptile primordium, while the root cap is differentiated at the tip of the seminal root. Scutellum expansion continues until day 10 when most of the embryo cell differentiation features are complete and vascular bundles are established (Itoh et al. 2005). The size of the embryo increases rapidly after day 10 by rapid cell division and cell expansion, and the embryo gains the capacity for germination by day 12. Morphological differentiation is discontinued thereafter while the embryo tissues mature physiologically. Embryo growth and development reach a climax on day 25, and the tissue shrinks in size during the next 5 days, while the grain becomes ready for harvest.

4.5.7 Endosperm Development

Endosperm growth coincides with that of the embryo. The endosperm undergoes a coenocytic stage at 1–2 days after fertilization. Two daughter nuclei are formed in the first division of the endosperm mother nucleus. They move away from each other and generate no cell wall between them. They divide rapidly and form a systematic multinucleate endosperm, which is in the coenocytic/nuclear stage. The nuclei organize themselves into a single layer covering the endosperm as a peripheral layer. The endosperm nuclei divide faster in the area near the proembryo and gradually decrease towards the antipodal end. The nuclei of the peripheral layer divide periclinal to form one additional layer.

The nuclei of the peripheral layers form cell walls, putting the endosperm in the cellular stage. Cell wall formation between the nuclei begins on day 3 and finishes on day 5 after fertilization. In this stage, the endosperm enters a period of cellular development, during which the peripheral layer works like a cambial tissue, driving repeated cell multiplication. Cell division is directed towards the central cavity of the embryo sac until complete filling. Cell division mostly occurs during the night period, and the process ends 9–10 days after fertilization; this cessation of cell division coincides with that of the embryo. Among the radial arrangement of endosperm cells in rice, the number of cell layers is greater on the dorsal side than on the ventral side. At the end of cell division, cells in the outermost cell layer differentiate into aleurone layers. On the dorsal side, there are 3–6, 1–2, and 1 aleurone layers on the dorsal, ventral, and lateral sides of the endosperm tissue, respectively. All the cells of the endosperm are parenchymatous and store starch at different levels. Cells at the centre are the oldest, and they accumulate more starch than peripheral cells.

Endosperm filling is characterized by the filling of starch and protein bodies. For starch accumulation, amyloplasts are formed on day 5 after fertilization, and protein body formation is initiated on day 9 after fertilization. Starch accumulation reaches a climax between days 10 and 20 after fertilization (Fu et al. 2013; Zhu et al. 2011). The endosperm reaches the maturation stage from day 21 onwards, and there is no gain in fresh weight from days 20 to 30. Most of the starchy endosperm and maternal tissues, such as vascular bundles, nucellar projections and nucellar epidermal cells, are degraded by the end of maturation (Wu et al. 2016a, b). As far as seed dry matter is concerned, the contribution of the embryo is negligible compared to that of the endosperm. The temporal accumulation of dry matter in the seed comprises three distinct phases in an S-shaped curve. Phase I begins soon after fertilization, during which the amount of dry matter is minimal. This stage is crucial for grain development because it fixes the final cell number of the endosperm. This phase lasts for 12 days after fertilization (Patel and Mohapatra 1996). Phase II is responsible for the linear increase in dry matter, during which cells in the endosperm expand and gain weight via the transport of assimilates from maternal tissue in an aqueous medium facilitated by the entry of water. This phase is characterized by endoreduplication, which leads to the formation of polyploid cells, increasing the strength of the endosperm for the storage of essential reserve food materials. In the final phase, very little dry matter growth occurs. The seeds lose water and mature while entering the dormant stage. The stages of embryo and endosperm development are shown in Fig. 4.19 and were adapted from An et al. (2020).

The temporal changes in the entry and exit of water from developing rice seeds are described by Mohapatra et al. (1993) in Fig. 4.18. The water content of the

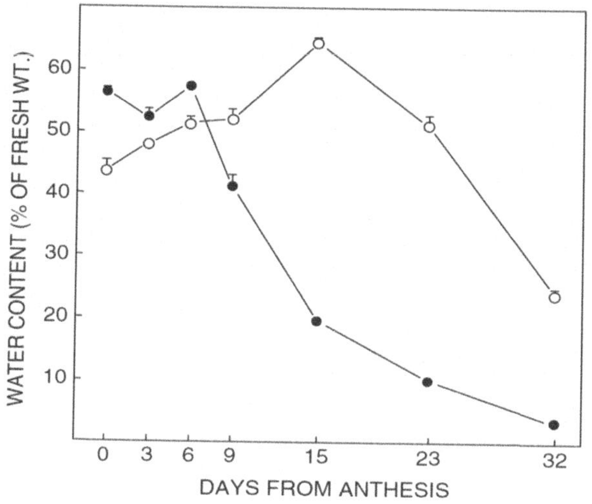

Fig. 4.18 Temporal changes in the water content of superior (good-quality high-density grain, closed symbol) and inferior (poor-quality low-density grain, open symbol) developing grains of rice (cultivar Kajalghara) panicles during the post-anthesis period. (Adapted from Mohapatra et al. 1993)

high-density or superior grains increases during the first week after anthesis and decreases sharply thereafter until maturity. In comparison, the water content increases slowly up to day 15 after anthesis and decreases thereafter in the inferior type grain; the rate of decrease is much slower than that of the superior grain.

4.5.8 Interaction Between the Embryo and Endosperm During Grain Filling

Double fertilization induces concomitant development of the embryo and endosperm, and these two tissues are intimately correlated with subsequent growth. The development of both tissues is temporally and intimately synchronized. Thus, both compartments should be closely coordinated and interact physically and chemically to develop high-density grains. The challenge is enormous in the drive to clarify the mechanisms of the flow of resources, possibly hormones and/or nutrients, between the two compartments and establish a model for inter-organ communication. Nevertheless, An et al. (2020) summarized the shreds of evidence from the literature for inter-organ exchange in the bidirectional flow of growth-controlling factors and combined them into a comprehensive module, as given below in Fig. 4.19.

This exploration of cross-talk between the embryo and endosperm was explored further by workers who found that the ratio of glucose to sucrose mediates endosperm differentiation and that the ratio was greater in the embryo and endosperm between days 5 and 10 after fertilization and decreased thereafter in a manner corresponding to the transition from differentiation to storage accumulation; moreover,

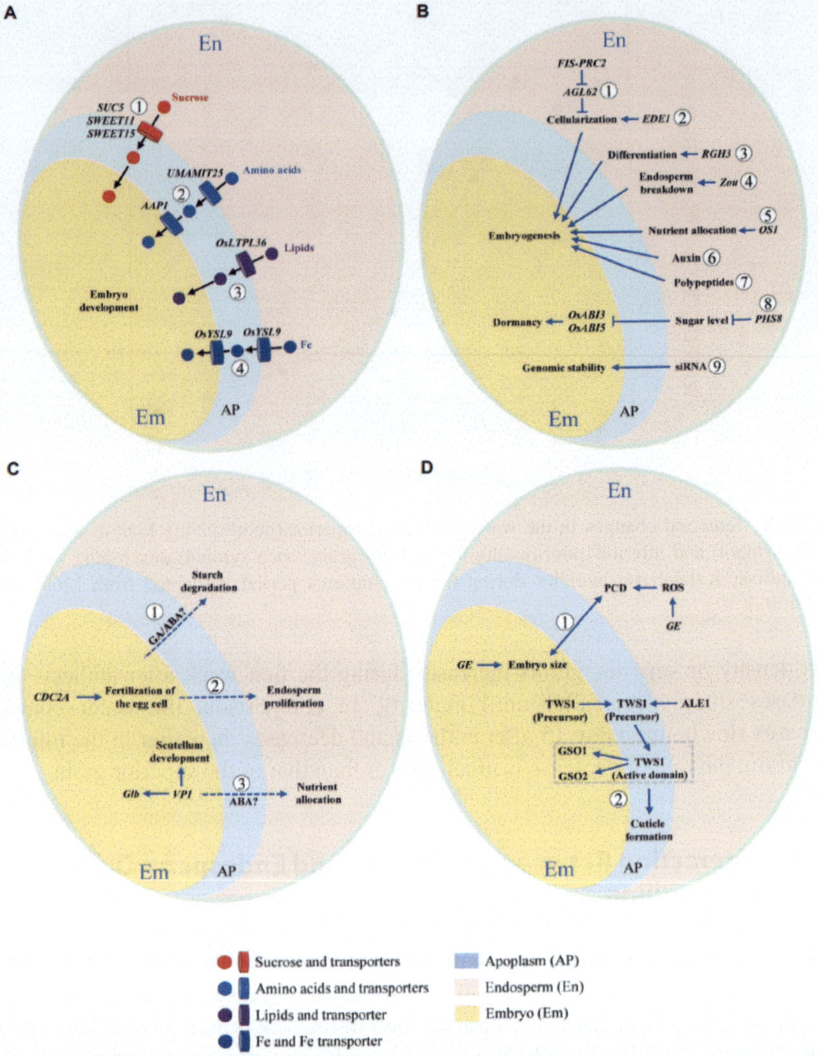

Fig. 4.19 Schematic illustration of embryo–endosperm interactions during rice grain filling. Concomitant development of the embryo and endosperm within the limited space of the seed hull requires bidirectional dialogues between the two compartments. These interactions can be summarized by the following four aspects. (**A**) Nutrient flow from the endosperm to the embryo via apoplastic transport pathway. Nutrients and their transporters are (1) Sucrose transporters, *SWEET11*, *SWEET15,* and *SUC5*,(2) amino acid transporters, *UMAMIT25* and *AAP1*, (3) Lipid transporters, *OsLTPL36*, and (4) Fe transporters, *OsYSL9*. (**B**) Dependence of embryo development on the endosperm. Genes exclusively expressed in or signalling released from the endosperm that regulate embryo development are shown. (1) *FIS-PRC2* and *AGL62*, (2) *EDE1*, (3) *RGH3*, (4) *Zou*, (5) *OS1*, (6) Auxin, (7) Poly-peptides, (8) *PHS8*, and (9) Small-interfering RNA (siRNA). (**C**) Regulation of endosperm development by the embryo. Genes expressed in or signalling released from embryos that modulate endosperm development are shown. (1) GA/ABA, (2) *CDC2A,* and (3) *VP1*. (**D**) Developmental coordination between the embryo and endosperm. Bidirectional dialogues are presented by signalling pathways mediated by (1) GE and (2) TWS1. The dotted line indicates a putative or unproven signalling pathway (Adapted from An et al. 2020 with permission)

the sugar contents were comparatively greater in the embryo. Sucrose synthase enzyme-encoding genes differ unequally between the embryo and endosperm; out of the five isoforms, two were over-expressed in the embryo, and the other three were over-expressed in the endosperm. The endosperm contains a greater starch concentration than the embryo, and accordingly, the expression of genes controlling starch synthesis also differs. In plants, the trehalose-6-phosphate (T6P) signal controls the availability of sucrose in plant cells via its target protein SnRK1. It was shown that T6P, SnRK1, gibberellins, and auxin signalling in embryos regulate endosperm developmental transition during the grain-filling stage (Tao et al. 2021).

The maternal tissues are composed of the pericarp, testa, and nucleus and are located in the outer layer. The endosperm consists of aleurone and starchy endosperm packed with starch granules and protein bodies, respectively. Generally, the aleurone has two or three layers on the dorsal side, whereas one is on the ventral side of the rice grain (Fig. 4.20). Opaque tissues are called chalky and generally occur on the ventral side and central parts of the endosperm. The embryo is composed of two structures: the scutellum and the embryonic axis (plumule, radicle, and embryonic axis) (An et al. 2020).

Grain:

- Rice grains are ripened ovaries with lemmas and paleae firmly adhered to them.
- The lemma and palea with other smaller components from the hull are removed from the shelling rice for consumption.
- The rice fruit is a caryopsis in which a single seed is fused with the wall of the ovary (pericarp).
- The seed consists of an endosperm and an embryo. The embryo is very small and is found on the ventral side of the caryopsis. It contains plumules (embryonic leaves) and radicles (roots).

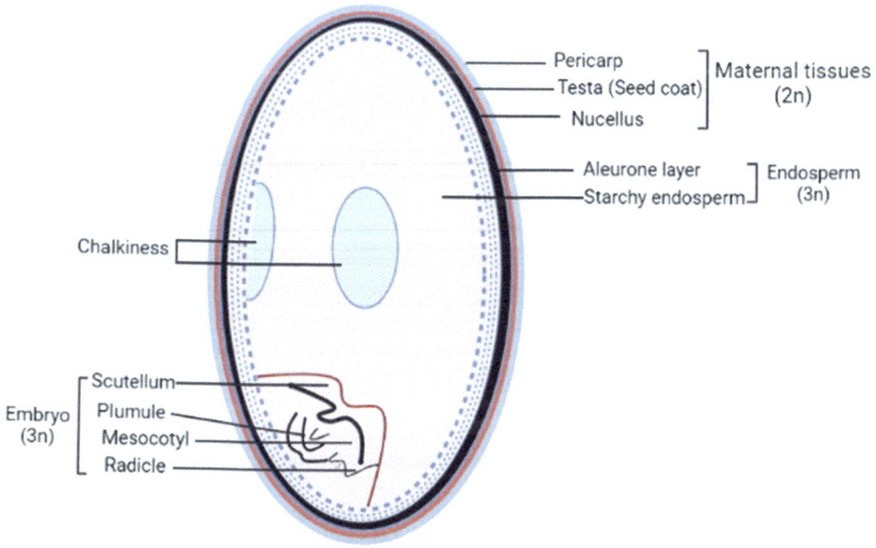

Fig. 4.20 The structure of mature rice grains. (Adapted from An et al. 2020)

Upon immersion in water or field water in sowing, the radicle grows as a root, and the plumule grows as a shoot.

References

An L, Tao Y, Chen H, He M, Xiao F, Li G, Ding Y, Liu Z (2020) Embryo-endosperm interaction and its agronomic relevance to rice quality. Front Plant Sci 11:587641. https://doi.org/10.3389/fpls.2020.587641

Aung W, Takashi S, Tanaka T, Matsui T (2020) Panicle inclination influences the pollination stability of rice (*Oryza sativa* L.). Plant Prod Sci 23:60–68. https://doi.org/10.1080/1343943X.2019.1698971

Bai X, Cao Z, Zhao L, Zhang J, Lv C, Li C, Xie J (2018) Rice heading stage automatic observation by multi-classifier cascade-based rice spike detection method. Agric For Meteorol 259:260–270

Chen S, Zhong W, Liu M, Xie Z, Wang H (2008) Pollen grain germination and pollen tube growth in pistil of rice. Ric Sci 15:125–130

Chen H, Xu N, Wu Q, Yu B, Chu Y, Li X, Huang J, Jin L (2018) OsMADS27 regulates root development in a NO_3^--dependent manner and modulates the salt tolerance in rice (*Oryza sativa* L.). Plant Sci 277:20–32

Chen J, Xu Y, Fei K, Wang R, He J, Fu L, Shao S, Li K, Zhu K, Zhang W, Wang Z, Yang J (2020) Physiological mechanism underlying the effect of high temperature during anthesis on spikelet-opening of photo-thermo-sensitive genic male sterile rice lines. Sci Rep 10:22101. https://doi.org/10.1038/s41598-020-59183-0

Coen HS, Meyerowitz EM (1991) The war of the whorls: genetic interactions controlling flower development. Nature 353:31–37

Cui R, Han J, Zhao S, Su K, Wu F, Du X, Xu Q, Chang K, Thei Ben G, Meng Z (2010) Functional conservation and diversification of class E floral homeotic genes in rice (*Oryza sativa*). Plant J 61:767–781

Das K (2018) The relative contribution of variation in endosperm starch biosynthesis enzyme activities during grain filling of rice cultivars with contrasting panicle architecture. PhD thesis, Sambalpur University, India

Desai SV, Balasubramanian VN, Fukatsu T, Ninomiya S, Guo W (2019) Automatic estimation of the heading date of paddy rice using deep learning. Plant Methods 15:76

Fu J, Xu YJ, Chen L, Yuan LM, Wang ZQ, Yang JC (2013) Changes in enzyme activities involved in starch synthesis and hormone concentrations in superior and inferior spikelets and their association with grain filling of super rice. Rice Sci 20:120–128. https://doi.org/10.1016/S1672-6308(13)60116-X

GRiSP (Global Rice Science Partnership) (2013) Rice Almanac, 4th edn. International Rice Research Institute, Los Banos

Huang S, Liang Z, Chen S, Sun H, Fan X, Wang C, Xu G, Zhang Y (2019) A transcription factor, OsMADS57, regulates long-distance nitrate transport and root elongation. Plant Physiol 180:882–895

Ikeda K, Sunohara H, Nagato Y (2004) Development course of inflorescence and spikelet in rice. Breed Sci 54:147–156

Irish V (2017) The ABC model of flower development. Primer 27:887–890

Itoh J-I, Nonomura K-I, Ikeda K, Yamaki S, Inukai Y, Yamagishi H, Kitano H, Nagato Y (2005) Rice plant development: from zygote to spikelet. Plant Cell Physiol 46(1):23–47

Julia C, Dingkuhn M (2012) Variation in time of day of anthesis in rice in different climatic environments. Eur J Agron 43:166–174

Kariali E, Sahoo SS, Panigrahi S, Pati D, Sahu BB, Dash SK, Mohapatra PK (2022) Inter-auricular distance: a morphological biomarker for ethylene production in the boot and coincidental grain loss of rice plant. Crop Environ 1:41–49

Kim SL, Lee S, Kim HJ, Nam HG, An G (2007) OsMADS51 is a short-day flowering promoter that functions upstream of Ehd1, OsMADS14, and Hd3a. Plant Physiol 145:1484–1494

Lancashire PD, Bleiholder H, Boom TVD, Langeluddeke P, Stauss R, Weber E, Witzenberger A (1991) A uniform decimal code for the growth stages of crops and weeds. Ann Appl Biol 119:561–601

Laza MRC, Peng S, Akita S, Saka H (2004) Effect of panicle size on grain yield of IRRI-released *Indica* rice cultivars in the wet season. Plant Prod Sci 7:271–276

Lee S, Kim J, Han JJ, Han MJ, An G (2004) Functional analyses of the flowering time gene *OsMADS50*, the putative SUPPRESSOR OF OVEREXPRESSION OF CO 1/AGAMOUS-LIKE 20 (SOC1/AGL20) orthologue in rice. Plant J 38:754–764

Lee DY, Lee J, Moon S, Park SY, An G (2007) The rice heterochronic gene *SUPERNUMERARY BRACT* regulates the transition from spikelet meristem to floral meristem. Plant J 49:64–78

Lee S, Choi SC, An G (2008a) Rice SVP-group MADS-box proteins, OsMADS22 and OsMADS55, are negative regulators of brassinosteroid responses. Plant J 54:93–105

Lee S, Woo Y-M, Ryu S-I, Shin Y-D, Kim WT, Park KY, Lee I-J, An G (2008b) Further characterization of a rice AGL12 group MADS-box gene, *OsMADS26*. Plant Physiol 147:156–168

Li S, Li W, Huang B, Cao X, Zhou X, Ye S, Li C, Gao F, Zou T, Xie K, Ren Y, Ai P, Tang Y, Li X, Deng Q, Wang S, Zheng A, Zhu J, Liu H, Wang L, Li P (2013) Natural variation in PTB1 regulates rice seed setting rate by controlling pollen tube growth. Nat Commun 4:2793. https://doi.org/10.1038/ncomms3793

Li C, Liu XJ, Yan Y, Alam MS, Liu Z, Yang ZK, Tao R, Yue EK, Duan MH, Xu JH (2022) OsLHY is involved in regulating flowering through the Hd1- and Ehd1- mediated pathways in rice (*Oryza sativa* L.). Plant Sci 315:111145. https://doi.org/10.1016/j.plantsci.2021.111145

Liu C, Teo ZWN, Bi Y, Song S, Xi W, Yang X, Yin Z, Yu HA (2013) Conserved genetic pathway determines inflorescence architecture in *Arabidopsis* and rice. Dev Cell 24:612–622

Liu L, Zou Z, Qian K, Xia C, He Y, Zeng H, Zhou X, Reimann M, Yin C (2017) Jasmonic acid deficiency leads to scattered floret opening time in cytoplasmic male sterile rice Zhenshan 97A. J Exp Bot 68:4613–4625

Matsui T, Omasa K, Horie T (1999) Mechanism of anther dehiscence in rice (*Oryza sativa* L.). Ann Bot 84:501–506

Meng F, Xiang D, Zhu J, Li Y, Nao C (2019) Molecular mechanisms of root development in rice. Rice 12:1. https://doi.org/10.1186/s12284-018-0262-x

Mohapatra PK, Patel R, Sahu SK (1993) Time of flowering affects grain quality and spikelet partitioning within rice panicle. Aust J Plant Physiol 20(2):231–241

Mohapatra PK, Panigrahi R, Turner N (2011) Physiology of spikelet development on the rice panicle: is the manipulation of apical dominance crucial for grain yield improvement? Adv Agron 110:330–359

Moon S, Jung KH (2020) The first step in the successful fertilization of rice and *Arabidopsis*: plants involve pollen longevity, adhesion, and hydration. Plan Theory 9:956. https://doi.org/10.3390/plants9080956

Nakagawa M, Shimamoto K, Kyozuka J (2002) Overexpression of RCN1 and RCN2, rice TERMINAL FLOWER 1/CENTRORADIALIS homologs, delays phase transition and alters panicle morphology in rice. Plant J 29:743–750

Panigrahi S (2023) Transcriptome dynamics and regulation of gene expression in relation to starch biosynthesis and grain filling in contrasting Recombinant Inbred Lines (RIL) of rice. PhD thesis, Sambalpur University, India, pp 200

Patel R, Mohapatra PK (1996) Assimilate partitioning within floret components of contrasting rice spikelets produces qualitatively different types of grains. Aust J Plant Physiol 23(1):85–92

Ren DY, Li YF, Zhao FN, Sang XC, Shi JQ, Wang N, Guo S, Ling Y, Zhang C, Yang Z, He G (2013) *MULTI-FLORET SPIKELET1*, which encodes an AP2/ERF protein, determines SPIKELET meristem fate and sterile lemma identity in rice. Plant Physiol 162:872–884

Ren D, Xu Q, Qiu Z, Cui Y, Zhou T, Zeng D, Longbiao Guo L, Qian Q (2019) *FON4* prevents the multi-floret spikelet in rice. Plant Biotechnol J 17(6):1007–1009

Sage TL, Sage RF (2009) The functional anatomy of rice leaves: implications for refixation of photorespiratory CO_2 and efforts to engineer C4 photosynthesis into rice. Plant Cell Physiol 50:756–772

Shah L, Sohail A, Ahmad R, Cheng S, Cao L, Wu W (2022) The roles of MADS-box genes from root growth to maturity in *Arabidopsis* and rice. Agronomy 12:582. https://doi.org/10.3390/agronomy12030582

Shi P, Zhu Y, Tang L, Chen J, Sun T, Cao W, Tian Y (2016) Differential effects of temperature and duration of heat stress during anthesis and grain filling stages in rice. Environ Exp Bot 132:28–41

Shi W, Li X, Schmidt RC, Struik PC, Yin X, Jagadish SVK (2018) Pollen germination and in vivo fertilization in response to high temperature during flowering in hybrid and inbred rice. Plant Cell Environ 41:1287–1297

Sugiyama S-H, Yasui Y, Ohmuri S, Tanaka W, Hirano H-Y (2019) Rice flower development revisited: regulation of carpel specification and flower meristem determinacy. Plant Cell Physiol 60:1284–1295

Tao Y, An L, Xiao F, Li G, Ding Y, Paul MJ, Liu Z (2021) Integration of embryo–endosperm interaction into a holistic and dynamic picture of seed development using a rice mutant with notched-belly kernels. Crop J 10:729. https://doi.org/10.1016/j.cj.2021.10.007

Thakur AK, Uphoff N, Antony E (2010) An assessment of physiological effects of system of rice intensification (SRI) practices compared with recommended rice cultivation practices in India. Exp Agric 46:77–98

Winter CM, Yamaguchi N, Wu MF, Wagner D (2015) Transcriptional programs are regulated by both LEAFY and APETALA1 at the time of flower formation. Physiol Plant 155:55–73

Won SY, Jung JA, Kim JS (2021) Genome-wide analysis of the MADS-box gene family in *chrysanthemum*. Comput Biol Chem 90:107424

Wu XB, Liu JX, Li DQ, Liu CM (2016a) Rice caryopsis development I: dynamic changes in different cell layers. J Integr Plant Biol 58:772–785. https://doi.org/10.1111/jipb.12440

Wu XB, Liu JX, Li DQ, Liu CM (2016b) Rice caryopsis development II: dynamic changes in the endosperm. J Integr Plant Biol 58:786–798. https://doi.org/10.1111/jipb.12488

Wu J, Yu C, Huang L, Gan Y (2021) A rice transcription factor, OsMADS57, positively regulates high salinity tolerance in transgenic *Arabidopsis thaliana* and *Oryza sativa* plants. Physiol Plant 173:1120–1135

Xiang X, Zhang P, Yu P, Zhang Y, Yang Z, Sun L, Wu W, Khan RM, Abbas A, Cheng S, Cao L (2019) LSSR1 facilitates the seed-setting rate by promoting fertilization in rice. Rice 12:31. https://doi.org/10.1186/s12284-019-0280-3

Xu X, Vergara BS (1986) Morphological changes in rice panicle development. A review of the literature. In: IRRI research paper series no. 117. International Rice Research Institute, Manila

Yamaguchi T, Nagasawa N, Kawasaki S, Matsuoka M, Nagato Y, Hirano H-Y (2004) The YABBY gene *DROOPING LEAF* regulates carpel specification and midrib development in *Oryza sativa*. Plant Cell 16:500–509

Yamaguchi T, Lee YD, Miyao A, Hirochika H, An G, Hirano H-Y (2006) Functional diversification of the two C-class genes, *OSMADS3* and *OSMADS58*, in *Oryza sativa*. Plant Cell 18:15–28

Yoshida S, Hayakawa Y (1970) Effect of mineral nutrition on tillering of rice. Soil Sci Plant Nutr 16: 180–191

Yoshida S (1981) Fundamentals of rice crop science. International Rice Research Institute, Manila

Yoshida H, Nagato Y (2011) Flower development in rice. J Exp Bot 62:4719–4730

You L, Yu L, Liang R, Sun R, Hu F, Lu X, Zhao J (2021) Identification and analysis of genes involved in double fertilization in rice. Int J Mol Sci 22(23):12850. https://doi.org/10.3390/ijms222312850

Yu C, Liu Y, Zhang A, Su S, Yan A, Huang L, Ali I, Liu Y, Forde BG, Gan Y (2015) MADS-box transcription factor OsMADS25 regulates root development through the affection of nitrate accumulation in rice. PLoS One 10:e0135196

Yujugullu O, Marelli S, Erten E, Sudret B, Hajnsek I (2017) Determining rice growth stage with X-band SAR: a meta-model-based inversion. Remote Sens (Basel) 9:460. https://doi.org/10.3390/rs9050460

Zahn LM, Feng B, Ma H (2006) Beyond the ABC model: regulation of floral homeotic genes. Adv Bot Res 44:163–207

Zeng Y, Zhang Y, Xiang J, Uphoff NT, Pan X, Zhu D (2017) Effects of low-temperature stress on spikelet-related parameters during anthesis in *indica–japonica* hybrid rice. Front Plant Sci 8:1350. https://doi.org/10.3389/fpls.2017.01350

Zhang H, Li Y (2019) The spikelet is a different "flower" in rice. Plant Cell. https://doi.org/10.1105/tpc.18.00682

Zhang G, Xu N, Chen H, Wang G, Huang J (2018) OsMADS25 regulates root system development via auxin signalling in rice. Plant J 95:1004–1022

Zhao S, Gu J, Zhao Y, Hassan M, Li Y, Ding W (2015) A method for estimating spikelet number per panicle: integrating image analysis and a 5-point calibration model. Sci Rep 5:16241. https://doi.org/10.1038/srep16241

Zhou S, Wang Y, Li W, Zhao Z, Ren Y, Wang Y, Gu S, Lin Q, Wang D, Jiang L, Su N, Zhang X, Liu L, Cheng Z, Lei C, Wang J, Guo X, Wu F, Takahashi H, Wang H, Wan J (2011) *Pollen semi-Sterility1* encodes a Kinesin-1–like protein important for male meiosis, anther dehiscence, and fertility in Rice. Plant Cell 23(1):111–129

Zhu GH, Ye NH, Yang JC, Peng XX, Zhang JH (2011) Regulation of expression of starch synthesis genes by ethylene and ABA in relation to the development of rice inferior and superior spikelets. J Exp Bot 62:3907–3916. https://doi.org/10.1093/jxb/err088

Tiller Development in Rice

5

Abstract

The rice plant has several distinct successive nodes on the culm. A node bears one leaf. On the axil of the leaf, there is a bud that forms a stem branch or tiller. Tiller development is characterized by three stages: (A) apical and axillary meristem initiation, (B) bud development, and (C) extension of bud growth to a tiller. The tillers generated on the leaf axils of the main culm are primary. In the same sequence, secondary and tertiary tillers are produced on primary and secondary tillers, respectively. Tiller production continues in the leaf axils until the end of the vegetative period. Tiller genesis occurs temporally in the main and other shoots of the plant, but maturity is almost synchronous. The tiller number varies with the variation in genotype and growth habitat. Fertilizing rice with NPK during the vegetative growth period promotes tillering, leading to high grain yield. However, heterogeneity in tiller yield associated with tiller formation and fertilizer application reveals that tiller asynchrony increases with the level of fertilization, mainly nitrogen. The positive effects of nitrogen on the rice population are the outcome of an increase in the grain yield of early-formed tillers. Thus, the primary tiller number is always critical for achieving high grain yield in rice.

Keywords

Rice tillers · Tiller dynamics · Tiller management · Fertilization · Phytomers

5.1 Rice Tiller

Rice plants are typical rhizomatous monocot grasses that have immense potential to generate tillers. At the germination stage, the plant is single-culmed but subsequently produces a variety of branches (tillers). Tiller production in rice is an important agronomic trait for grain yield determination. According to Li et al. (2003), the

© The Author(s), under exclusive license to Springer Nature Singapore Pte Ltd. 2025

P. K. Mohapatra et al., *Tillering Behavior of Rice Plant*,
https://doi.org/10.1007/978-981-97-5235-5_5

tiller is a specialized grain-bearing branch of the main shoot that is formed on the unelongated basal internode of the stem during the vegetative period and becomes independent of the main shoot during growth using its capacity for own adventitious root formation. Rice plants have several distinct successive nodes on the culm. Each node has one leaf on the unelongated basal part of the culm that produces tillers, whereas the elongated upper part remains tillerless. On the axil of the leaf, there is a bud. This axillary bud subsequently develops into a stem branch or tiller. Tillers develop on the nodes of mother shoots and subsequently from early-stage tillers, especially in the basal region of rice hills. The tillering pattern in rice during the early tillering stage is described in Figs. 5.1 and 5.4.

The shoot architecture of rice plants, including tiller formation, is determined by the activity and determinacy of the shoot apical meristem and axillary meristem. The event of tillering is a three-stage process (Hussein et al. 2014). First, an axillary meristem bud is initiated by the establishment of a stem cell population in the leaf axil. In the second stage, leaf primordia are produced from the axillary meristem to support axillary bud development. In the third stage, bud outgrowth occurs, after which the growing bud extends into a tiller. Theoretically, each stem node has the potential to produce a leaf, tiller, and root. However, tillers and roots develop later than leaves. Plants enter the tillering stage from the pre-tillering stage at the 3–4 leaf stage when tiller initiation occurs. During the pre-tillering stage, both seminal and secondary or lateral roots develop further, the first four leaves appear distinctly, and the plant becomes ready to show its first tiller. Tillers are formed at the base of the plant inside the stem surrounded by the leaves. The tillering stage is numbered

Fig. 5.1 Tillering pattern in rice. *M* mother shoot, *P1* first primary tiller, *P2* second primary tiller, *P3* third primary tiller, *P4* fourth primary tiller, *P1S1* first secondary tiller emerged from P1 and so on. (Adapted from Kariali 2014)

according to the number of tillers. Tillering stage '1' refers to the appearance of first tiller (Fig. 5.2).

Katayama (1951) proposed the synchronous tiller development theory for cereals, including rice, in which the initiation of the *N*th tiller coincides with the emergence of the (N + 3)th leaf. Accordingly, the first tiller initiation begins in rice seedlings at the five-leaf stage (four true leaves + prophyll). At this stage, the culm nodes are compressed, and the internode length is less than 0.04 inch. The first tiller originates between the main stem and the second leaf from the base of the plant. The second tiller develops from the axillary position between the third leaf and the stem when the sixth leaf appears on the plant. Subsequent tiller initiation on the stem continues in the same order. On average, modern rice cultivars produce 20–25 tillers, 14–15 of which become fertile, while the rest are unproductive. Normally, one tiller bud arises from the axil of each leaf on the culm, but only those on the unelongated compact culm possess the potential to develop into tillers bearing productive panicles, whereas the rest of the tillers borne on the nodes above become unproductive or sterile. Some of the tillers do not have the capacity for outgrowth and remain

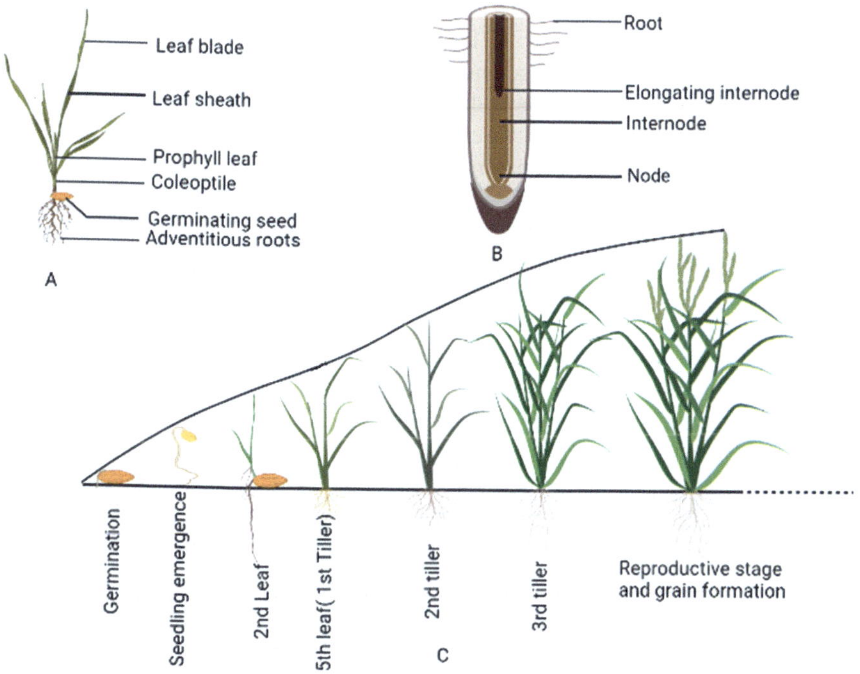

Fig. 5.2 Cumulative leaf number and tiller formation. (**A**) Seedling stage or pre-tillering stage. (**B**) Vertical hypothetical section of a stem showing compact and elongated nodes. (**C**) Germination to first tiller formation under non-stress conditions on a 120-day-long plant. Tiller formation is a strict synchronization process. At the 3–4 leaf stage, tillering initiates. The plants enter the tillering stage at the seedling stage. The first tiller emerges at the fifth leaf stage and is visible from the outside. These processes continue until the beginning of the elongation of the internode. (Adapted from Counce et al. 2000; Moldenhauer et al. 2001)

confined inside the leaf sheath. Hence, the final tiller number of a rice plant is determined by the number of tiller buds capable of outgrowth. Tiller initiation generally finishes at the internodal elongation and/or panicle initiation stage. There is a time gap between maximum tillering and panicle initiation, mainly in medium- to long-duration genotypes (maturity duration of 120 days or more). The time gap between maximum tillering and panicle initiation is termed the 'vegetative lag phase'. The vegetative phase is inseparable from the reproductive phase at 105 days or before. The cessation of tiller initiation at this stage marks the end of the vegetative period and the beginning of the reproductive period.

5.2 Tillering Capacity

The phyllochron is defined as the period during which leaves on the main stem of a plant appear. In rice, leaves emerge sequentially on the main stem one after another, maintaining a time gap. A phytomer is defined as 'one of the individual structural units that in serial arrangement make up the body of a plant' (Webster's Dictionary). A typical phytomer in rice contains a node with a bounding internode and leaf and an axillary bud between the stem and leaf attached to the node. The number of tillers in rice plants is a trait that is specific mainly to the cultivar (Mohanan and Mini 2008; Pawar et al. 2016; Yoshida and Hayakawa 1970). The number of tillers is a key component of grain yield because it determines the panicle number. In addition to variations in cultivar, environmental factors such as hill spacing, allelopathic weed competition, fertility, damage from insects and pests, and depth of surface water in the field significantly influence tillering capacity. The tillering ability of rice plants is calibrated with the number of phyllochrons during growth, and the plant completes the process before it reaches the reproductive period (Nemoto et al. 1995; Stoop et al. 2002). The phyllochron is a period or cycle of plant growth in which one or more phytomers (units of tiller, leaf, or root) are generated from meristematic tissue (Fig. 5.3). This number increases during the vegetative growth period. Depending on the extent of favourable growth conditions, the length of phyllochrons can vary from 4 to 10 days; soil and ambient temperatures, duration of sunlight period, plant spacing, nutrient availability, nature of soil compaction, soil water regime, and level of aeration are the key factors determining length (Thakur et al. 2016). These workers reported that the tiller number per plant increases beyond 200 under the most favourable conditions when more than 12 or 13 phyllochrons were completed before panicle initiation. Increasing tiller numbers can lead to greater grain yield realization, which is debatable. Song et al. (2009) reported that grain yield was significantly lower under direct seeding than under transplanting conditions despite the greater number of tillers in the former situation.

The synchronism of leaf and tiller emergence relative to the position and to the main stem developmental stage of rice has been elucidated further by Jaffuel and Dauzat (2005). The tiller and main shoot leaf phyllochrons are not identical at either the vegetative or reproductive stage. However, leaf emergence on the main shoot and tiller development are closely linked, suggesting a clear synchrony between leaf

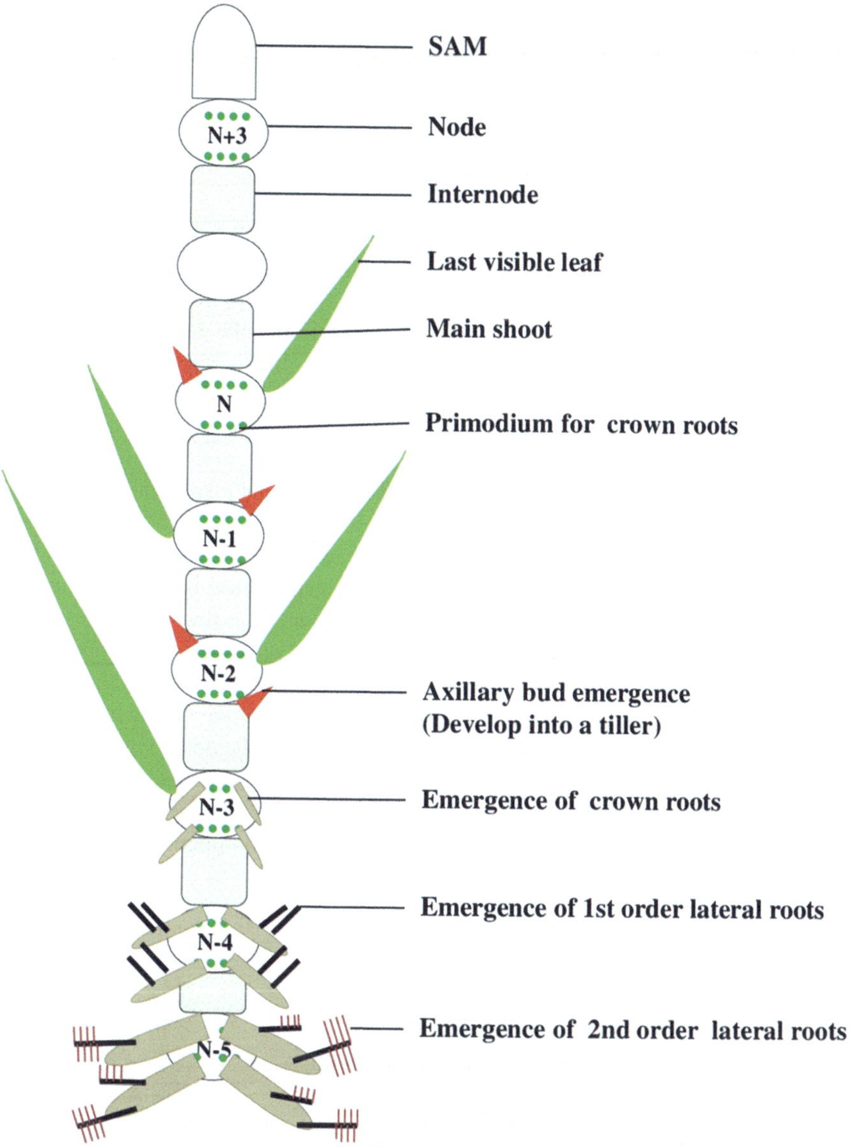

Fig. 5.3 A phytomer of rice is made up of an axillary bud, a leaf and a node. N: node. (Adapted from Gonin et al. 2019)

emergence rates of the main shoot and tillers where the final leaf number of a tiller becomes dependent on the main shoot leaf number. Jaffuel and Dauzat (2005) examined the morphological and developmental dynamics of each axis from the third leaf stage up to the heading; they showed a clear decrease in the number of phytomers per axis with increasing branching order and rank and synchronous emergence of leaves on the main stem and tillers up to the flowering stage. Thus, axillary bud development in tillers depends on their topological location and plant developmental stage.

5.3 Varietal Difference in Tillering

The tillering potential of rice is an important attribute of rice cultivation. The tillering capacities of the cultivars are divided into three types: high tillering (26 to 35 tillers), medium tillering (10 to 25 tillers), and low tillering (0 to 2 tillers). Tillering ability is genetically controlled (Pawar et al. 2016), and the farmer chooses the genotype most suitable for his field. A high-tillering ability can increase the plasticity of plants in adapting to thin or dense stands in the field, whereas low-tillering ability leads to a lack of this essential attribute and failure of such adaptation. The varietal difference in the tillering ability of two contrasting (high- and low-tillering genotypes) rice cultivars are presented in Fig. 5.4. Under favourable conditions, modern high-yielding semi-dwarf rice produces as many as 20–25 tillers, among which only 14–15 produce grain-bearing panicles, while the rest become unproductive. In early-maturing varieties, the vegetative growth period is short, and plants produce a relatively small number of leaves, no-tillers or a few tillers. However, varieties with lower tiller numbers have the advantage of producing a greater proportion of heavier grains (Padmaja Rao 1987). The correlation between the number of leaves on the main shoot and the number of tillers is positive (Fukushima 2019). A cultivar with more leaves on the main shoot can sustain the development of more tillers by enhancing the provision of photosynthetic assimilates, whereas low-tillering cultivars lack this advantage. Nuruzzaman et al. (2000) explored the morphological dynamics of rice cultivars and found that tiller numbers and correlations

Fig. 5.4 Varietal difference in tillering of two different rice cultivars. Cultivar Lalat showing high-tillering behaviour (**A**) and cultivar IR6844-29-2-1-3-1-2 showing low-tillering feature (**B**). (Source: PhD thesis of E. Kariali, 2008, Sambalpur University, India)

between plant height and specific leaf area and between specific leaf area and maximum tiller number were strongly negative and positive, respectively. The number of tillers per unit area is lower for the panicle number type (PNT) than for the panicle weight type (PWT). A comparison between PNT and PWT revealed that tiller leaf emergence and tiller bud formation are relatively slow in the latter, i.e. in the PWT (Fukushima 2019). The PWT promoted main shoot development and suppressed tiller formation in contrast to the PNT, in which main shoot development was suppressed, and promoted tiller and rachis branch development to a greater extent. Varieties with short plant heights produce shorter and thinner leaves, which marginalizes the competition for dry matter and nitrogen assimilates between the main shoot and tiller and among tillers. Hence, tiller production efficiency would be greater in high-tillering rice cultivars for the same amount of dry matter production and nitrogen absorption because of the ability of the tillers to claim more assimilates. In addition, high tiller production in favourable environments permits rice plants to close their canopies and maximize the absorption of incident radiation, contributing to an increase in plant biomass. However, it is not true that a larger number of tillers will always improve yield. Rice cultivars with a high tiller number can exhibit greater inconsistency in terms of mobilizing assimilates and nutrients among the tillers generated, leading to variations in grain development and yield between tillers (Yoshida 1981). There is a clear gap between the number of tillers generated at the vegetative stage and their survival at the maturity stage because many of the late-stage tillers decimate during senescence curtailing potential yield benefits (Kariali and Mohapatra 2007).

5.4 Types of Tillers

Tillers in rice plants are initiated in a specific chronological sequence. The tillering potential of rice plants is a variable trait whose expression mainly depends on duration and morphology (Mohanan and Mini 2008). During development, plants produce a variety of tillers, primary, secondary, and tertiary (Mohapatra et al. 2011) (Fig. 5.5), and they emerge at different stages of growth. Because tiller origin is linked to leaf emergence on the culm, tiller type denotes the category of culm from which it originates. Primary tillers arise from the leaf base of the lower stem nodes of the main shoot. After a few primary tillers are formed, secondary tillers are initiated on the primary tillers. Tillering in short-duration traditional rice is limited to primary tillers only, but it continues in some other late-duration cultivars up to the level of secondary tiller production. In comparison, modern semi-dwarf rice plants with a long duration of vegetative growth exhibit profound tillering, which includes the initiation of tertiary tillers from secondary tillers. According to the conventional concept of plant ideotype (Donald 1968), a high tillering and branching height, tall stature, and large leaves in lax orientation plants embody the great advantage of grain yield because of its ability to serve as a biological model with the expectation of performing predictably within a defined environment. This ideotype approach

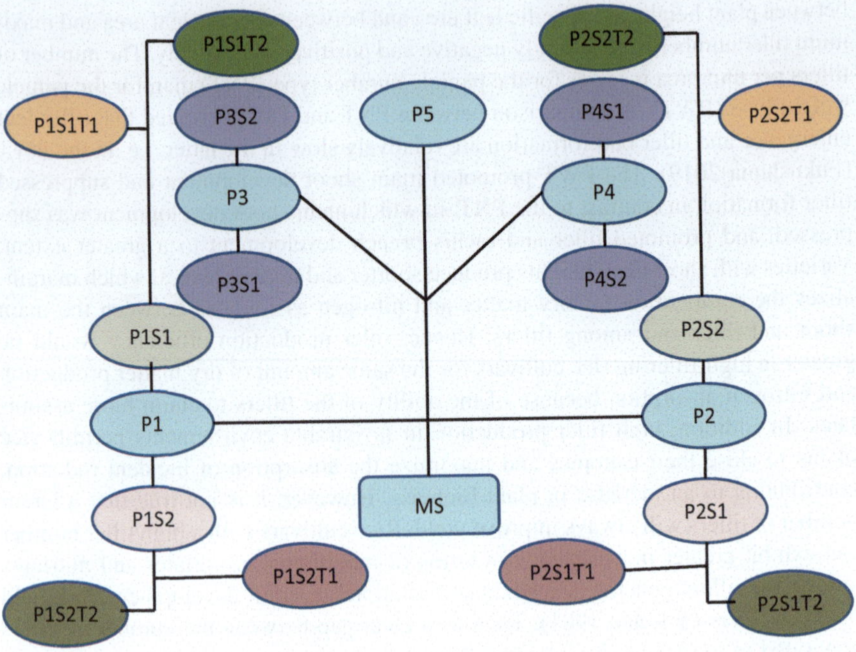

Fig. 5.5 Tillering pattern of *Oryza sativa* plants. MS, P, S, and T represent the main shoot, primary, secondary, and tertiary tiller, respectively. *P1* first primary tiller, *P5* fifth primary tiller, *P1S1* first primary tiller emerged on P1, *P2S1* first secondary tiller of P2, *P1S1T1* first tertiary tiller of P1S1, and so on (Adapted from Mohapatra et al. 2011)

was construed to have the potential to define the plant traits of high-yielding cultivars (Sedgley 1991), and preference in adoption was given for profuse-tillering rice.

5.5 Disadvantage of High-Tillering Rice

Padmaja Rao (1987) reported that genotypes with lower tiller numbers produce greater proportions of high-density grains. The growth and development of ear-bearing culms in a mono- or bi-tiller plant stand are mostly synchronized, which contributes to uniformity in grain growth in the absence of asymmetric competition for assimilate distribution. In contrast, excessive tiller formation precludes grain production in many tillers and causes reduced leaf area and reduced photosynthetic efficiency via mutual shading. In addition, tillers in the upper nodes emerge late but synchronize maturity with early-stage dominant tillers, which ultimately reduces the period of effective growth and size of these tillers. Thus, asynchrony in age, size, and spatial organization exacerbate stiff competition between individual tillers for resources available in the environment, such as nutrients, water, and irradiance for survival, although theoretically, these organs possess the ability to exist independently. According to Kariali et al. (2012), tiller emergence on rice culms is staggered and temporally spaced, but maturity is synchronous; the duration of growth of a tiller on a upper node decreases, and tiller emergence contributes to decreased biomass and grain yield (Table 5.1). Therefore, it is necessary to rationalize tiller number and achieve synchrony in flowering, maturity, and panicle size to improve yield (Zhao et al. 2020; Pawar et al. 2016; Khush 2000). Tillers with large internodal vascular bundles transport metabolites more efficiently and thus increase grain yield. The main shoot (culm) has greater and larger vascular bundles than newly formed primary tillers. Among the primary tillers, old tillers possess greater and larger vascular bundles than new tillers (Lee et al. 1992; Li et al. 2022). Due to mitochondrial impairment, the *dvb1* mutant has thinner and shorter panicles and internodes and narrower leaves. The yield of the *dvb1* mutant is severely lower than that of the wild type (Zhang et al. 2021).

5.6 Management of Tiller Production

5.6.1 Ratoon Tillers

Ratoon cropping is defined as the regrowth of a rice plant from the stubble left after harvest. This sequential cropping process helps augur crop intensification in the temporal dimension and prevents intercrop competition. In ratoon or stubble crops, the growth of the underground part of rice stubble is encouraged for successful cultivation. Ratoon cropping has advantages over traditional cropping. The cost of cultivation is reduced significantly because no investment in the resources necessary for fresh cultivation occurs. Second, the crop matures approximately 30 days earlier, ameliorating the burden of crop management for a longer duration. However, there are several deficiencies in the use of this crop compared with the main crop. There is greater asynchrony in growth and development among ratoon tillers than among those of the main crop, and there is a significant increase in the number of unproductive tillers in the latter, which curtails overall grain yield (Oad et al. 2002). The latter type of tiller senesces early and produces no grains or few grains.

The morphology of rice ratoon plants differs prominently from that of the main crop. The overall grain yield is lower because of the reductions in plant height and effective tiller number (Bahar and De Data 1977; Balasubramanian et al. 1970). However, the number of tillers produced during ratoon cropping is always greater than that produced during main cropping, and many of these tillers are perfectly unproductive (Samson 1980). These unproductive tillers and buds continue metabolic activities at the cost of grain filling (Vergara et al. 1988). There is wide variation between cultivars in terms of their genetic potential for yield levels because the ability of plants to shoot formation during tiller production is related positively to the duration of the main crop. A ratoon crop of a late-maturing rice cultivar produces more tillers than a late-maturing cultivar and results in a greater yield. In general, the yield of the ratoon crop was approximately 60% lower than that of the main crop, while there was a 40% reduction in crop duration. In addition, cultivar

Table 5.1 Morphological features of the panicles of individual tillers of two *indica* rice genotypes, Lalat and MGD-106

Name of rice cultivar	Name of tiller	Panicle length (cm)	Panicle weight (g)	No. of primary branches	No. of spikelets/ panicle	No. of filled grains/ panicle	No. of unfilled grains/ panicle	Percentage of Filled grains	Percentage of Unfilled grains	Total filled grain weight (g)	Total unfilled grain weight (g)	1000 grain weight (g)
Lalat	M	28.26 a	3.82 a	12.4 a	171.00 a	162.66 a	08.75 e	95.13 a	04.87 d	3.48 a	0.035 b	21.86 b
	P	26.26 b	3.39 a	12.2 a	168.66 a	157.00 a	11.66 e	93.08 a	06.91 d	3.21 a	0.058 b	20.81 b
	S	26.12 b	2.26 b	12.0 a	164.00 a	145.12 b	18.21 d	88.42 b	11.57 d	2.90 a	0.107 b	20.66 b
	T	21.66 c	1.83 b	08.66 b	090.33 c	067.23 d	26.00 d	74.18 c	25.82 c	1.46 c	0.142 b	18.53 b
MGD-106	M	19.75 d	2.98 a	11.5 a	135.33 b	083.15 c	52.33 c	61.26 d	38.73 b	2.28 b	0.302 a	25.46 a
	P	19.00 e	2.19 B	09.66 b	134.33 b	073.75 c	61.12 b	54.59 e	45.41 b	2.11 b	0.363 a	24.13 a
	S	17.37 f	1.87 b	09.2 b	128.33 b	046.82 e	87.26 a	32.24 f	67.76 a	1.34 c	0.475 a	22.26 a
LSD (5%)		0.647	1.104	1.713	22.041	14.8	7.841	3.422	13.83	0.789	0.194	3.355

The total number of tillers at maturity in Lalat was 29.0 ± 1.732, and that in MGD-106 was 12.667 ± 1.528. ±values represent the standard deviation (SD) of three replicates. *M* mother tiller, *P* primary tiller, *S* secondary tiller, *T* tertiary tiller (Kariali et al. 2012)

Means followed by common letters within a column shows no significant difference in Duncan's multiple range test (DMRT) differences also exist concerning the generation of ratoon tillers from different nodes of the stubble. All nodes of the stubble may bear ratoon tillers, but tiller emergence is sometimes confined to the lower nodes or any specific node of the stubble (Krishnamurthy 1988). Vergara et al. (1988) reported that stubble height determines the number of tiller buds available for rejuvenation; some cultivars produce tillers from high nodes, while others produce tillers from lower nodes. The basal ratoon density is not affected by cutting height. Furthermore, ratoon tillers that develop early do not compete with tillers that develop later; instead, they support their growth by providing photosynthetic assimilates.

In addition, the cutting height at harvest has a significant effect on the ratooning potential because many dead buds are present on the culm close to the ground, and lower cutting produces many missing hills and poor tillering. The cutting height of stubble is correlated with regrowth potential or ratoon potential, nodal and basal tiller numbers, uniform growth and maturity, ratoon rating and ability, missing hills, and filled spikelets (Oad et al. 2002). Although higher cuttings permit tiller emergence from the upper part of the stubble (Vergara et al. 1988), effective tillers grow only from the stubble base, and others are less productive (Evatt 1958). The other field manipulation practices that influence ratooning ability include ploughing depth, transplanting time, spacing, crop establishment, and water and fertilizer management. These factors influence the population of the main crop and affect the ratooning ability. A high population of main crops is conducive to increasing the tiller number per unit area, which is positively correlated with the ratoon potential. Ratooning in a rice intergrade grown in the botanical garden of the School of Life Sciences, Sambalpur University, India is presented in Fig. 5.6.

5.6.2 Physiology of Ratoon Tillering

Vergara et al. (1988) reported that stubble height determines the number of tiller buds available for rejuvenation; some cultivars produce tillers from high nodes, while others produce tillers from lower nodes only. The basal ratoon density was not affected by cutting height. Furthermore, ratoon tillers that develop early do not compete with tillers that develop later; instead, they support their growth by providing photosynthetic assimilates. Ratoon tiller occurrence depends on the pool of assimilates in stubble and roots that remain after the main crop is harvested (Cuevas-Perez 1980; Ichii and Sumi 1983). Because rice culms are storehouses for food reserves, thick culms could possess more assimilates than thin culms, exhibiting greater expression of ratoon plant potential. Irrespective of culm thickness, healthy tillers arise from the lower nodes of the stubble, possess better ratoon yield potential (Chauhan et al. 1985), and sustain development at the carbohydrate level retained in the stubble. Variation in carbohydrate levels affects the number of ratoon tillers. Krishnamurthy (1988) suggested that a low C: N ratio is favourable for supporting tiller emergence. Rice ratooning depends on the ability of dormant buds on the stubble of the main crop to remain viable. The buds may be at different stages of development, and this difference could be manifested in the variation in their

Fig. 5.6 Production of ratoon tillers after crop harvest in an intergrade between wild rice (*Oryza rufipogon)* and the traditional rice cultivar *Kalabati*. (Source: photo gallery of Preetam K. Senapati and E. Kariali)

regenerative growth. Ratoon tillers develop from the upper nodes of stubbles, grow rapidly, have fewer leaves, and mature early (Parashar 1970). However, panicles of ratoon tillers from lower nodes bear more grain and contribute more to grain yield (Sun et al. 1988).

The C:N ratio in the tillers of the upper nodes is high, and the plants behave like old plants; in comparison, tillers from the lower nodes possess a low C:N ratio and exhibit vigour similar to young plants (Iso 1954). In rice, the ratoon crop grows with only water added to the field, and the soil is kept saturated. However, for the promotion of early and profuse ratooning, it is necessary to apply fertilizer in the field immediately after the main crop harvest. Nitrogen fertilizer is considered the most important nutrient in the growth and yield of rice (Oad et al. 2002). In short, a high N concentration in the stubble of rice plants correlates positively with profuse tillering (Garcia 1981). Thus, efficient and economic ratooning can be achieved by developing and adopting rice varieties that embody profuse ratooning capabilities and applying appropriate N levels in the field. A study in China has shown that rice yield doubles (from 6.7 to 12.3 tons per hectare) when the two factors are combined (Hart 2018). For P and K fertilization, direct treatments were not effective at increasing the RRY (ratoon yield potential), but the RRY improved when fertilization was applied to the main crop (Vergara et al. 1988). Thus, the main crop environment has a substantial influence on rice ratoon tillering. Low irradiance levels during the reproductive and ripening stages decrease ratoon tillering, but the incidence of low temperature during the reproductive phase encourages ratoon tillering. The application of nitrogen fertilizer is considered the most important single nutrient element for the growth and yield of ratoons.

5.7 Effects of N:P:K Fertilization on Rice Tillering

Fertilizing rice during the vegetative growth period promotes tillering, leading to high grain yield. For optimum growth and yield, appropriate doses of fertilizers are necessary. Farmers provide balanced macronutrient fertilization because it directly improves yield. The N:P:K ratio in inorganic fertilizers is the percentage of the product contained in the volume of nitrogen, phosphorus and potassium. The application of N:P:K at an appropriate balance has been successful in improving rice yield by increasing plant height and tiller number (Fidiyawati et al. 2022). In principle, the overall nutrient balance should be such that it enhances plant growth and yield. In rice, the NPK ratio varies significantly with the growth duration of the plants, and the recommendations for duration groups are listed below. Like inorganic fertilizers, the application of a balanced combination of N, P, and K in organic fertilizers such as poultry manure, cow dung, and compost also results in a greater accumulation of elements in leaves, grains, panicles, and leaf sheaths, which ultimately increase the yield of rice (Moe et al. 2019). Thus, applying fertilizers, either organic or inorganic, in a correct balance between N, P, and K (Khem et al. 2018) is beneficial for rice cultivation, but the recommended dose should be cultivar- and field-based (Paiman et al. 2021).

Short duration	NPK@120:40:40 kg ha^{-1}
Medium duration	NPK@150:50:60 kg ha^{-1}
Long duration	NPK@150:50:80 kg ha^{-1}

5.7.1 Role of Individual Elements

5.7.1.1 Nitrogen

Nitrogen in rice crops functions as a primary substance for chlorophyll and protein synthesis. Rice yield is improved by a faster rate of growth, extension of leaf area, and increase in tiller number and grain setting. The number of tillers on plants is dynamic and adjustable by culture (Mohapatra et al. 2011), especially for the nitrogen status of the soil (Zhong et al. 2002). Nitrogen supplementation increases the level of accumulation in plants and contributes positively to the vegetative growth of rice plants by favouring shoot branching to increase tiller numbers (Liu et al. 2011). Kamiji et al. (2011) showed that nitrogen applications that increase plant nitrogen at the panicle development stage are conducive to increasing spikelet numbers, which is favourable for grain yield. N availability in plant cells induces interactions between elements and plant hormones, such as auxin, cytokinin, and strigolactone, which ultimately determine tiller bud formation in rice (Xu et al. 2015). Under nitrogen deficiency, rice plants exhibit fewer tillers, a stunted appearance, and yellowish leaves; these plants then begin to die from the tip and move downwards to the leaf base. In contrast, excess application of the element induces detrimental effects, weakening straw stability, and lodging of the shoot to the ground. In China, N application above the recommended dose causes/contributes to crop loss, environmental pollution, enhanced costs of farming, and, above all, increases in global warming (Peng et al. 2010).

5.7.1.2 Effects of Nitrogen on Different Tiller Types

In rice cultivation, the quality and quantity of tillers determine the number and structure of panicles for a commensurate yield contribution, and an optimized level of nitrogen application is essential for ensuring an appropriate level of yield. Although both tiller generation and survival are factors determining the ultimate tiller number, the effects of nitrogen on yield are enhanced by increasing the quality of effective tillers without reducing their number. Zhou et al. (2022) demonstrated the mechanism of improving tillering quality for yield gain in rice, where the optimized application of nitrogen increased the quality of tillers while not disturbing their number. A decrease in the emergence rate of secondary tillers augments the percentage of ear-bearing tillers and dry matter and nitrogen accumulation per tiller. In ear-bearing tillers, a decrease in branch degeneration and an increase in secondary spikelet survival contribute to an increase in grain yield. Nitrogen application improved the grain yield of panicles on main shoots, primary tillers and secondary tillers. Of course, among the tiller types, nitrogen effects are more discernible on primary tillers than on secondary tillers (Zhou et al. 2022, Fig. 5.7), and such effects could be used as an index for the optimization of nitrogen supplementation in the field. In a similar study, Wang et al. (2016) assessed the heterogeneity in tiller yield associated with tiller formation and nitrogen fertilizer and concluded that tiller asynchrony increases with the level of nitrogen fertilization and that the positive effects of nitrogen on the rice population are the outcome of an increase in the grain yield of early-formed tillers. Thus, the primary tiller number is always critical for achieving high grain yield in rice (Huang et al. 2020). Considering the optimum use of nitrogen fertilizers in a particular environment, there must be a balance between rice production and the environmental risk of the fertilizer. In the rice-growing habitat of the Yangtze River basin, in China, no yield benefits accrued when the nitrogen input rate exceeded 130 or 170 kg ha^{-1} (Wang et al. 2019). Wang et al. (2019) recommended nitrogen application rates between 130 and 170 kg ha^{-1} using fertilizer rates of 90–0-40 and 90–40–40 kg ha^{-1} nitrogen staggered at basal-tillering-panicle initiation stages, respectively, to achieve optimum utilization of the element with no sacrifice of yield. A model representing the effects of optimized N application on rice yield is presented in Fig. 5.8.

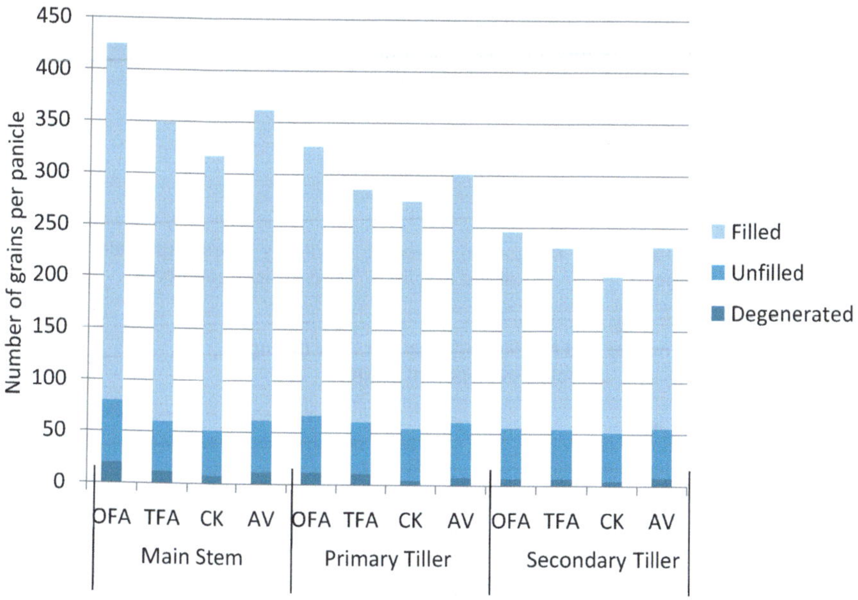

Fig. 5.7 Differentiation and development of tiller spikelets under different nitrogen treatments. *AV* average value, *CK* no nitrogen application, *TFA* traditional nitrogen fertilizer application, *OFA* optimized nitrogen fertilizer application. (Zhou et al. 2022 with permission of the authors)

5.7.2 Phosphorous (P)

Phosphorous participates in various metabolic activities in any life system. It is the second most common major element after nitrogen for its impact on the productivity of rice. A reduced availability of phosphorous decreases biomass formation, leading to a reduction in grain yield. Root hair development is impaired due to the lack of sufficient P, which impacts the overall nutrient balance in the whole plant system (Vejchasarn et al. 2016). Plant height and tiller formation greatly decrease due to low P availability. Vejchasarn et al. (2016) reported 42% and 41% reductions in shoot biomass and tiller numbers, respectively, in plots with low P availability. P easily binds to different soil chemical and biological elements. Moreover, this phenomenon hinders plant absorption efficiently. The additional application of phosphorous is needed to overcome such situations to obtain a reasonable yield. P acquisition from P-deficient soil differs among rice genotypes (Alam et al. 2009). However, additional application of P to soil is still the best way to overcome P deficiency. To date, greater P-acquisition capacity in high-yielding plant backgrounds has not been achieved. In contrast to N, the excess application of P does not improve or decrease vital functions or the appearance of plants; however, P pollution is heavily present in water bodies. Algal blooms are very common in rice fields and adjacent water bodies due to the excess application of P in agricultural fields. The tiller number per unit area was significantly greater in the plot where 72 kg of P_2O was

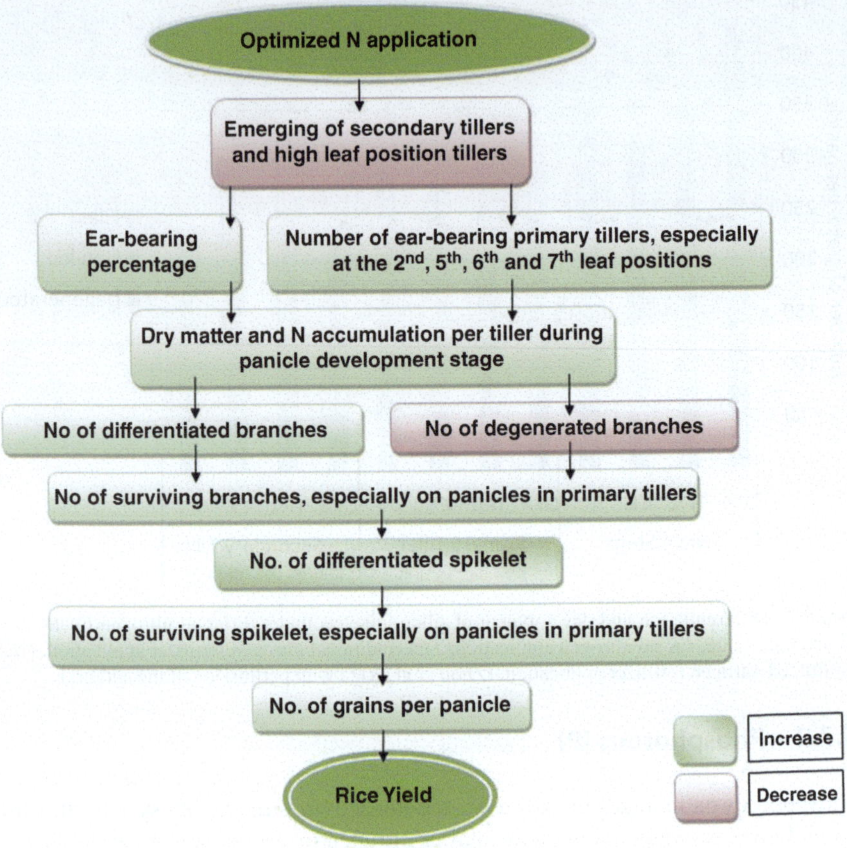

Fig. 5.8 The pathway for optimizing nitrogen for increasing rice grain yield. (Zhou et al. 2022 with permission of the authors)

applied than in the plot where no additional P_2O was applied. Moreover, additional P application did not improve the tiller number per unit of land.

5.7.3 Potassium (K)

Potassium is the third most common essential element for augmenting rice growth. K is needed for the activation of different enzymes, the regulation of cellular ionic balance and turgor, transpiration, and metabolite transport. Studies have shown that the application of K-fertilizer in K-deficient soil improves plant vigour and root growth, prevents lodging, enhances resistance to pests and diseases, and counteracts the deleterious effects of sodium in saline soil. Grain yield production is greatly improved by the additional application of K to K-deficient soil. The number of tillers, number of filled grains, and 1000-grain weight significantly improved in response to K, which led to greater yields. Islam et al. (2015) reported that the application of a low dose of glyphosate potassium salt (20 kg ha^{-1}) improved plant height, tiller number, panicle number, panicle length, and grain yield. They applied K at 0, 20, 40, 60, and 80 kg ha^{-1} and found linear improvement in these traits. At 80 kg ha^{-1} K, the improvements in plant height, tiller number, panicle number, panicle length, and grain yield were 19.5%, 25.5%, 21.4%, 14.5%, and 78.5%, respectively.

References

Alam MM, Hasanuzzaman M, Nahar K (2009) Tiller dynamics of three irrigated rice varieties under varying phosphorus levels. Am Eur J Agron 2:89–94

Bahar FA, De Data SK (1977) Prospects of increasing total rice production through ratoonig. Agron J 69:536–540

Balasubramanian B, Morachan YB, Kliappa R (1970) Studies on rationing in rice. I. Growth attributes and yield. Madras Agric J 57:565–570

Chauhan JS, Vergara BS, Lopez FSS (1985) Rice ratooning. IRRI Research Paper Series 102, p 19

Counce PA, Keisling TC, Mitchell AJ (2000) A uniform, objective, and adaptive system for expressing rice development. Crop Sci 40:436–443

Cuevas-Perez FE (1980) Inheritance and associations of six agronomic traits and stem-base carbohydrate concentrations on ratooning ability in rice *Oryza sativa* L. PhD thesis, Oregon State University, p 102

Das K (2018) The relative contribution of variation in endosperm starch biosynthesis enzyme activities during grain filling of rice cultivars with contrasting panicle architecture. PhD thesis, Sambalpur University, India

Donald CM (1968) The breeding of crop ideotypes. Euphytica 17:385–403

Evatt NS (1958) Stubble rice production tests. Rice J 61(6):18–19

Fidiyawati E, Latifah E, Krismawati A, Sihombing D, Setyorini D, Bakar A, Sugiono (2022) Effectivity of inorganic fertilizer NPK (15-15-6) to growth and yield of lowland rice (*Oryza sativa* L.) on alfisol soil. IOP Conf Ser Earth Environ Sci 980:012014. https://doi.org/10.1088/1755-1315/980/1/012014

Fukushima A (2019) Varietal differences in tiller and panicle development determine the total number of spikelets per unit area in rice. Plant Prod Sci 22:192–201

Garcia RN (1981) The effects of growth duration and different levels of light intensity on the ratooning ability of rice. BS thesis, University of the Philippines at Los Baños, Philippines

Gonin M, Bergougnoux V, Nguyen TD, Gantet P, Champion A (2019) What makes adventitious roots? Plan Theory 8:240. https://doi.org/10.3390/plants8070240

Hart N (2018) Rice fields in China double yields by Ratooning. International development correspondent. August 2018

Huang M, Shan S, Cao J, Fang S, Tian A, Liu Y, Cao F, Yin X, Zou Y (2020) Primary-tiller panicle number is critical to achieving high grain yields in machine-transplanted hybrid rice. Sci Rep 10:2811. https://doi.org/10.1038/s41598-020-59751-4

Hussein A, Tavakol E, Horner DS, Munoz-Amatriain H, Muehlbauer GJ, Rossini L (2014) Genetics of tillering in rice and barley. Plant Genome 7:1–20. https://doi.org/10.3835/plantgenome2013.10.0032

Ichii M, Sumi Y (1983) Effect of food reserves on the ratoon growth of rice plant. Jpn J Crop Sci 52(1):15–21

Islam A, Biswas JC, Sirajul Karim AJM, Salmapervin MST, Saleque MA (2015) Effects of potassium fertilization on growth and yield of wetland rice in Grey terrace soils of Bangladesh. Res Crop Ecophysiol 10:64–82

Iso E (1954) Ratoon culture of horai varieties. In: Rice and crops in its rotation in subtropical zones. Japan-FAO Association, Tokyo, pp 197–200

Jaffuel S, Dauzat J (2005) Synchronism of leaf and tiller emergence relative to the position and main stem developmental stage in a rice cultivar. Ann Bot 95:401–4012

Kamiji Y, Yoshida H, Palta JA, Sakuratani T, Shiraiwa T (2011) N applications that increase plant N during panicle development are highly effective in increasing the spikelet number in rice. Field Crop Res 122(3):242–247. https://doi.org/10.1016/j.fcr.2011.03.016

Kariali E (2008) Hierarchy in Tiller Development and Physiology of Unequal Partitioning of Assimilates between Tillers and It's Inheritance in Rice. Ph D thesis, Sambalpur University, India

Kariali E (2014) Environmental factors undermine genetic expression of tiller dynamics in wild rice *Oryza nivara* and *Oryza rufipogon*. Am J Plant Sc 5: 2617–2622

Kariali E, Mohapatra PK (2007) Hormonal regulation of tiller dynamics in differentially-tillering rice cultivars. Plant Growth Regul 53:215–223

Kariali E, Sarangi S, Panigrahi R, Panda BB, Mohapatra PK (2012) Variation in senescence pattern of different classes of rice tillers and its effect on panicle biomass growth and grain yield. Am J Plant Sci 3:1047–1057. https://doi.org/10.4236/ajps.2012.38125

Katayama T (1951) Studies on tillering of rice, wheat and barley. Yokendo, Tokyo

Khem B, Hirai Y, Yamakawa T, Mori Y, Inoue E, Okayasu T, Mitsuoka M (2018) NPK balances in whole-crop rice cultivation under different application methods of manure and chemical fertilizer. J JASS 34(4):87–99

Khush GS (2000) New plant type of rice for increasing the genetic yield potential. In: Nanda JS (ed) Rice breeding and genetics. Science Publishers, Enfield, NH, pp 1–99

Krishnamurthy K (1988) Rice ratooning as an alternative to double cropping in tropical Asia. In: Rice ratooing. International Rice Research Institute, Manila, pp 3–15

Lee DJ, Vergara BS, Zamora OB, Kim BK, Chae JC (1992) Development of vascular bundle in the peduncle of different tillers and its relationship to panicle characteristics in rice. Kor J Crop Sci 37:155–165

Li X, Qian Q, Fu Z, Wang Y, Xiong G, Zeng D, Wang X, Liu X, Teng S, Hiroshi F, Yuan M, Luo D, Han B, Li J (2003) Control of tillering in rice. Nature 422:618–621. https://doi.org/10.1038/nature01518

Li G, Chen X, Zhou C, Yang Z, Zhang C, Huang Z, Pan W, Xu K (2022) Vascular bundle characteristics of different rice varieties treated with nitrogen fertilizers and Its relation to stem assimilate allocation and grain yield. Agriculture 12:779. https://doi.org/10.3390/agriculture12060779

Liu Y, Ding YF, Wang QS, Meng DX, Wang SH (2011) Effects of nitrogen and 6-benzylaminopurine on rice tiller bud growth and changes in endogenous hormones and nitrogen. Crop Sci 51(2):786–792

Moe K, Htwe AZ, Thu TTP, Kajihara Y, Yamakawa T (2019) Effects on NPK status growth dry matter and yield of rice (*Oryza sativa*) by organic fertilizers applied in field condition. Agriculture 9:109. https://doi.org/10.3390/agriculture9050109

Mohanan KV, Mini CB (2008) Relative contribution of rice tillers of different status towards yield. Int J Plant Breed Genet 2:9–12

Mohapatra PK, Panda BB, Kariali E (2011) Plasticity of tiller dynamics in wild rice *Oryza rufipogon* Griff.: a strategy for resilience in sub-optimal environments. Int J Agron 2011:1–9

Moldenhauer KEWC, Counce P, Hardke J (2001) Rice growth and development. In: Rice production handbook, pp 7–14. https://www.uaex.uada.edu/publications/pdf/mp192/chapter-2-word.pdf

Nemoto K, Morita S, Baba T (1995) Shoot and root development in rice related to the phyllochron. Crop Sci 35:24–29

Nuruzzaman M, Yamamoto Y, Nitta Y, Yoshida T, Miyazaki A (2000) Varietal differences in tillering ability of fourteen *japonica* and *indica* rice varieties. Soil Sci Plant Nutr 46:381–391

Oad FC, Cruz PS, Memon N, Oad NL, Hassan JU (2002) Rice ratooning management. Pak J Appl Sci 2:29–35

Padmaja Rao S (1987) High-density grain among primary and secondary tillers of short and long duration rices. IRRN 12(4):12

Paiman A, Kusumastuti CT, Gunawan S, Adriani F (2021) Maximizing the rice yield (*Oryza sativa* L.) using NPK fertilizer. Open Agric J 15:33–38

Parashar CRK (1970) Paddy ratoons. World Crops 22(3):145–147

Pawar SY, Radhakrishnan VV, Mohanan KV (2016) The importance of optimum tillering in rice–an overview. Indian J Biol Sci 2:125–127

Peng S, Buresh RJ, Huang J, Zhong ZY, Yang J, Wang G, Liu Y, Hu R, Tang Q, Cui K, Zhang F, Dobermann A (2010) Improving nitrogen fertilization in rice by site-specific N management. Agron Sustain Dev 30:649–656

Samson BI (1980) Rice ratooning effects of varietal type and some cultural management practices. MSc Thesis, University of the Philippines at Los Banos

Sedgley RH (1991) An appraisal of the Donald ideotype after 21 years. Field Crop Res 26:93–112

Song C, Cai S-g, Xin C, Zhang G-p (2009) Genotypic differences in growth and physiological responses to transplanting and direct seeding cultivation in rice. Ric Sci 16:143–150

Stoop WA, Uphoff N, Kassam A (2002) A review of agricultural research issue raised by the system of rice intensification (SRI) from Madagascar: opportunities for improving the system for resource-poor farmers. Agr Syst 71:249–274

Sun XH, Zhang JG, Liang YJ (1988) Ratooning with rice hybrids. In: Rice ratooning. IRRI, Manila, pp 155–161

Thakur A, Uphoff NT, Stoop WA (2016) Scientific underpinnings of the System of Rice Intensification (SRI): what is known so far? Adv Agron 135:147–179

Vejchasarn P, Lynch JP, Brown KM (2016) Genetic variability in phosphorus responses of rice root phenotypes. Rice 9:29. https://doi.org/10.1186/s12284-016-0102-9

Vergara BS, Lopez FSS, Chuahan JS (1988) Morphology and physiology of ratoon rice. In: Rice ratooing. International Rice Research Institute, Manila, pp 31–40

Wang Y, Ren T, Lu J, Ming R, Li P, Hussain S, Cong R, Li X (2016) Heterogeneity in rice tillers yields associated with tillers formation and nitrogen fertilizer. Agron J 108:1717–1725

Wang J, Penghao F, Wang F, Fahad S, Mohapatra PK, Chen Y, Zhang C, Peng S, Cui K, Nie L, Huang J (2019) Optimizing nitrogen management to balance rice yield and environmental risk in the Yangtze River's middle reaches. Environ Sci Pollut Res 26(11). https://doi.org/10.1007/s11356-018-3943-5

Xu J, Zha M, Li Y, Ding Y, Chen L, Ding C, Wang S (2015) The interaction between nitrogen availability and auxin, cytokinin, and strigolactone in the control of shoot branching in rice (*Oryza sativa* L.). Plant Cell Rep 34(9):1647–1662. https://doi.org/10.1007/s00299-015-1815-8

Yoshida S, Hayakawa Y (1970) Effect of mineral nutrition on tillering of rice. Soil Sci Plant Nutr 16: 180–191

Yoshida S (1981) Fundamentals of rice crop science. International Rice Research Institute, Manila

Zhang L, Feng P, Deng Y, Yin W, Wan Y, Lei T, He G, Wang N (2021) Decreased vascular bundle affects mitochondrial and plant development in rice. Rice 14:13. https://doi.org/10.1186/s12284-021-00454-3

Zhao S, Jang S, Lee YK, Kim D-G, Jin Z, Koh H-L (2020) Genetic basis of tiller dynamics of rice revealed by genome-wide association studies. Plants 9:1695. https://doi.org/10.3390/plants9121695

Zhong X, Peng S, Sheehy JE, Visperas RM, Liu H (2002) Relationship between tillering and leaf area index: Quantifying critical leaf area index for tillering in rice. J Agric Sci 138(3):269–279. https://doi.org/10.1017/S0021859601001903

Zhou W, Yan F, Chen Y, Ren W (2022) Optimized nitrogen application increases rice yield by improving the quality of tillers. Plant Prod Sci 25:311–319. https://doi.org/10.1080/1343943X.2022.2061538

Heterogeneity in Tiller Development

6

Abstract

The tiller number is associated with grain yield. Old tillers contribute to yield as much as main shoots do, but late tillers waste investment in primary production, although some of them remobilize reserve carbon. Late tillers synchronize with early tillers at maturity, resulting in less biomass and yield, and are sometimes unproductive altogether. The competence for accruing yield benefits declines progressively with age. It happens so because the initiation of tillers is staggered and spaced in chronological order, and the later-formed tillers on upper internodes have little time for primary production supportive of grain yield. The hierarchy of tiller development becomes evident as grain yield becomes poorer for each successive tiller. High tillering was a breeding priority for semi-dwarf rice over traditional rice. But, augmenting the tiller number, as has been done in profuse-tillering semi-dwarf rice, by adopting an ideotype approach was not a preferential choice for advancement of yield potential. Therefore, the pyramiding of potentially useful traits in super or NPT rice invoked the preclusion of unproductive tillers amounting to low tiller numbers per plant.

Keywords

Tiller heterogeneity · Senescence · NPK effects · Deficiency symptoms

6.1 Tiller Heterogeneity

According to Yoshida (1981), panicle number (effective tillers), spikelets per panicle, grain filling percentage, and grain weight are the key parameters for rice yield estimation, and these features collectively affect final grain yield while also altering plant architecture. Incidentally, each component of grain yield is positively correlated with the final grain yield, revealing an additive effect of one with the other. In

© The Author(s), under exclusive license to Springer Nature Singapore Pte Ltd. 2025

P. K. Mohapatra et al., *Tillering Behavior of Rice Plant*,
https://doi.org/10.1007/978-981-97-5235-5_6

contrast, an increase in any component in practice does not always induce a positive cognitive influence on another component. In rice, there are strong restrictions and compensation among the four parameters (Tong et al. 2011), precluding a synergistic effect on grain yield. Hence, there is a need for coordinated developments among the components for obtaining higher yields, and foremost attention is needed for increasing grain number by improving the effective tiller number.

Tiller development in rice is asynchronous. Tiller bud development in rice is also asynchronous. Individual tillers develop in an acropetal succession from the culm base upwards. Although the initiation of tillers is staggered and spaced in chronological order (Mohanan and Pavithran 2007), the maturity of tillers is nearly synchronous (Kariali et al. 2012), and the later-formed tillers on upper internodes have little time for primary production supportive of grain yield. Early emerging tillers from lower culm nodes of the main stem or tiller result in high biomass and yield at maturity. However, secondary and tertiary tillers appearing at the juvenile stage are less productive (Kariali et al. 2012) (Table 6.1 and Fig. 6.1).

The hierarchy of tiller development becomes evident as grain yield becomes poorer for each successive tiller (Padmaja Rao 1991; Sahu et al. 2004). The latter-formed tertiary tillers of rice plants are either sterile or scarcely produce grains suitable for humans. Thus, augmenting the tiller number, as has been done in profuse-tillering semi-dwarf rice by adopting an ideotype approach (Donald 1968), becomes irrelevant in innovative breeding for the enhancement of rice yield potential. In contrast, selection for new plant types with few tillers, preferably originating from basal culm nodes, is a profound breeding priority. Here, early tiller dominance during growth is encouraged to preclude the production of late tillers by genotype modification (Mohapatra and Kariali 2008), and resources essential for grain yield are conserved. Such manipulation of tiller number is also possible by environmental factors (Yoshida 1981; De Datta 1981). Tillers growing early exhibit a greater sink–source ratio, spikelet number, leaf area per tiller, sink capacity, and filled grain percentage (Choi and Kwon 1985), because they have more vascular bundles in the culm (Hayashi 1976) for better transport of assimilates and other growth regulating materials.

6.2 Effects of Tiller Productivity, Senescence, and Survival on Rice Yield

In rice antiquity, tiller senescence is highly important. Through evolution, rice has adapted to uneven and uncertain climatic conditions, mostly in the tropics, where weather conditions, primarily patterns of rainfall, are often erratic and unstable. Erratic rains invite incidences of flooding and drought during the growth period, infringing on the completion of the life cycle and jeopardizing plant survival. For survival in pernicious environments, plants adopt reproductive strategies, such as producing many more tillers than they did at maturity; these strategies salvage the residual units of tillers to complete their life cycle and provide seeds for propagation into the next season. Because tiller emergence in multi-tillered rice is asynchronous

Table 6.1 Morphological features of the panicles of individual tillers of two *indica* rice cultivars, Lalat and MGD-106

Name of rice cultivar	Name of tiller	Panicle length (cm)	Panicle weight (g)	No. of primary branches	No. of spikelets/ panicle	No. of filled grains/ panicle	No. of unfilled grains/ panicle	Percentage of Filled grains	Percentage of Unfilled grains	Total filled grain weight (g)	Total unfilled grain weight (g)	1000 grain weight (g)
Lalat	M	28.26 a	3.82 a	12.4 a	171.00 a	162.66 a	08.75 e	95.13 a	04.87 d	3.48 a	0.035 b	21.86 b
	P	26.26 b	3.39 a	12.2 a	168.66 a	157.00 a	11.66 e	93.08 a	06.91 d	3.21 a	0.058 b	20.81 b
	S	26.12 b	2.26 b	12.0 a	164.00 a	145.12 b	18.21 d	88.42 b	11.57 d	2.90 a	0.107 b	20.66 b
	T	21.66 c	1.83 b	08.66 b	090.33 c	067.23 d	26.00 d	74.18 c	25.82 c	1.46 c	0.142 b	18.53 b
MGD-106	M	19.75 d	2.98 a	11.5 a	135.33 b	083.15 c	52.33 c	61.26 d	38.73 b	2.28 b	0.302 a	25.46 a
	P	19.00 e	2.19 B	09.66 b	134.33 b	073.75 c	61.12 b	54.59 e	45.41 b	2.11 b	0.363 a	24.13 a
	S	17.37 f	1.87 b	09.2 b	128.33 b	046.82 e	87.26 a	32.24 f	67.76 a	1.34 c	0.475 a	22.26 a
LSD (5%)		0.647	1.104	1.713	22.041	14.8	7.841	3.422	13.83	0.789	0.194	3.355

The total number of tillers at maturity was 29.0 ± 1.732 in Lalat and 12.667 ± 1.528 in MGD-106. The ± values represent the SDs of three replicates. *M* mother, *P* primary, *S* secondary, *T* tertiary tillers (Source: Kariali et al. 2012)

Means followed by common letters within a column shows no significant difference in Duncan's multiple range test (DMRT)

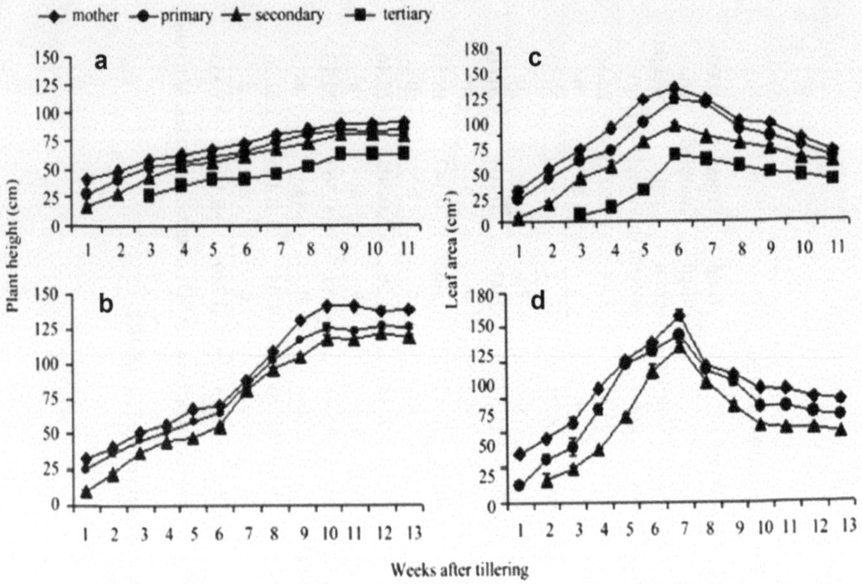

Fig. 6.1 Plant height (**a**) and (**b**) and leaf area (**c**) and (**d**) of different types of tillers of high-tillering Lalat (top) and medium-tillering MGD-106 (bottom) rice cultivars. The data represent the means of three replicates. The F values between the tillers and sampling occasions were as follows: (**a**) {158.23*** and 120.857*** ($n1$ = 3; $n2$ = 10)}; (**b**) {97.129***and 456.673*** ($n1$ = 2; $n2$ = 12)}; (**c**) {78.566*** and 36.985*** ($n1$ = 3; $n2$ = 10)}; and (**d**) {77.664*** and 81.345*** ($n1$ = 2; $n2$ = 12)}. The data were analyzed via ANOVA (analysis of variance) in the Microsoft Excel computer program (double factor ANOVA without replication). ***$P \geq 0.001$ (Kariali et al. 2012)

(Wu et al. 1998), poorly developed inferior tillers are primary target sites for decimation. Tiller senescence usually occurs in late-formed tillers between the end of active tillering and the beginning of the reproductive phase (Moldenhauer and Slaton 2001) or sometime later. Thus, the primary feature of failure in tiller survival is grossly related to tiller dominance, which is an integral part of crop improvement. This observation has been confirmed by Davidson and Chevalier (1990), who reported that late-emerged tillers, which are unsuccessful competitors, die much earlier than complete longevity.

Because maturation among tillers is synchronous (Mohapatra and Kariali 2008), later-initiated tillers have a shorter life cycle than tillers initiated earlier, and they generate less biomass and grain yield. The impact of tiller dynamics on productivity, senescence, and survival on grain yield has been investigated in two different tillering cultivars, Lalat (with high tillering) and MGD-106 (with moderate tillering), by Kariali et al. (2012) to ascertain how the tiller production capacity influences dry matter partitioning among various categories of tillers. The progress of senescence in the tillers was assessed by measuring the gradual decrease in plant height and leaf area (Fig. 6.1), photosynthetic pigments (Fig. 6.2) and total nitrogen and protein concentrations (Fig. 6.3) and the increase in lipid peroxidation and

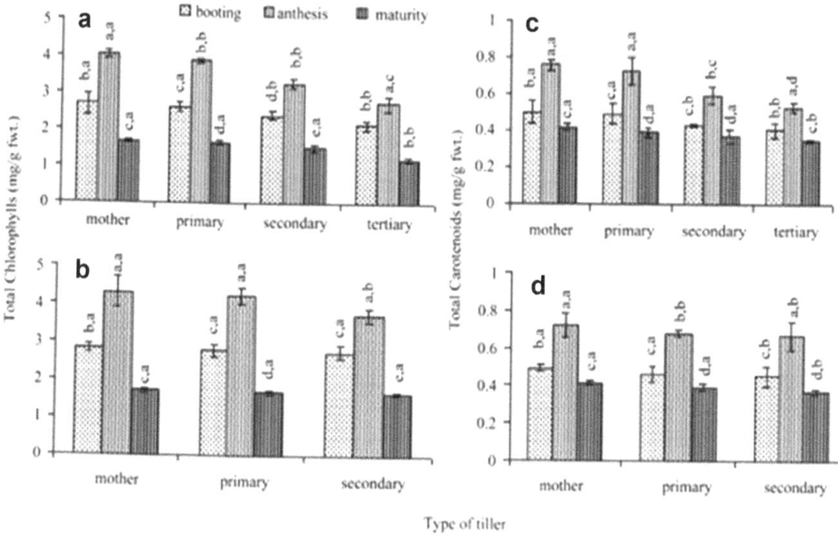

Fig. 6.2 Total chlorophyll (**a**) and (**b**) and carotenoid (**c**) and (**d**) concentrations in the flag leaves of different types of tillers of high-tillering Lalat (top) and medium tillering MGD-106 (bottom) rice cultivars at booting, anthesis, and maturity. The data represent the means of three replicates. The small letters above the vertical bars {lower ones: v × ds × r, the interaction between varieties (v), ds (developmental stages) and r (replicates) and upper ones "v × t × r" interaction between varieties (v), t (tillers) and replicates (r)}. Common letters with "v × ds × r" (CV = 0.769, LSD = 0.294 at $P \geq 0.05$) and "v × t × r" (CV = 0.905, LSD = 0.387 at $P \geq 0.05$) for (**a**) and (**b**) and "v × ds × r" (CV = 0.029, LSD = 0.044 at $P \geq 0.05$) and "v × t × r" (CV = 0.126, LSD = 0.051 at $P \geq 0.05$) for (**c**) and (**d**) indicate no significant difference according to Duncan's multiple range test (DMRT) (Kariali et al. 2012)

peroxidase activities (Fig. 6.4) in the flag leaves. These physiological changes affect leaf photosynthesis, and the effects are more pernicious for late-formed tillers than for tillers formed early as clearly indicated from the fluorescence measurement of flag leaves of different tillers (Fig. 6.5). It was concluded that the staggered initiation of tillers over time makes the flag leaves of late-emerged tillers more vulnerable to photooxidative stress (Kariali et al. 2012), and as a consequence, both source and sink activities decline. Older tillers enjoy physiological advantages over new tillers in grain filling and the bias against development-limiting yield capacity of newer tillers decreases with the increase of tiller number.

In rice plants, each tiller develops independently and flag leaf photosynthesis becomes the main source of carbon assimilates for the sustenance of each tiller. Thus, a tiller emerging late bereft of optimum leaf area contributes to poor survival and decreases survival early by becoming vulnerable to photooxidative stress. Subject to stress, both source and sink activities are limited, resulting in no or less grain yield. Hence, an increase in the tiller number in semi-dwarf rice is often not beneficial for high grain yield. Mohapatra and Kariali (2008) estimated that a threshold biomass of approximately 2.5 g/tiller is needed for 50% grain filling in the high-tillering *indica* rice cultivar Lalat and this feature is fully cultivar-specific.

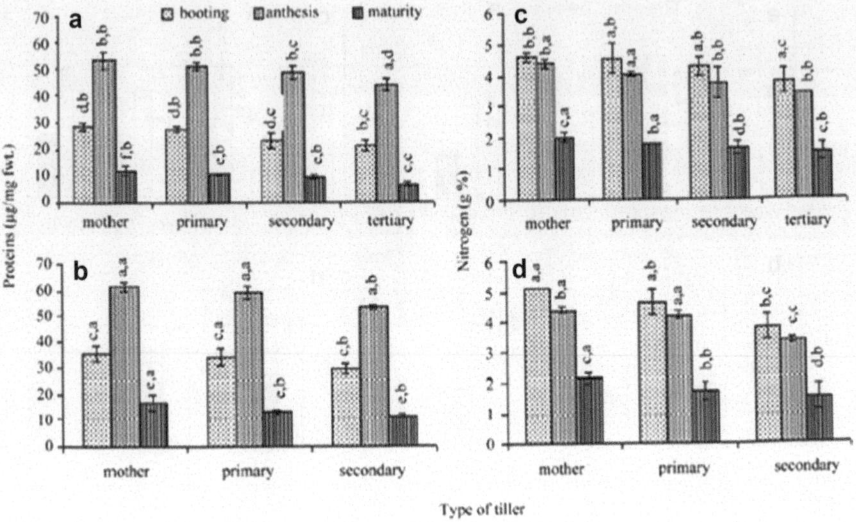

Fig. 6.3 Protein (**a**) and (**b**) and nitrogen (**c**) and (**d**) concentrations in the flag leaves of different types of tillers of high-tillering Lalat (top) and medium tillering MGD-106 (bottom) rice cultivars at booting, anthesis, and maturity. The data are presented as the means of three replicates. The small letters above the vertical bars {lower ones: v × ds × r, the interaction between varieties (v), ds (developmental stages) and r (replicates) and upper ones "v × t × r" interaction between varieties (v), t (tillers) and replicates (r)}. Common letters with "v × ds × r" (CV = 0.628, LSD = 3.213 at $P \geq 0.05$) and "v × t × r" (CV = 0.736, LSD = 3.71 at $P \geq 0.05$) for (**a**) and (**b**) and "v × ds × r" (CV = 0.081, LSD = 0.335 at $P \geq 0.05$) and "v × t × r" (CV = 0.905, LSD = 0.387 at $P \geq 0.05$) for (**c**) and (**d**) indicate no significant difference according to Duncan's multiple range test (DMRT) (Kariali et al. 2012)

6.3 N, P, and K Effects on Rice Tillering

Adequate tissue levels of mineral elements are needed for optimum growth. Carbon, hydrogen, and oxygen are acquired from water by absorption in the roots and carbon dioxide assimilation in the leaves. Among the other elements, seven are macronutrients and nine are microelements and are absorbed by roots from the soil (Table 6.2).

The elements are classified as either macro- or micro-nutrients according to their concentration in the tissue. However, this classification system is unjustified in consideration of the physiological activities of the elements, and an alternative function-based classification has been proposed by Mengel and Kirkby (1987). The function-based classification has the merit of defining the essentiality of the element for participation in various biochemical activities, which highlights its relative contribution to the plant's life cycle. The essential elements of this system can be classified into four groups (Taiz et al. 2015).

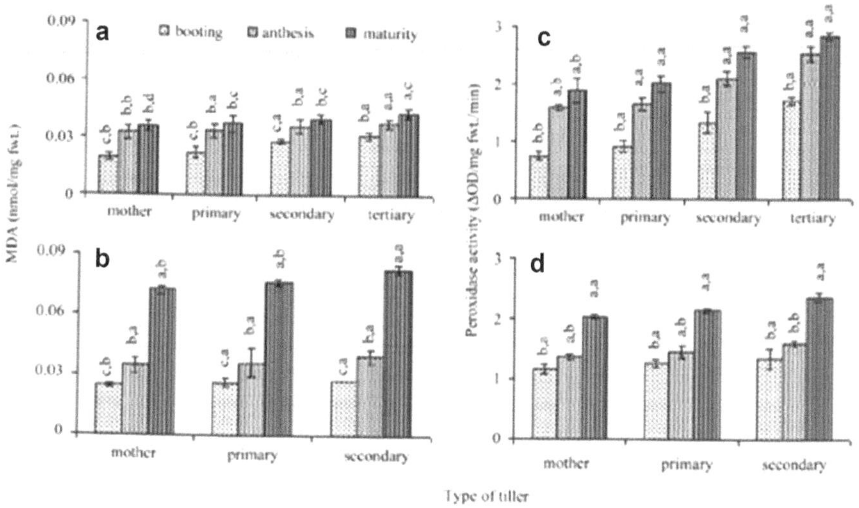

Fig. 6.4 Lipid peroxidation (**a**) and (**b**) and peroxidase activity (**c**) and (**d**) in the flag leaves of different types of tillers of high-tillering Lalat (top) and medium tillering MGD-106 (bottom) rice cultivars at booting, anthesis, and maturity. The data are presented as the means of three replicates. The small letters above the vertical bars {lower ones: v × ds × r, the interaction between varieties (v), ds (developmental stages) and r (replicates) and upper ones "v × t × r" interaction between varieties (v), t (tillers) and replicates (r)}. Common letters with "v × ds × r" (CV = 0.005, LSD = 0.329 at $P \geq 0.05$) and "v × t × r" (CV = 0.234, LSD = 0.006 at $P \geq 0.05$) for (**a**) and (**b**) and "v × ds × r" (CV = 0.556, LSD = 0.774 at $P \geq 0.05$) and "v × t × r" (CV = 0.578, LSD = 0.894 at $P \geq 0.05$) for (**c**) and (**d**) indicate no significant difference according to Duncan's multiple range test (DMRT) (Kariali et al. 2012)

6.3.1 Individual Contributions of N, P, and K

Because rice soils are often deficient in nitrogen (N), phosphorus (P), and potassium (K), the addition of these elements to the field improves their absorption into the tissue for growth and grain yield. Fertilizer treatments become effective when applied in the deficiency zone of the growth curve drawn about variations in the concentration of the element in the tissue. Figure 6.6 describes the effects of varying concentrations of nutrients on plant growth and yield while identifying the three zones, i.e. the deficiency zone, adequate zone, and toxic zone. For maximum tillering of rice plant N, P and K levels in leaf blades are required at 5%, 0.2% and K 1.5% respectively, whereas no tillers are formed when the respective nutrient concentration decline to 2%, 0.03% and 0.5% (Yoshida and Hayakawa 1970).

The chemical fertilizers applied to rice fields contain inorganic salts of the macroelements N, P, and K. These fertilizers are in high demand during the growth and development of rice crops. The fertilizers are given in a balanced proportion to stabilize the soil medium and sustain rice growth. For a balanced ratio, the farmer is advised to test the soil by chemical analyses before cultivation. For example, a 12:10:09 ratio between the three elements on the package label of fertilizer bags indicates the effective percentage of $N:P_2O_5:K_2O$. Although the proper balance of these elements is conducive to

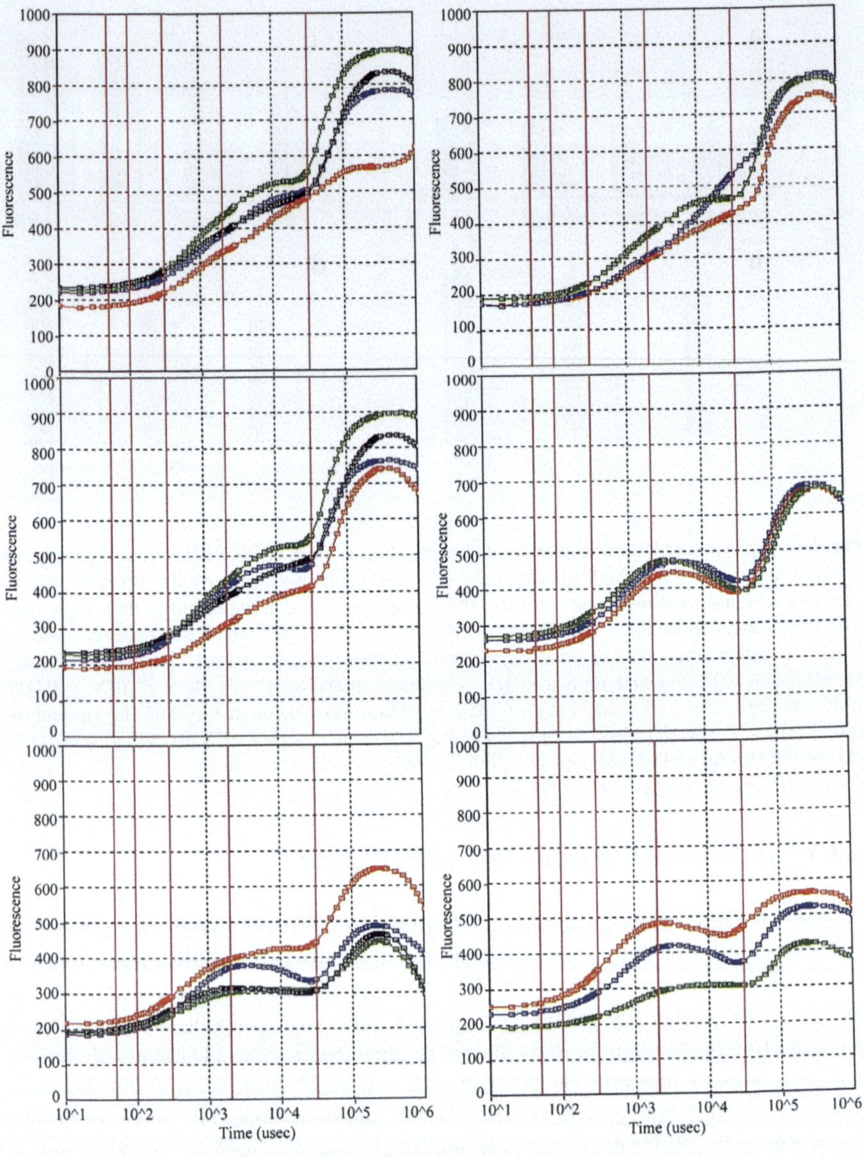

Fig. 6.5 Fluorescence (O-J-I-P) curves of the mother (red), primary (blue), secondary (green), and tertiary (black) tillers of high-tillering Lalat (left) and medium-tillering MGD-106 (right) rice cultivars at booting (top), anthesis (middle), and maturity (bottom) (Kariali et al. 2012)

achieving optimum yields mainly by promoting tiller emergence, their roles in cell metabolism are not identical. Plants absorb elements in ionic form, where potassium is taken up as a cation, phosphorus is absorbed as an oxyanion phosphate (PO_4^{3-}), and nitrogen enters as either an anion nitrate (NO_3^-) or a cation ammonium (NH_4^+).

Table 6.2 Adequate tissue levels of elements that may be needed by plants (Source: Taiz et al. 2015)

Adequate tissue levels of elements that may be needed by the plants			
Name of the element	Chemical symbol	Concentration in dry matter (% or ppm)[a]	Relative number of atoms concerning molybdenum
A. Source: Water or carbon dioxide assimilation			
Carbon	C	45	40,000,000
Hydrogen	H	6	60,000,000
Oxygen	O	45	30,000,000
B. Source: Micronutrients absorbed from the soil			
Boron	B	20	2000
Chlorine	Cl	100	3000
Copper	Cu	6	100
Iron	Fe	100	100
Manganese	Mn	50	1000
Molybdenum	Mo	0.1	1
Nickel	Ni	0.1	2
Sodium	Na	10	400
Zinc	Zn	20	300
C. Source: Macronutrients absorbed from the soil			
Calcium	Ca	0.5	125,000
Magnesium	Mg	0.2	80,000
Nitrogen	N	1.5	1,000,000
Phosphorus	P	0.2	60,000
Potassium	K	1.0	250,000
Silicon	Si	0.1	30,000
Sulphur	S	0.1	30,000

6.3.2 Nitrogen

In any type of soil, the demand for nitrogen is greater than that for the other two elements P and K, for crop production. The plants require nitrogen in greatest demand. Rice plants absorb nitrogen from the soil medium in the form of nitrate and ammonium, and both of these ions are highly mobile. Organic nitrogen in the form of amino acids is less mobile. Nitrogen is mainly taken up in the form of nitrate because ammonium taken up at high concentrations becomes toxic. For nitrate uptake, there are transporters encoded by the *NRT1* and *NRT2* gene families. Rice plants, grown under floodwater submergence access nitrogen as an ammonium cation. There are high and low-affinity NH_4^+ transporters encoded by AMT family genes (Maathuis 2009). Nitrate reduction occurs in plastids/chloroplasts via the participation of two enzymes nitrate reductase and nitrite reductase (Fig. 6.7).

Nitrogen is necessary for the development of various organs, photosynthesis, protein and nucleic acid synthesis, and distribution within cells, which in turn reflects rice yield and quality (Maathuis 2009). The most important function of N is to provide the amino group of amino acids, which subsequently builds up various proteins. Nitrogen is also a constituent of the ring structures of purines and pyrimidines, providing infrastructure for DNA and RNA, which by their right, control

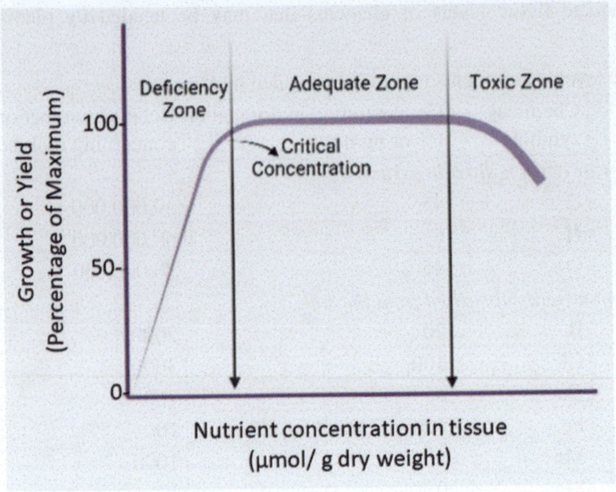

Fig. 6.6 Relationships between yield or growth and nutrient content of plant tissue. The critical concentration is the concentration below which growth is reduced. (Adapted from Plant Physiology, Taiz et al. 2015)

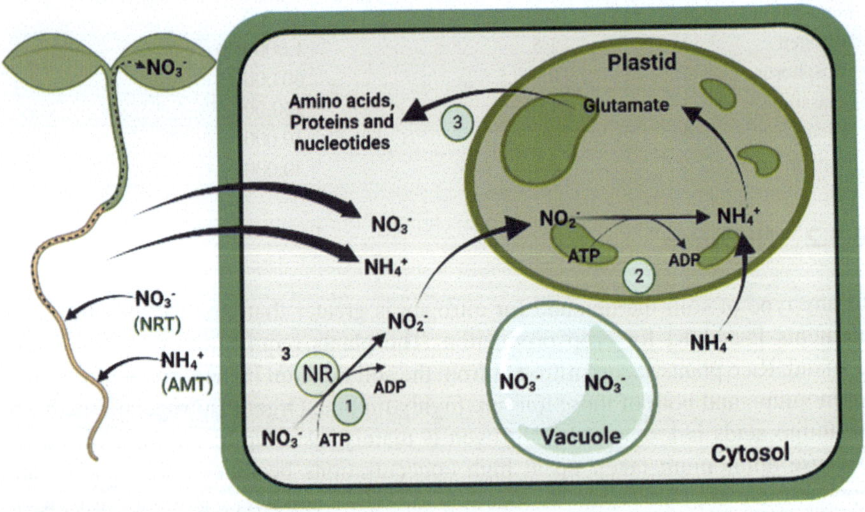

Fig. 6.7 Nitrogen acquisition primarily occurs in the form of inorganic NO_3^- and NH_4^+. NO_3^- uptake is mediated by high-affinity H^+-coupled symporters from the NRT family, whereas NH_4^+ uptake is mediated by AMT transporters that function as ion channels or possibly NH3:H^+ symporters. Most NO_3^- reduction and assimilation occur in the shoot. The first reductive step (1) occurs in the cytoplasm by the enzyme nitrate reductase (NR) and nitrite is further reduced to $NH4^+$ (ammonium ion) in chloroplasts (2). Reduced N is assimilated into the amino acid glutamate, which in turn is used in the synthesis of proteins and nucleotides in the cytoplasm (3). Further, surplus N is typically stored in the central vacuole as NO_3^-. (Adapted from Maathuis (2009)

Table 6.3 Classification of plant mineral nutrients according to their biochemical functions (Taiz et al. 2015)

Classification of nutrients based on their mode of biochemical functions	
Mineral nutrients	Functions
Group 1	*Nutrient elements that constitute carbon compounds*
N	Constituents of proteins, nucleic acid, nucleotides, amides, amino acids, coenzymes, hexosamines, etc.
S	Constituents of cystine, cysteine, methionine and coenzyme A, lipoic acid, thiamine pyrophosphate, glutathione, biotin, adenosine-5′-phosphosulphate, and 3-phosphoadenosine
Group 2	*Nutrients that act as energy storage or structural components*
B	Complexes with mannitol, mannan, polymannuronic acid, and other constituents of cell walls. Involved in cell elongation and nucleic acid metabolism
P	Component of sugar phosphates, nucleic acids, nucleotides, coenzymes, phospholipids, phytic acid, etc. plays an important role in ATP-mediated reactions
Si	Important component of the cell wall and is mostly deposited in the form of amorphous silica. It provides cell wall mechanical properties, rigidity, and elasticity and is involved in lodging resistance
Group 3	*Nutrients that remain in ionic form*
Ca	Important components of the middle lamella of cell walls and act as cofactors for some enzymes involved in ATP hydrolysis and phospholipids catabolism. It also acts as a second messenger in the plant's metabolic regulation
Cl	It is needed for the photosynthetic reaction involved in O_2 evolution in oxygen evolving complex
K	Acts as a cofactor for many important enzymes and important cations in maintaining cell turgor, cell electroneutrality, and plays an important role in stomatal opening/closing
Mg	Required by many enzymes involved in phosphate transfer reactions and acts as a major constituent of the chlorophyll molecule
Zn	Constituent of some dehydrogenases like alcohol dehydrogenase, glutamic dehydrogenase, and carbonic anhydrase
Na	Involved with the regeneration of phosphoenolpyruvate in both C_4 and CAM pathways in plants and often acts as a substitute for potassium in many plant functions
Group 4	*Nutrients involved in redox reactions*
Cu	Component of many enzymes such as ascorbic acid oxidase, tyrosinase, monoamine oxidase, uricase, cytochrome oxidase, phenolase, laccase, and plastocyanin
Fe	Constituent of cytochromes and nonheme iron proteins involved in photosynthesis, N_2 fixation, and respiration
Mo	Constituent of nitrogenase, nitrate reductase, and xanthine dehydrogenase
Ni	Constituent of urease. Constituent of hydrogenases in N_2-fixing bacteria
Mn	Required for the activity of some important enzymes such as dehydrogenases, decarboxylases, kinases, oxidases, and peroxidases. Involved in other cation-activated enzymes and acts as an important component of photosynthetic O_2 evolution in the oxygen evolving complex

various biological functions of plant cells (Table 6.3). The element is highly mobile in plant tissues. During maturity, the concentration of nitrogen in the leaf blades and leaf sheath declines progressively, while that in the stems increases at tillering before a fall. However, at maturity, there is no difference in the concentration of nitrogen between these organs. This phenomenon is indicative of the remobilization of nitrogen in plant organs with the progression of ageing when root uptake decreases and demand increases due to partitioning for reproductive growth. Symptoms of deficiency appear first on older leaves and spread to younger leaves before any adverse effects on yield. Old leaves become chlorotic or light green under N deficiency and die prematurely in rice plants. The N-deficient rice leaves are narrow, short, erect, and yellowish. Nitrogen deficiency is common in all rice-growing soils of Asia where high-yielding semi-dwarf rice is grown. This deficiency leads to poor tiller growth and a loss of yield.

6.3.3 Nitrogen and Tiller Heterogeneity

A study showed that N, P, and K fertilizers can increase panicle grain number by 31.4, 23.9, and 48.2%, respectively, and effective tiller number by 55.1, 29.2, and 6.7%, respectively (Ye et al. 2019). Compared to those in the zero-application treatment, the increases in P and K levels in the growing medium had significant positive effects on plant growth and yield attributes, such as the number of grains per panicle and the number of panicles, but these impacts were far below the values associated with nitrogen application effects. Thus, nitrogen treatments are highly essential for improving the effective tiller number and thereby overall grain yield in rice (Ndille et al. 2021), but the response is not identical among different categories of tillers. In this context, Wang et al. (2017) classified rice tillers into three different categories based on their production capacity: superior, medium, and inferior. The production capacity was calculated as the ratio of the grain yield of individual tillers divided by the grain yield of the optimally producing tillers. On this scale, tillers were denoted as inferior (ratio < 50%), medium (ratio between 50% and 75%) or superior (>75%), and their response to nitrogen supplementation was tested with the specific objectives of (1) quantifying the proportion of the three types of tillers under different N levels and (2) comparing the yield components of the three types of tillers under different N levels. The tillers responded differently to nitrogen application, while the heterogeneity in tiller development increased (Wang et al. 2017). The quantitative proportions and yield contributions decreased in superior tillers but increased in inferior tillers, while medium tillers were not responsive to increasing nitrogen supply (Wang et al. 2017). Under high-nitrogen conditions, the number of spikelets per panicle increased, but the grain filling percentage and grain weight of medium and inferior tillers decreased. Therefore, enhancing the grain filling and weight of inferior tillers could be an effective approach for further improving the per-acre yields of rice.

6.3.4 Phosphorus

Approximately 90% of soil phosphorus is fixed and not accessible to roots, causing the soil to be deficient in this element. The residual 10%, comprising the labile fraction, exchanges with the soil solution and contributes to the uptake of inorganic P (PO_4^{3-}) by roots. The process of release is very slow and unsustainable for ongoing plant growth and ultimately exacerbates P deficiency in the soil, devastating crop growth. Therefore, in rice agricultural settings, farmers must replenish the soil with a large number of P fertilizers during the initial period of field preparation, without which the chances of remunerative yield are minimal. The massive use of P fertilizers in agriculture indicates that the P level in rice ecosystems is suboptimal for crop growth. Normally, rock phosphates are the major source of P fertilizers.

Phosphorus (such as phosphate-PO_4^{3-}) is an important macronutrient that plays a central role in all major metabolic processes of plant cells, including respiration and photosynthesis and sustains energy transduction for the maintenance of life. Sugar phosphate is an integral component of the sugar-phosphate intermediates of respiration and photosynthesis and the phospholipid content of the cell membrane preserves cellular integrity. Most prominently, it is a component of nucleotides used in energy metabolism, such as ATP, and in DNA and RNA. In addition, ATP is actively involved in the energy-dependent cellular processes of biosynthesis, ion pumping, and mechanical work.

Like nitrogen, phosphorus is highly mobile, and deficiency symptoms appear first in older organs. In phosphorus-deficient rice plants, phosphorus use efficiency was shown to be linked to tissue-specific biomass and P allocation patterns, where nascent leaves and roots became major sinks and old leaves were the sources of P translocation (Irfan et al. 2020). This deficiency disrupts cell metabolism. Physiologically, P deficiency is first reflected in a sharp decline in leaf photosynthesis because sugar phosphates are involved in many intermediate steps of the carbon fixation process. The characteristic symptoms are stunted growth of young plants and dark green colouration of leaves, often bearing necrotic spots. The leaves may accumulate anthocyanin pigments, but unlike plants deficient in N, they are not chlorotic. Plaxton and Tran (2011) reported that within species, plants can limit their ability to acclimatize to extended periods of P deprivation by upregulating a complex array of morphological, physiological, and biochemical adaptations, collectively called the Pi starvation response (PSR). PSR arises because of the induction of many Pi-starvation-inducible genes. These genes encode enzymes that reorient intrinsic phosphate use while maximizing external phosphate acquisition. A suite of miRNAs has been implicated in the control of Pi homeostasis in cells in addition to pre- or post-transcriptional modification of starvation-induced gene expression. Although the genetic or epigenetic regulation of gene expression has not been pinpointed, Plaxton and Tran (2011) suggested a model for various adaptive metabolic processes that elucidates the mechanisms involved in the acclimatization of plants to Pi deficiency (Figs. 6.8 and 6.9).

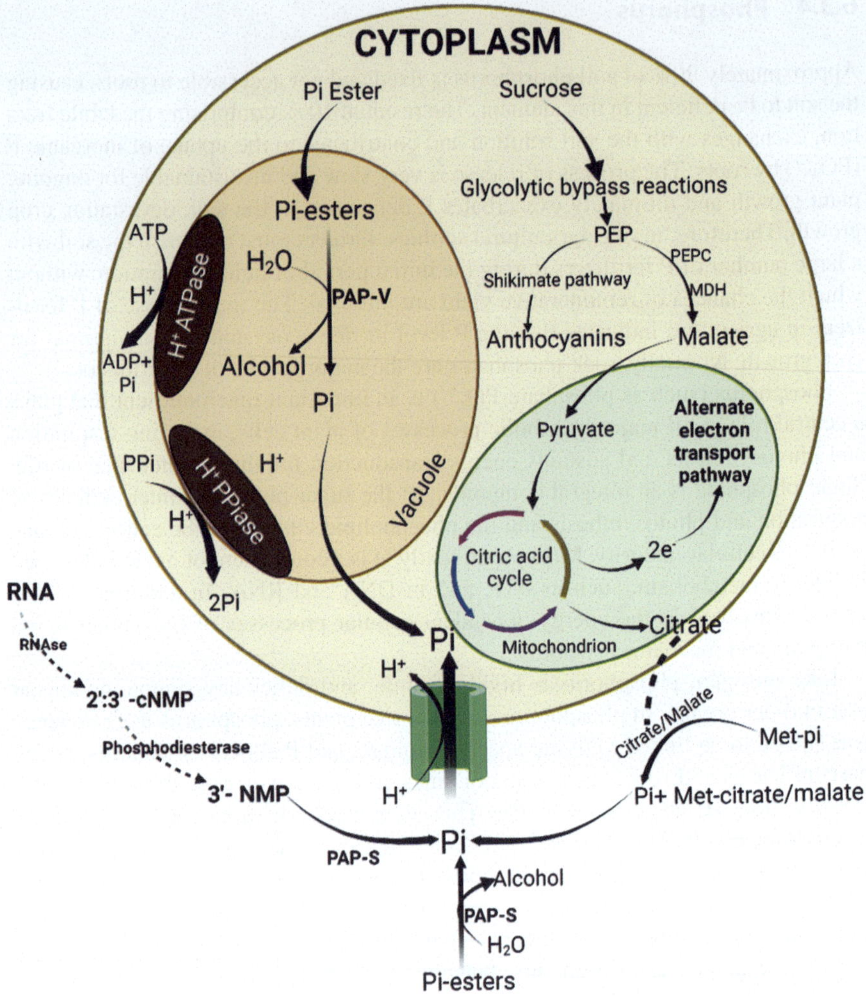

Fig. 6.8 Adaptive metabolic process during Pi deficiency in plants. Alternative pathways of cytosolic glycolysis, mitochondrial electron flow, and vacuolar H^+ pumping facilitate respiration and maintenance of vacuolar pH by Pi in plant cells as they negate the dependence on adenylates and Pi. The levels of adenylates and Pi become markedly depressed during severe Pi starvation. Large amounts of organic acids like citrate/malate are produced by PEPC, malate dehydrogenase (MDH), and citrate synthase (CS) which may also be excreted by roots to increase the availability of: (1) mineral-bound Pi and (2) organic P and its amenability to hydrolysis by secreted PAPs. (Adapted from Plaxton and Tran 2011)

6.3.5 Phosphorus Absorption

The soil medium is composed of solid, liquid, and gaseous phases. The liquid phase constitutes the soil solution, which contains all the dissolved nutrients and provides a medium in which ions move to the root surface. Phosphorus available as calcium phosphate is attached to the solid phase, and it is sparingly soluble in the aqueous phase and

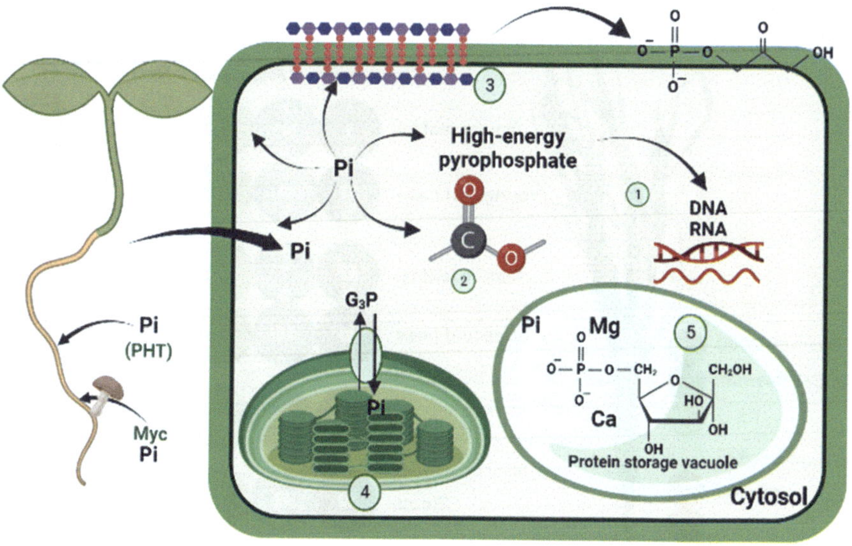

Fig. 6.9 Phosphorus is often a growth limiting factor, and many plants are associated with mycorrhizal fungi (Myc), which improve P nutrition in return for reduced carbon. P is absorbed by the plants directly as inorganic PO_4^{3-} (Pi) through H^+- coupled high-affinity transporters. Cellular Pi is essential for cellular energy homeostasis as it readily forms high-energy pyrophosphate (1) and ester (2) bonds. Phosphate groups form the lipophilic component of many membrane lipids (3) and are therefore essential for membrane composition and integrity. Chloroplast P, necessary for the synthesis of high-energy bonds in photosynthesis, enters the chloroplast (4) in exchange for glyceraldehyde-3-phosphate (G3P). In stored tissues (5), P is sequestered in the protein storage vacuole (PSV) as phytate together with minerals. (Adapted from Maathuis 2009)

absorbed very slowly by the roots. Hence, plant species adapt to phosphorus deficiency by inducing mycorrhizal symbiosis, which helps them acquire scarce minerals. Fungal hyphae contribute up to 80% of plant phosphorus at the high cost of carbon assimilation. To alleviate P deficiency, plants produce protoid roots that expand the invaded soil volume and extrude large amounts of chelators in the form of organic acids to help release soil particle-bound phosphates and transport them to the liquid phase.

6.3.6 P Deficiency in Rice Tillers

In addition to nitrogen, phosphorus has been a major driver of the optimum productivity of rice crops. The suboptimal availability of the element in arable lands has forced researchers to identify resilient genotypes and reduce the reliance on costly synthetic fertilizers. In this context, Irfan et al. (2020) compared the utilization efficiency and internal remobilization of phosphorus among six rice cultivars. The rice cultivars PS2 (P-efficient) and Super Basmati (P-inefficient) exhibited greater P uptake, utilization efficiency, and internal remobilization than did the other cultivars grown in the absence of external phosphorus supplementation. In comparison to Super Basmati, the PS-2 cultivar had greater plant height, larger panicles, more tillers, and greater grain yield under P-deficient conditions because of its ability to

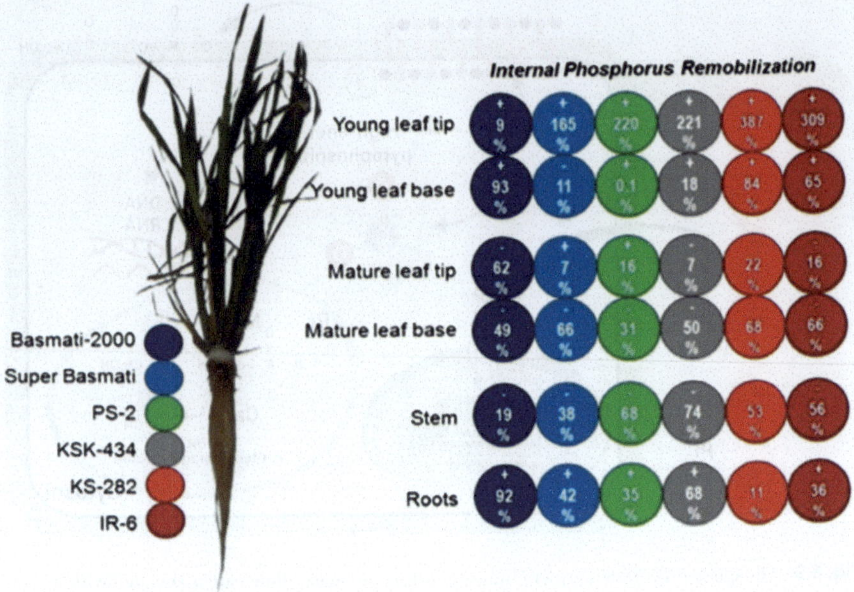

Fig. 6.10 Internal phosphorus remobilization among various plant tissues of six rice cultivars at 50 days following transplanting after feeding for 20 days with an adequate P concentration (200 μM Pi) and subsequent P omission for the next 30 days. Positive signs indicate P accumulation, while negative signs indicate P remobilization in specific plant tissues/organs in response to P omission. (Adapted from Irfan et al. 2020 with permission)

take up more P (Figs. 6.10 and 6.11). P deficiency is also a factor responsible for the redistribution of nitrogen in plant tissues. The concentration of nitrogen in the stem decreases at the tillering stage, restricting tiller production to a level much lower than the optimum level (Takahashi et al. 1956). This relationship also revealed the close relationship between stem nitrogen concentration and tiller number in rice (Fig. 6.12).

The tolerance to low phosphorus among rice varieties was tested by characterizing physiological and morphological traits associated with phosphorus use efficiency by Deng et al. (2020). With a two-by-two factorial design, they cultivated two rice cultivars, one with strong tolerance and one with weak tolerance to phosphorus at two levels of phosphorus, comprising normal and half of the normal levels, respectively. A reduction in the P concentration in the growth media improved the internal use efficiency, harvest index, and translocation efficiency of P in both cultivars, but these effects were lower for the strongly tolerant cultivars than for the weakly tolerant cultivars. P deficiency reduced the tiller number (Deng et al. 2020), shoot biomass, leaf area index, leaf photosynthesis, and root diameter, irrespective of cultivar tolerance. Low-phosphorus conditions increased the number of productive tillers, the root biomass, the root-to-shoot ratio, the root bleeding rate, and the root acid phosphatase activity. Under low P, the length, oxidation activity, and total

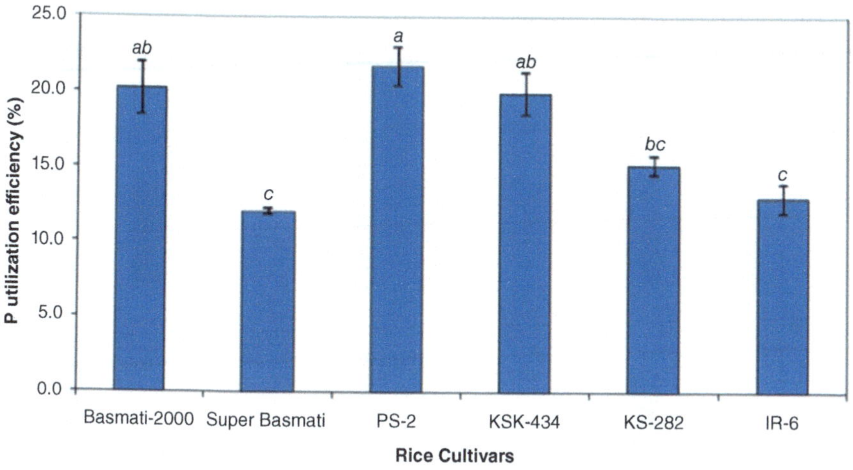

Fig. 6.11 Phosphorus utilization efficiency (PUTE) of six rice cultivars calculated at two P levels under hydroponic conditions. The values are the means of three replicates ($n = 3$) and are presented with standard errors. Bars not sharing identical letter(s) are significantly different from each other (LSD test, $P \leq 0.05$). The LSD value for PUTE at $P \leq 0.05$ was 5.53. (Adapted from Irfan et al. 2020 with permission)

and active absorption surface area increased significantly in the strongly tolerant cultivars, but the responses of the low-tolerant cultivars were the opposite. The results suggested that the varietal resiliency associated with P deficiency is closely correlated with increased growth and activity in roots (Deng et al. 2020).

6.3.7 Potassium

Potassium (K^+) is the most abundant cation in plants and constitutes 2–5% of the plant's dry weight (Marschner 2012). For rice plants, potassium (K^+) is the most limiting nutrient after nitrogen and phosphorus and needs to be applied in a balanced amount with other nutrients for optimum growth. In plant cells, K^+ remains soluble in the ionic form in cytoplasmic or vacuolar fluids and is not assimilated in any organic matter or cellular constituents. The cation charge allows fast and efficient control of the electrical balance and osmotic potential in plant cells. Potassium, together with organic solutes, becomes the most important osmolyte for establishing cell turgor and driving pressure-dependent processes such as stomatal movement, cell elongation, and long-distance phloem transport. The osmoregulatory functions also involve the exchange of materials with the extracellular environment. K^+ activates many enzymes involved in respiration and photosynthesis. Depending on the appropriate duration of fertilization, potassium helps to improve rice yield and quality. K^+ fertilizer is applied at a rate of 37.5 kg of muriate of potash, 50% higher than the recommended dose, at the time of heading and has the potential to achieve the highest head rice yield (Atapattu et al. 2018). According to the Rice

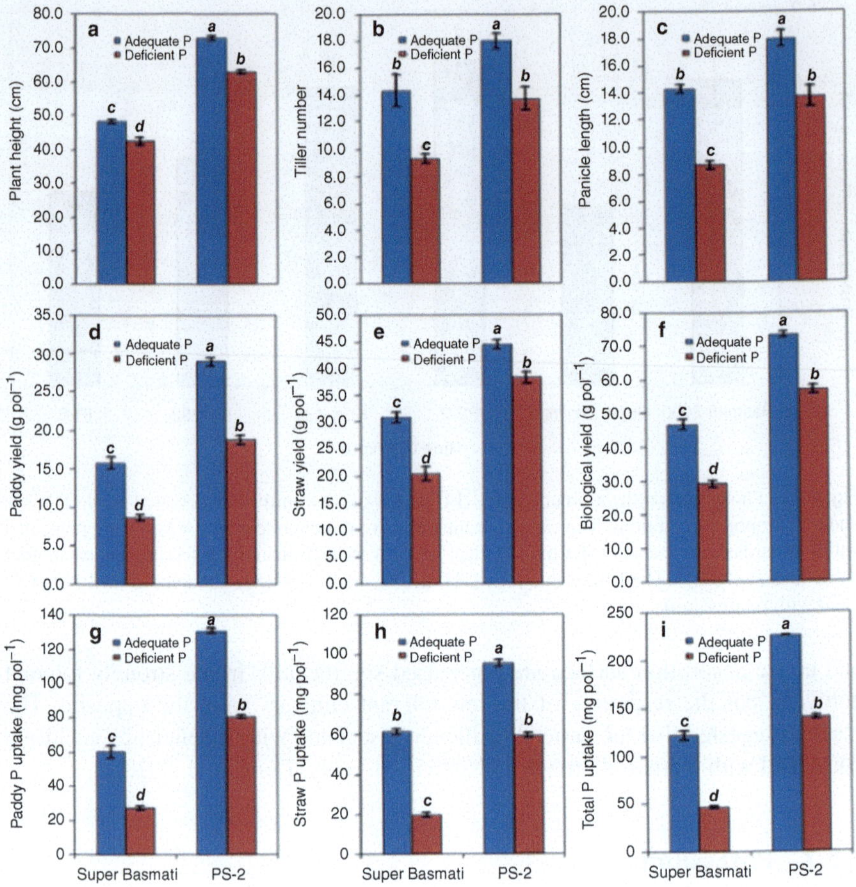

Fig. 6.12 Plant height (**a**), tiller number (**b**), panicle length (**c**), paddy yield (**d**), straw yield (**e**), biological yield (**f**), paddy P uptake (**g**), straw P uptake (**h**), and total P uptake (**i**) of two rice cultivars. The values are the means of three replicates ($n = 3$) and are presented with standard errors. Bars not sharing identical letter(s) are significantly different from each other at two P levels (LSD test, $P \leq 0.05$). (Adapted from Irfan et al. 2020 with permission)

Knowledge Bank, rice plants absorb high amounts of K^+ from the medium of growth. Rice straw is rich in K^+ and contains 14.0 kg K^+ or 16.8 kg K_2O per ton of straw.

6.3.8 Potassium Uptake

Like those involved in carbon autotrophy, plants are also potassium autotrophs. A special mechanism is needed to extract potassium from the growing medium, collect it from the soil, and ensure its distribution within the system. A set of channels and transporters are used for the transfer of K^+ across several membranes in the cell.

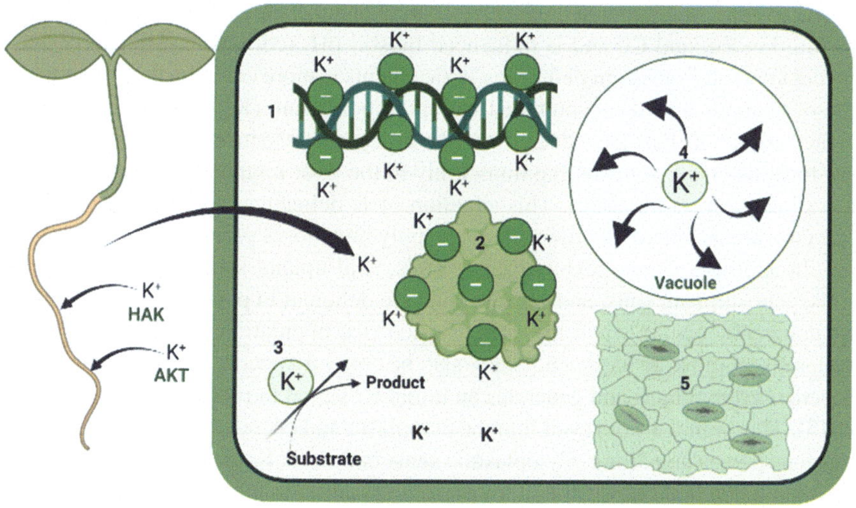

Fig. 6.13 Potassium acquisition occurs through high-affinity H^+-coupled symporters (HAKs) and low-affinity ion channels (AKT). Inside plants, the chaotropic properties of K^+ make it an ideal counterion because of its negative charges on nucleic acids (1) and proteins (2). In addition, K^+ activates specific enzymes (3) by acting as cofactors in enzymatic reactions between substrates and products. As the main cation in vacuoles (4), K^+ generates turgor to provide structure and drive cell expansion, plant growth, and plant movement, such as the regulation of stomatal apertures (5). (Adapted from Maathius 2009)

Energy for transport is provided by the splitting of ATP during K^+-ATPase activity, which generates an electrical gradient and a proton gradient across the membrane. Proton-coupled K^+-transporters harness gradients for the absorption of potassium. In addition, potassium can use an electrical gradient for entry into the cell via potassium channels. Potassium channels are also used for the exit of K^+ by acting in concert with anion channels (Dreyer et al. 2019). The mode of uptake and function of K^+ have been described below by Maathuis (2009) (Fig. 6.13).

6.4 Deficiency Symptoms

Like N and P, K is highly mobile in plant cells. K^+ omission induces the remobilization of the element from older to younger organs. Hence, deficiency symptoms appear first in older leaves and subsequently move towards younger leaves. The deficiency symptoms include stunted stem growth with short internodes. The plant becomes weak and slender and bears many pronounced yellow- or orange-coloured leaves. The leaves develop marginal chlorosis that results in the formation of necrotic lesions at the leaf tip first and spreads towards the leaf base. Plants bearing weak stems become highly susceptible to wind action, bend to the ground, and remain lodged on surface water in the field. Potassium availability in the soil is influenced by several physicochemical factors. Sinclair et al. (1990) reported that

K^+ availability is dependent on the clay content and presence of microelements such as Mn, Fe, Zn, and Cu and is influenced by soil pH, redox potential, and organic matter content. Potassium deficiency in rice plants is more common under excessive use of N and P fertilizers combined with poor potassium (K^+) application. K^+ deficiency also contributes to a decrease in the nitrogen concentration in plants after the ear-forming stage, contributing immensely to the poor regulation of ineffective tillers (Takahashi et al. 1956). This situation of K deficiency appears intensively in rice crop areas, where ambitious farmers apply high doses of nitrogen and phosphorus. In addition, coarse-textured/sandy soils, acid upland soils, degraded lowland soils, acid sulphate soils, and organic soils are deficient in potassium. Additionally, K^+ deficiency occurs when the direct-seeded rice population is large and the root system is shallow. K^+ deficiency may also be cultivar-specific. K^+ deficiency stress is perceived by plants and generates an initial K^+ signal in root cells (Wang and Wu 2013). This signal is conveyed into the cytoplasm and encoded by Ca^{2+} and reactive oxygen species signalling. Cytoplasmic sensors decode K^+-deficiency-induced signalling and regulate downstream transcriptional and post-transcriptional responses, eventually producing a series of adaptive responses in both physiological and morphological alterations for survival in K^+ deficiency.

6.5 Mechanism of K^+ Deficiency Perception

K^+-deficiency signals are first perceived by the root epidermis plasma membrane and transduced into the interior of the cytoplasm. Perception inside cells induces a series of physiological and biochemical reactions, which include both long- and short-term responses. Short-term responses occur within hours of exposure of the roots to the soil medium and involve several transduction signalling components, including the membrane potential, reactive oxygen species (ROS), and plant hormones. In the early stage of infection, K^+-deficiency-sensing K^+ inside the cell may not be affected, but prolonged exposure leads to a decrease in K^+ in the cytoplasm and initiates a long-term response. In summary, Wang and Wu (2013) proposed a model of K^+ transport and signalling pathway in *Arabidopsis* in response to K^+ deficiency (Fig. 6.14).

6.6 Mechanism of K^+ Transport

K^+ deficiency activates H^+-ATPase, especially the AHA2 activity, resulting in plasma membrane hyper-polarization and extracellular acidification. The hyperpolarized membrane potential (ψ) activates the inwards K^+ channel like AKT1 and Ca^{2+}-permeable channels whereas extracellular acidification energizes K^+ uptake via a K^+ transporter protein like HAK5. Further, K^+ deficiency induces ROS accumulation in root cells via RHD2 and RCI3, which in turn regulate the transcription of K^+-deficiency-responsive genes like *HAK5*. Higher levels of ROS also

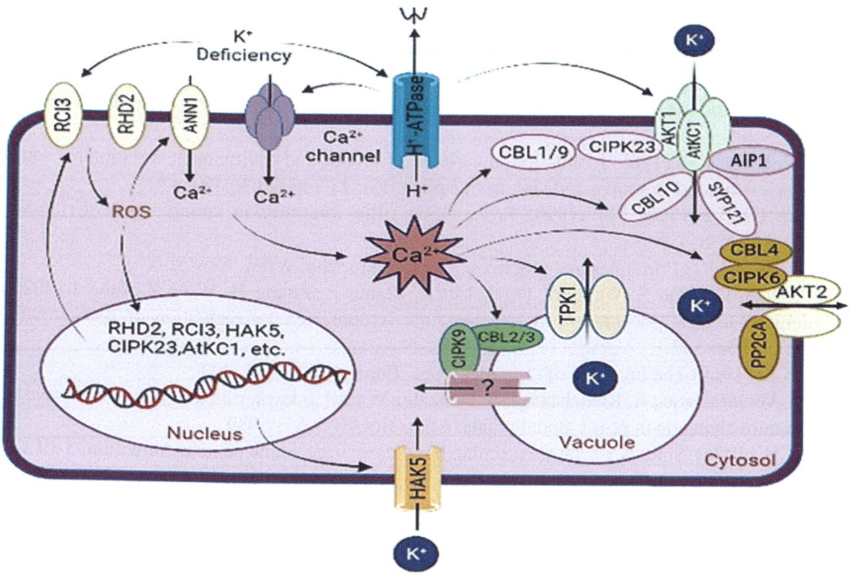

Fig. 6.14 K^+ transport and signalling pathways in *Arabidopsis* in response to K^+ deficiency (Adapted from Wang and Wu 2013)

activate the Ca^{2+}-permeable ANN1 channels that mediate Ca^{2+} influx into the cytoplasm. The activation of Ca^{2+}-permeable channels by membrane hyperpolarization and ROS results in the increase of Ca^{2+} in the cytosol and this Ca^{2+} signal can be perceived and transduced downstream by specific Ca^{2+} sensors like CBL proteins. CBL1 (and/or CBL9) recruits the cytoplasm-localized kinase CIPK23 to the plasma membrane, where CIPK23 activates AKT1-mediated K^+ uptake through phosphorylation, and AKT1 activity can also be regulated by CBL10, AIP1 (a protein phosphatase), AtKC1 (a K^+ channel subunit), and SYP121 (a SNARE protein). The CBL4-CIPK6 complex enhances AKT2-mediated K^+ flux by facilitating the translocation of the AKT2 from the endoplasmic reticulum to the plasma membrane which is kinase interaction-dependent but independent of phosphorylation. At the vacuolar level, CBL3/CBL2 interacts with CIPK9 and causes the activation of an unknown vacuole-localized K^+ channel leading to the release of K^+ from the vacuole to the cytoplasm. Apart from this, the tonoplast channel TPK1 also mediates K^+ efflux from the vacuole into the cytoplasm (Wang and Wu 2013).

References

Atapattu A, Prasantha BDR, Amartunga KSP, Marambe B (2018) Increased rate of potassium fertilizer at the time of heading enhances the quality of direct seeded rice. Chem Biol Technol Agric 5:22

Choi HC, Kwon KW (1985) Evaluation of varietal difference and environment variation for some characters related to source and sink in rice plant. Kor J Crop Sci 30:460–470

Davidson DJ, Chevalier PM (1990) Pre-anthesis tiller mortality in spring wheat. Crop Sci 30(4):832–836

De Datta SK (1981) Principles and practices of rice production. Wiley, New York

Deng Y, Men C, Qiao S, Wang W, Gu J, Liu L, Zhang Z, Zhang H, Wang Z, Yang J (2020) Tolerance to low phosphorus in rice varieties is conferred by regulation of root growth. Crop J 8:534–547

Donald CM (1968) The breeding of crop ideotypes. Euphytica 17:385–403

Dreyer I, Vergara-Jaque A, Riedelsberger J, Gonzalez W (2019) Exploring the fundamental role of potassium channels in novel model plants. J Exp Bot 70(5985):5989

Hayashi H (1976) Studies on large vascular bundles of rice plant panicle formation. I-III (in Japanese with English summary). Proc Crop Sci Soc Japan 42:322–342

Irfan M, Aziz T, Maqsood MA, Bilal HM, Siddique KHM, Xu M (2020) Phosphorus (P) use efficiency in rice is linked to tissue-specific biomass and P allocation patterns. Sci Rep 10:4278. https://doi.org/10.1038/s41598-020-61147-3

Kariali E, Sarangi S, Panigrahi R, Panda BB, Mohapatra PK (2012) Variation in senescence pattern of different classes of rice tillers and its effect on panicle biomass growth and grain yield. Am J Plant Sci 3:1047–1057

Maathuis FJ (2009) Physiological functions of mineral macronutrients. Curr Opin Plant Biol 12:250–258

Marschner P (2012) Marschner's mineral nutrition of higher plants, 3rd edn. Academic Press, San Diego

Mengel K, Kirkby EA (1987) Principles of plant nutrition. International Potash Institute, Worblaufen-Bern

Mohanan KV, Pavithran K (2007) Chronology of tiller emergence and tiller orientation in rice (*Oryza sativa* L.). Oryza 44:307–310

Mohapatra PK, Kariali E (2008) Time of emergence determines the pattern of dominance of rice tillers. Aust J Crop Sci 1(2):53–62

Moldenhauer K, Slaton N (2001) Rice growth and development. In: Rice production handbook, pp 7–14

Ndille CN, Lena EM, Mupeta I, Nkegngafac NJ (2021) Effect of the amount and timing of ammonium sulfate single top-dressing application on growth and yield of Akitakomachi rice (*Oryza sativa* L.). Asian J Agric Food Sci 9:187–201

Padmaja Rao S (1991) Breaking the yield barriers in rice through enhancing high-density grains. In: Proceedings of the International Conference on Plant Physiology. Banaras Hindu University, Varanasi, pp 307–316

Plaxton WC, Tran HT (2011) Metabolic adaptations of phosphate-starved plants. Plant Physiol 156:1006–1015

Sahu KC, Kariali E, Mohapatra PK (2004) Tiller dominance in rice is dependent on assimilate concentration of the panicle during grain development. Indian J Plant Physiol 9:402–406

Sinclair AH, Mackiedawson LA, Linehan DJ (1990) Micronutrient inflow rates and mobilization into soil solution in the root zone of winter-wheat (*Triticum-aestivum* l). Plant and Soil 122:143–146

Taiz L, Zeiger E, Moller IM, Murphy A (2015) Plant physiology and development, 6th edn. Sinauer Associates, Sunderland, MA, pp 119–142

Takahashi N, Okajima H, Takagi S, Honda T (1956) The mechanism of tiller development in the rice plant. Jpn J Crop Sci 25:74–75. https://doi.org/10.1626/jcs.25.73

Tong HH, Chen L, Li WP, Mei HW, Xing YZ, Yu XQ, Xu XY, Zhang SQ, Luo LJ (2011) Identification and characterization of quantitative trait loci for grain yield and its components under different nitrogen fertilization levels in rice (*Oryza sativa* L.). Mol Breed 28:495–509

Wang Y, Wu WH (2013) Potassium transport and signalling in higher plants. Annu Rev Plant Biol 64:451–476

Wang Y, Lu JW, Ren T, Hussain S, Jia Q, Zhang JL, Yousaf M, Li XK (2017) Influence of tiller heterogeneity on yield components of rice grown under different nitrogen regimes. Int J Plant Prod 11(3):437–452

Wu WG, Wilson LT, Am MC (1998) Contribution of rice tillers to dry matter and grain yield. Agron J 90:317–323

Xie X, Yoneyama K, Yoneyama K (2010) The strigolactone story, Annu. Rev. Phytopathol. 48, 93–117

Ye T, Li Y, Zhang J, Hou W, Zhou W, Lu J, Xing Y, Li X (2019) Nitrogen, phosphorus, and potassium fertilization affects the flowering time of rice (*Oryza sativa* L.). Glob Ecol Conserv 20:e00753. https://doi.org/10.1016/j.gecco.2019.e00753

Yoshida S, Hayakawa Y (1970) Effect of mineral nutrition on tillering of rice. Soil Sci Plant Nutr 16: 180–191

Yoshida S (1981) Climatic environment and its influence. In: Fundamentals of rice crop science. The International Rice Research Institute, Los Banos

7 Plasticity of Tiller Dynamics

Abstract

The plasticity of tiller dynamics in rice has been evident in the course of its adaptation to a wide spectrum of environmental conditions where the magnitude of the diversity of physical environments differs sharply and dictates irreversible changes in morphological features. The extraordinary genotypic permissibility of phenotypic plasticity has always perpetuated the changeover of tiller dynamics during rice antiquity and subsequent crop domestication. Rice crops grown under low soil moisture conditions develop ephemeral attributes by having a short life span, low biomass yield, and, above all, a mono-tillering feature. Conversely, rice plants grown in deep water extend their vegetative growth period and accumulate greater amounts of above-ground biomass at the cost of a poor harvest index. In addition to deep-water regions, plants with medium to lowland topologies harbour profusely tillering rice cultivars. In each situation, resilience under aberrant environmental conditions changes tiller dynamics resulting in the origination of new ecotypes or landraces pertinent to the niche.

Keywords

Tiller dynamics · Tillering ability · Heat stress · Cold stress · Water depth · Injury symptoms

7.1 Introduction

Rice is a tillering crop. The tiller number is the most important factor affecting rice plant architecture and grain yield because of its relationship with the final number of productive panicles. Because of its significant impact on grain yield, the rice tillering pattern has become a model system for branching in monocot plants, especially among cereal plants. Currently, tiller number has been a primary phenotyping

© The Author(s), under exclusive license to Springer Nature Singapore Pte Ltd. 2025
P. K. Mohapatra et al., *Tillering Behavior of Rice Plant*,
https://doi.org/10.1007/978-981-97-5235-5_7

indicator for screening cultivars for crop breeding. A tiller provides nutrients to the panicle for spikelet development and grain yield. On average, modern rice varieties produce 15–25 tillers, 10–15 of which become productive, while the others are either sterile or unproductive. Examples of tiller types of a high-yielding rice cv. Lalat and its phenological features are given in Table 7.1 (Mohapatra and Kariali 2008).

7.1.1 Tiller Dynamics of Rice Under Contrasting Environments

It has been surmised that tillering in rice species is cultivar-dependent because of the wide variation in genetic potential between cultivars for tiller production and survival. However, the tiller number of rice plants is dynamic and adaptable to the growing environment. The plasticity of the tiller dynamics of rice has been evident in the course of its adaptation to a wide spectrum of environmental conditions, where the magnitude of the diversity of physical environments differs sharply and dictates irreversible changes in plant phenotype. Environmental factors such as spacing, level of irradiation, nutrient supply, depth of soil, field water regime, and cultural practices influence tillering and tiller development in rice. Environmental factors modulate phenotypic expression by controlling several intrinsic physiological mechanisms, such as nutrient assimilation, the capacity for assimilate production, and partitioning and cellular hormone homeostasis. Hence, the genetic limitation of tillering behaviour becomes subjugated to the environment in a rice habitat (Mohapatra et al. 2011).

The plasticity of the tiller dynamics of wild rice plants of *Oryza rufipogon* has been used as a model system to illustrate and highlight the impact of environmental factors on the genetic potential of rice for tillering. The species is amenable to changes in tillering behaviour in response to environmental perturbation. Kariali et al. (2008) tested the pattern of tiller dynamics of *O. rufipogon* in deep-water marshlands and dry land ecotypes (often called *O. nivara*) in response to changes in natural habitat and well-managed growing conditions in cemented pots. The deep-water and dry land ecotypes had four and one tillers each in the natural habitat, and the preference was for the primary type of tiller. In contrast, the abundance of resources under pot conditions allowed profuse tiller production of both ecotypes, and many of them were non-productive tertiary tillers. The number of total tillers produced was lower in the dry land ecotype than in the deep-water ecotype grown under uniform pot conditions, which implies that the genetic potential for tillering is not entirely overwhelmed by environmental variation. Changes in tiller dynamics concerning various factors have been discussed.

7.1.2 Cultivar-Specific Tillering

In rice-growing habitats, the effect of environmental factors could be most decisive for tillering and tiller development. However, in the same environment, rice species differ markedly in terms of tillering ability, tiller performance, and contribution to

Table 7.1 Morphological and phenological features of the semi-dwarf rice cultivar Lalat grown in pots in 2001–2002

Type of tiller	Days from transplanting					Length of panicle (cm)	Number of grains		Weight of fertile grains (g)	% yield contribution (g)
	TE	PI	BT	AN	MA		Fertile	Sterile		
M	0	64	77	81	121	27.7 ± 0.5	157.3 ± 08.1	21.3 ± 3.2	3.47 ± 0.2	7.4
P1	16	66	77	81	121	27.2 ± 0.3	150.0 ± 01.5	22.7 ± 2.5	3.10 ± 0.4	6.6
P2	19	68	79	83	121	26.8 ± 0.3	141.0 ± 08.9	25.7 ± 2.9	2.90 ± 0.1	6.2
P3	22	70	79	83	121	26.5 ± 0.0	134.0 ± 10.2	26.0 ± 2.7	2.81 ± 0.5	5.9
P4	22	70	79	83	121	25.7 ± 0.6	125.0 ± 05.7	29.3 ± 4.9	2.63 ± 0.1	5.6
P5	26	74	81	85	123	25.3 ± 0.6	118.7 ± 15.6	31.7 ± 0.6	2.48 ± 0.1	5.3
P1S1	22	70	79	83	121	26.0 ± 0.1	129.3 ± 12.3	27.0 ± 3.5	2.76 ± 0.2	5.9
P1S2	26	74	81	85	123	25.7 ± 0.3	114.7 ± 08.5	35.0 ± 7.9	2.40 ± 0.1	5.1
P1S3	30	77	83	87	123	24.3 ± 0.8	107.3 ± 07.2	37.0 ± 8.7	1.90 ± 0.1	4.0
P2S1	26	74	81	85	123	25.3 ± 1.1	117.0 ± 02.7	32.7 ± 4.5	2.41 ± 0.0	5.1
P2S2	30	77	83	87	123	23.5 ± 0.8	107.3 ± 03.5	38.0 ± 2.7	1.96 ± 0.5	4.2
P2S3	32	78	85	90	124	22.2 ± 1.0	085.0 ± 03.6	42.3 ± 2.5	1.64 ± 0.3	3.5
P3S1	30	76	81	85	123	23.5 ± 0.5	102.0 ± 15.7	34.3 ± 4.0	1.00 ± 0.2	2.1
P3S2	34	79	85	90	124	19.5 ± 1.1	076.0 ± 04.4	36.3 ± 4.0	0.76 ± 0.5	1.6
P3S3	40	79	85	90	124	14.6 ± 0.5	057.3 ± 04.5	36.3 ± 3.1	0.55 ± 0.2	1.2
P4S1	34	78	83	87	123	20.0 ± 0.1	066.0 ± 07.9	35.0 ± 2.0	0.81 ± 0.1	1.7
P5S1	42	79	87	92	124	17.1 ± 0.4	050.3 ± 03.5	38.3 ± 1.5	0.51 ± 0.2	1.1
P1S1T1	67	97	103	110	124	14.0 ± 0.3	009.3 ± 02.5	43.4 ± 2.1	0.07 ± 0.0	0.2

TE tiller emergence, *PI* visible panicle initiation, *BT* booting, *AN* 50% anthesis, *MA* maturity. Date of sowing and transplanting: 10.12.2001 and 17.01.2002
Total grain weight: 47.2 ± 4.3 g (±SD, $n = 3$)

grain yield, and these features are considered varietal characteristics (Yoshida and Hayakawa 1970). Both breeders and rice farmers have been interested in identifying genotypic differences in tillering to determine the importance of this agronomic trait in adapting rice crops to target environments. Accordingly, rice cultivars have been classified into low, medium, and high-tillering types depending on the number of tillers produced during their life cycle. A high-tillering attribute is suitable for cultivation under the most favourable growing conditions; however, low tillering could be the desired choice for stress-prone environments. For example, high-tillering cultivars are preferred for achieving optimum yields in low-elevation irrigated systems, whereas genotypes must have fewer tillers for wet direct seeding conditions, where the seed rate is high to ensure good competitiveness against weeds. Similarly, in drought-prone rice fields where plants encounter episodic shortages of moisture during crop growth, desirable genotypes produce few but vigorous tillers, which minimize water utilization while storing reserves to cope with the adverse effects of moisture deficiency. Research on this topic has led to the development of new cultivar lines with low-tillering and a high productive tiller ratio for enhancing crop adaptation to direct seeding conditions (Khush 2000).

The uniculm tillering species bear a single tiller, a unique characteristic closely related to plant type, and this attribute is recalcitrant to modification irrespective of any variation in environmental factors. Strong apical dominance precludes the growth of any subordinate lateral branch or tiller in the culm of monoculm rice. Compared to that of conventional tillering lines, the grain yield of uniculm lines is poorer due to the low grain number per panicle. Although cultivation is not attractive for gross yield, uniculm character can be beneficial for synchronization in plant stands and avoid variation in grain quality. Achieving uniform maturity by maintaining an appropriate plant density per unit area has been considered crucial for obtaining optimum yield and grain quality in direct-seeded cultural system rice (Gravois and Helms 1992). In addition, tillering attributes also significantly influence the morphological characteristics of rice plants. The tiller number per plant is positively correlated with the leaf emergence rate and mean specific leaf area and negatively correlated with plant length (Nuruzzaman et al. 2000). It is implied that high-tillering character improves leaf number and area per plant at the cost of reducing plant height.

Since the number of tillers per plant is closely related to the number of panicles per plant, a high-tillering attribute is considered desirable in rice production (Miller et al. 1991). An increase in tiller number per plant could be a potent factor for improving rice yield potential, but this benefit comes at the cost of several undesirable traits. When the plant density increases, there is a matching negative effect on the number of filled grains per panicle because of the compensatory nature of the two important agronomic traits. The tiller number increases in rice plants due to the initiation of a greater number of late-stage tillers, such as secondary and tertiary tillers (Mohanan and Mini 2008), and these tillers do not contribute much to grain yield. The excessive tillers initiated at later stages of plant development increase differences in grain maturity; within individual plants, grain starch synthesis and grain quality are affected by the prevailing climatic factors, especially high temperature, at later grain filling stages. Hence, the greater the number of tillers per plant is, the greater the variation in grain quality within a plant because plants with more tillers are more likely to be subjected to higher temperatures during the grain maturation period (Wang et al. 2007). Wang et al. (2007) reported that grain yield declines both quantitatively and qualitatively, with later-initiated tillers exhibiting a lower chalky rice percentage and amylose and protein contents than early-initiated tillers. The other disadvantages of profuse tillering rice include a reduction in leaf area and a decrease in photosynthetic efficiency caused by mutual shading. Conversely, the low-tillering trait underrates biomass production and a deficiency in grain-filling capacity and carbohydrate production (Peng et al. 1999) are not ideal choices for identifying plant ideotypes.

Therefore, the initiation of these tillers above the resource capacity of the plant is corrected by enhanced senescence (Mohapatra et al. 2011). According to Lafarge et al. (2004), tiller senescence occurs since the number of tillers produced during the vegetative stage is greater than the number of tillers that become productive at maturity. The carbon assimilates invested in senescent tillers are wasted because of poor remobilization. The timescale tillering strategy does not differ between high-tillering and low-tillering rice cultivars until the mid-tillering stage, but tiller emergence in the latter stops thereafter. Cessation of tiller emergence at this crucial stage precludes the waste of carbon assimilates for the growth of undesirable tillers, which is otherwise used for improving grain yield in productive tillers. Thus, the yield advantage of reduced tillering rice is induced by its greater ability to control tiller emergence. This strategy has been adopted for the development of new moderate tillering plant types and hybrid rice varieties that have enlarged the prospect of yield gain over high-tillering semi-dwarf rice. Compared to those of profuse tillering cultivars, heavy panicle cultivars possess relatively wider shoot apical meristems (Fig. 7.1), which promote main shoot development and suppress tiller development. As a result, the panicle number per plant decreases, but the spikelet number per panicle increases with the prospect of a higher yield (Fukushima 2019). Hence, extending the period of stem elongation without compromising the flowering date effectively reduces the tillering phase to induce a faster phyllochron to compensate for tiller or panicle numbers (Dreccer et al. 2013).

7.1.3 Tillering Ability

The number of growth-phyllochrons completed by the rice plant before entry into the reproductive phase is considered the tillering ability (Nemato et al. 1995; Stoop et al. 2002). It is an important agronomic trait because it determines grain yield. Since the number of tillers per plant is closely associated with the number of panicles per plant, tillering ability is considered desirable for high yield. Wu et al. (1998) characterized the contribution of cultivar-specific tillering ability to dry matter accumulation, yield components, and grain yield. They found a greater yield of the rice cultivar Teqing (918 g m^{-2}) than that of the lower-yielding Gulfmont (791 g m^{-2}) and Rosemont (729 g m^{-2}) plants, largely because of the greater tillering ability,

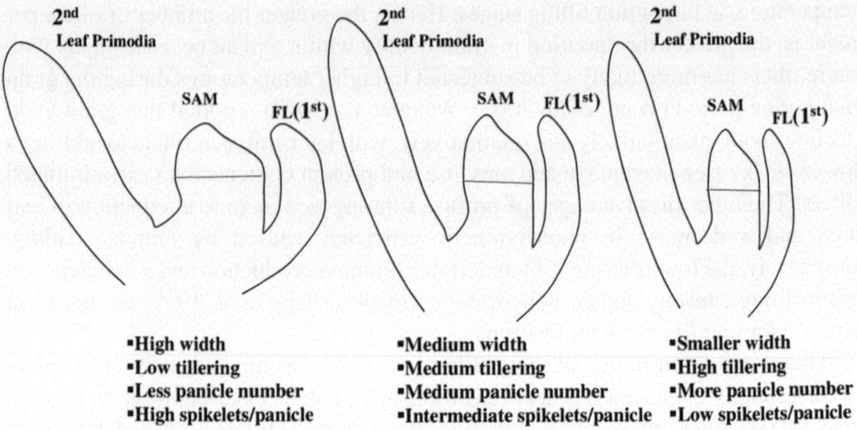

Fig. 7.1 Schematic representation of longitudinal sections of the width of the shoot apical meristem (SAM) in low, medium, and high-tillering rice genotypes at the panicle initiation stage and its relationship with tiller number during the active tillering stage and panicle number and spikelet number at the time of grain maturity. The blue line indicates the width of the SAM and the FL represents the flag leaf (Fukushima 2019)

greater spikelet density, and longer maturation period of the former, which results in greater use of the relatively long growing season length. Therefore, the yield potential of a rice cultivar is characterized by its tillering capacity (Wu et al. 1998). The ability to produce several tillers and tiller types differs among rice varieties (Afza et al. 2020), and the array of differences provides choices for breeders while selecting breeding materials. However, rice plants that produce a large number of tillers exhibit high inconsistency in the partitioning of carbon assimilates and nutrients among tillers, leading to significant variation in grain yield (Miller et al. 1991). Hence, the variation in grain development and yield among tillers varies among cultivars concerning tillering ability (Wang et al. 2007). When the tiller number increases, the yield contribution of all categories of tillers, including the main stem, decreases significantly. This feature has been evident in the comparison of panicle growth between the low-tillering rice cultivar Xiushui 11 and its profuse tillering mutant. The dry mass weight and increase in panicle weight of each tiller are higher in the former than their respective counterparts in the latter because the sink size and harvest index are low in profuse tillering rice. Minimizing unproductive tillers is important for yield, as has been the case for China's super rice plants (Peng et al. 2008).

Irrespective of varietal differences and environmental conditions, rice plants produce tiller buds on leaf axils during the vegetative period, but the growth of these tiller buds is determined by the genetic factors of the varieties and growing conditions (Kakizaki 1987). In the post-initiation stage, tiller growth becomes dependent on the plant's nutritional status, primarily the amount of nitrogen absorbed from the soil and the carbohydrates produced during leaf photosynthesis during the tiller bud growth stage (Tanaka and Garcia 1965). High-yielding semi-dwarf rice plants are fertilizer-responsive, and given the optimum fertility conditions, they exhibit maximum tillering capacity for weed competition. Compared to that, in low-tillering cultivars, the distribution of dry matter produced by tillers is greater in high-tillering varieties (Honda 1977; Yamamoto et al. 1995), and the increase in tillering ability is associated with the provision of fertilizers. Nuruzzaman et al. (2000) reported that the tillering ability of a rice variety is intricately dependent on the amount of nitrogen absorbed from the growth medium and the dry matter produced by leaf photosynthesis until the maximum tillering stage and the efficiency of these materials in tiller production. The role of nitrogen in fertilizer application is also critical for tiller survival because the percentage of productive tillers is correlated with the amount of nitrogen absorbed between the maximum tillering stage and heading (Nuruzzaman et al. 2000). The extent of competition for nutrients between tillers determines the quantity of non-productive tillers. The non-productive tillers unsuccessfully compete for nutrients and carbohydrates from the mother shoot or elongating stem after the panicle formation stage and degenerate. The nitrogen flow efficiency in late-stage tillers is very low, and this low nitrogen flow efficiency is responsible for the poor nitrogen harvest index and insufficient assimilate supply, leading to low productivity (Wang et al. 2017). The number of effective tillers in rice is also influenced by root morphology and physiology. An increase in the number of roots and the root-to-shoot ratio enhances the water use efficiency of plants. According to Yang et al. (2012), in terms of root oxidation activity, the ultrastructure of root tip cells, root-sourced cytokinins, and ABA are the most crucial factors for root activity and are closely related to shoot growth and yield formation, such as panicle and spikelet number. High root oxidation activity and high ABA and cytokinin contents improve grain filling.

7.2 Plant Density and Tillering

7.2.1 Plant Density Effects on Tillering

Like nitrogen input, plant density is an important factor influencing tiller production in rice. The number of plants per unit area is an important index for crop growth monitoring and final grain yield. Population density is closely associated with cultivated area statistics and field management to maximize land use. An increase in plant density can suppress weed growth (Chauhan and Abugho 2013) and improve nitrogen use and grain yield (Zheng et al. 2020). The number of rice plants surviving in the field is a key index for environment-targeted rice breeding because it is often related to damage caused by typhoons (Blanc and Strobl 2016) and floodwater (Kwak et al. 2015). Hence, digital imagery and computer vision technology are used for plant counting from the transplantation to tillering stages with accuracy instead of manual counting (Liu et al. 2020).

The effects of plant density on tillering and grain yield have been intensively studied in transplanted flooded rice fields of many rice-growing nations. A higher plant density induces the initiation of fewer tillers per plant, higher tiller density in the crop, and a smaller number of grains per panicle. The grain yield increases proportionately with increasing plant density within the range of 5–100 plants m^{-2} and not beyond. A very high planting density increases the panicle number per unit area but does not invite yield gain because of the trade-off between tiller density and grains per panicle (Fukushima et al. 2011; Hayashi et al. 2006).

7.2.2 Plant Density and Tiller Appearance

Clerget et al. (2016) reported that the leaf appearance rate and panicle initiation on the main stem were faster at higher plant densities and that the number of out-growing tillers was linked to the number of emerged leaves. The onset of stem elongation at a distance of 20 cm regulates tiller density by cessation of tiller outgrowth, and this process occurs earlier at shorter plant spacing due to the increase in root density in the media. The effects of planting density on tillering have been studied intensively in transplanted lowland rice. A high planting density increases the tiller density per unit area but induces only a few tillers per plant, each bearing small numbers of grains. Several workers have reported an increase in grain yield with increasing planting density in the range of 5–100 plants m^{-2} (Hayashi et al. 2006; Nakano et al. 2012; San-Oh et al. 2008). No yield benefit occurs when the planting density increases to more than 100 plants m^{-2} because of the balance between tiller density and grain number per panicle in rice (Fukushima et al. 2011; Jones and Snyder 1987; Ottis and Talbert 2005). The physiological factors that influence tillering concerning plant density have been investigated by Dreccer et al. (2013). The stem water soluble carbohydrate concentration is plant density-dependent and could be an emergent property of tillering; a high stem water soluble carbohydrate concentration is related to low tillering and vice versa irrespective of variation in the environment and genotype. The availability of carbon assimilates for new tillers against their demand has become a key factor behind the final tiller number of rice plants (Dingkuhn et al. 2006). In addition, a common threshold of the irradiance red: far red ratio (0.39 to 0.44 with a standard error of difference = 0.55) coincided with the maximum tiller number per plant across plant densities and genotypes. This calibration method has been effectively applied in crop simulation modelling as a cut-off rule for tillering (Dreccer et al. 2013). The third physiological mechanism that influences the tiller dynamics of cereals is hormonal homeostasis, where root-derived cytokinin promotes tiller growth, whereas auxin moves basipetally from the stem apex and strigolactone moves acropetally from roots to suppress shoot branching. Dreccer et al. (2013) proposed a diagrammatic model showing the main processes affecting and linking tillering and stem water-soluble carbohydrates at the crop level (Fig. 7.2).

7.3 Tiller Dynamics Concerning Environmental Factors

7.3.1 Temperature

Rice plants thrive well in hot and humid climates in the tropics and subtropics where the ambient temperature fluctuates widely in the environment. Because of its origin in warm climates, the plant is a low-temperature sensitive crop; crop growth is

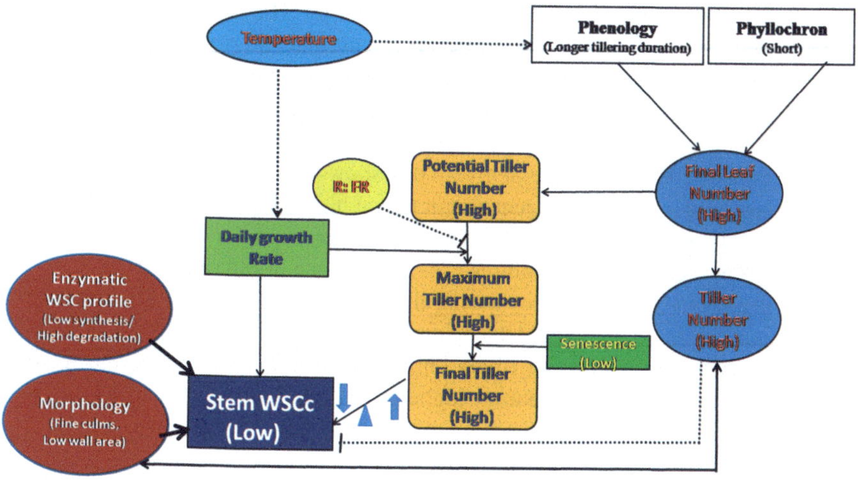

Fig. 7.2 Schematic pathway with the main processes affecting and linking tillering and WSCc (water-soluble carbohydrate concentration) at the crop level in wheat. (Adapted from Dreccer et al. 2013)

severely impacted at temperatures less than 15 °C. Additionally, rice plant growth is severely inhibited when plants are exposed to extremely low or high temperatures. The growth response of rice plants to temperature variation is unique because it affects dry weight increases more than varietal differences (Yoshida 1973). Rice plants are sensitive to temperature at different times during their life cycle, but the magnitude of the response varies widely with growth stage. The critically low and high temperatures, below 20 °C and above 30 °C, respectively, vary from one growth stage to another (Krishnan et al. 2011).

Moreover, all culture types of rice are not sensitive to temperature. In addition to genetic factors, the tillering behaviour of rice plants also depends partially on environmental factors, especially radiation, temperature, and nutritional conditions (Hanada 1993). Similarly, Yoshida (1973) suggested that the tillering behaviour of rice is related to temperature, light intensity, and carbohydrate metabolism. The projected increase in global temperature due to climate change is expected to reach 5 °C greater than the ambient temperature by the turn of the century (Tollefson 2020), suggesting the involvement of heat stress effects in rice production. The changing climate with increasing temperatures has posed a major challenge to rice production in Asia, where rice is the most important staple food. The adverse effects of stress can occur through changes in various metabolic processes in plants at all stages of development (Mittler and Blumwald 2010).

It was reported that temperature effects are discernible in all aspects of rice plant life, such as germination, rooting, tillering, sterility, anthesis, and ripening (Yoshida 1973). According to Yoshida (1981), 25–30 and 35 °C are the critical optimum and high-temperature responses, respectively, for the response of rice seedlings to temperature. Hanada (1993) suggested that a temperature of approximately 32 °C is optimal for vegetative growth and high-tillering ability. The tiller number decreases when the field is irrigated with cold water. Incident light and plant age influence the rice tillering response to temperature. Based on age, low temperature is more beneficial for tiller number than high temperature is. A comparison of the tillering response of rice plants at three different average temperatures, 30.5, 24.5, and 18.5 °C, under 32 klx irradiance at a 12-h day showed that tillering ability at lower temperatures was exceedingly greater than that at high temperatures, and the difference occurred due to the genesis of secondary and tertiary tillers. However, tillering becomes greater at high temperatures (25–30 °C) than at intermediate (19–24 °C) and low (12–17 °C) temperatures during the same period after seeding. However, the tiller number increases with decreasing temperature during which low temperature promotes the production of tillers on taller nodes, i.e. secondary and tertiary tillers. Lower temperatures encourage higher-order tiller production, although these plants are less productive.

7.3.2 Field Water Temperature

The water temperature around the basal part of the stem in rice is the critical factor that determines tiller dynamics. Its effects on tillering are more discernible than the water temperature around the roots. A temperature of 16 °C is more effective at the tillering of lower nodes than a temperature of 31 °C. This promotive effect of low temperature on tillering, however, becomes evident when rice nursery seedlings are exposed to low temperature during the second half of growth but not in the first half of the nursery growth period, when high temperature is conducive to greater tillering. Overall, in the first half stage of growth, a high air temperature is suitable for tillering, but tiller emergence requires a low temperature in the second stage of seedling growth.

7.3.3 Day and Night Temperatures

The difference in tiller development between day and night temperatures significantly affects rice plants. More tillers emerge when the difference is substantial or the night air temperature is low. As reviewed by Hanada (1993), several examples of adverse effects of high night temperatures on rice tiller production have been described. A rise in night temperature of 7 °C on average for a month critically impacts tiller emergence by reducing the number of tillers and panicles. The number of tillers or panicles is strongly affected by temperature variation when plants are exposed during the active tillering period. Conversely, a combination of high day temperature (36 °C) and low night temperature (16 °C) is most suitable for increasing tiller number, and a day temperature of 31 °C and night temperature of 21 °C are favourable for increasing panicle number. It is also suggested that the maximum number of tillers, although secondary or tertiary, are obtained under combinations of extremely different day and night temperatures, such as 36 °C day and 16 °C night temperatures. In

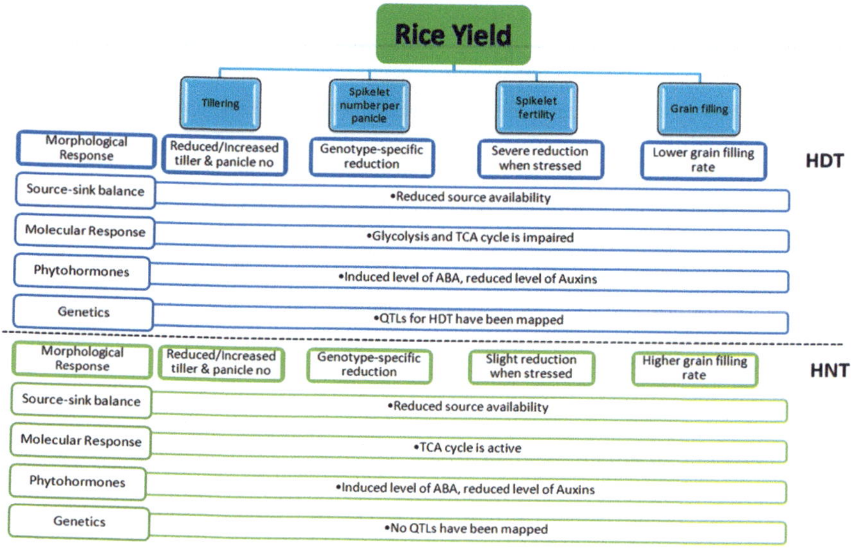

Fig. 7.3 Differential response of yield components and underlying molecular-physiological (source-sink balance, molecular response, phytohormones) and genetic mechanisms between high day temperature (HDT) and high night temperature (HNT) conditions. (Adapted from Xu et al. 2020)

most of the rice-growing regions of India, crop cultivation is most suitable on day/night temperatures of 28/22 °C, but a yield decrease of approximately 7–8% occurs for each 1 °C increase in the daytime maximum/night-time minimum at temperatures ranging between 28/21 and 34/27 °C (Dubey et al. 2018). An adverse effect on tillering is the prime cause of yield decline. Hanada (1993) summarized the effects of day and night temperature on rice tillering by various researchers and concluded that episodes of high temperature during the day are more suitable for tiller emergence than those at night. According to Yoshida (1973), carbon assimilated from daytime photosynthesis provides energy for tiller development during the active period of tillering, and photosynthetic activity does not vary within a narrow range of daytime temperatures to infringe on the carbon supply and restrict tillering. However, an increase in night-time temperature becomes critical for tiller development due to the loss of carbon assimilates caused by the heightened impact of temperature on respiration. A more comprehensive and updated account of the effects of high day temperature (HDT), high night temperature (HNT), and combined high day and night temperature (HDNT) has been provided by Xu et al. (2020). High temperature, irrespective of the duration of exposure, is detrimental to various yield components of rice plants, such as tiller number, spikelet number/fertility, and grain weight. The underlying physiological, molecular, and genetic mechanisms controlling the high-temperature response are explained in Fig. 7.3 (Xu et al. 2020).

According to Xu et al. (2020), HDT had much greater effects than HNT for yield component determination, especially on spikelet fertility. During grain filling, HDT caused a reduction both in the filling rate and growth duration of grains, while HNT had contrasting effects, namely, a faster filling rate and shorter filling duration. In terms of source-sink balance, HNT enhanced dark respiration. On the other hand, HDT inhibited photosynthesis. Both of the deleterious effects contributed to the limited availability of carbon reserves for sink tissue. Morphologically, impaired sink activities were observed under both HNT and HDT and there were no differences in phytohormone responses. However genetic studies identified some differences between HDT and HNT in the expression of quantitative trait loci (QTLs) conferring yield components and quality. At the biochemical level, differences also occurred in the responses to starch accumulation, heat shock proteins (HSPs) synthesis, and activities of antioxidants. Photosynthetic and tricarboxylic acid (TCA) cycle activities also differed between HDT and HNT conditions.

7.3.4 Effects of High Temperature

It is expected that global warming will occur faster than originally thought (Xu et al. 2018) and that its adverse impact will jeopardize the food security of rice-growing nations sooner than it was once expected. In Southeast Asia, rice provides 76% of food calories to consumers (Fitzgerald et al. 2009), and this region is expected to experience the fastest warming of the globe during climate change (IPCC 2014). High temperature refers to an episode of temperature above the threshold level that causes irreversible damage to crop growth and development. The temperature increase caused by climate change in recent years has significantly affected rice production by compromising various grain yield components of the plant, including tiller number (Ezin et al. 2022). In addition to studies on the effects of high temperatures on rice tillering, Kumar et al. (2021) examined the response of yield and yield-related traits to elevated temperatures in 11 rice cultivars grown inside field-mounted temperature gradient tunnels in the Bhabar region of the Indian Himalayas in the Kharif season in 2012 and 2013. Irrespective of cultivar differences, exposure to elevated temperature significantly impacted growth, yield, and yield-related traits, including tiller number. This simulated model experiment infers that changes in tiller dynamics could be a potential determinant of resiliency under climate change scenarios.

7.4 Morphological and Physiological Symptoms of Heat Stress

Prolonged exposure to elevated temperature is harmful to the growth and development of rice plants and different parts of the life cycle, which includes stages ranging from germination to grain ripening. A comprehensive account of the effects of heat stress on the vegetative, reproductive, and grain-filling stages of rice plants (Fig. 7.4) has been provided by a recent review (Xu et al. 2021).

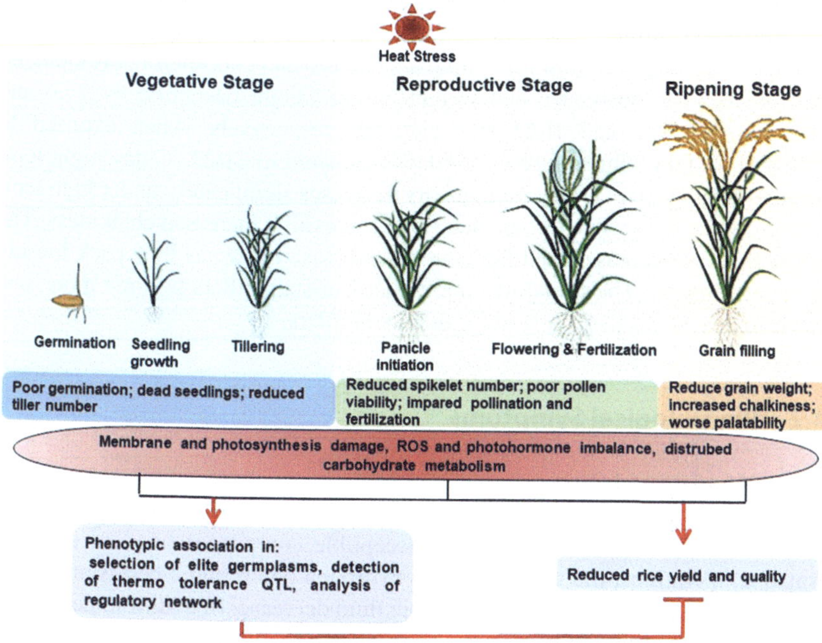

Fig. 7.4 Morphological and physiological characteristics of rice under heat stress at different growth stages. (Adapted from Xu et al. 2021)

7.4.1 Morphological Symptoms

Heat stress affects the seed germination rate and seedling vigour. The favourable temperature range for seedling growth is 25–28 °C, whereas exposure to 42–45 °C can cause stress. Exposed to the stress, seedlings lose water, exhibit symptoms of injury in the form of withered and yellowish leaves, and impaired shoot and root growth. Heat stress reduces the tiller number, and tillers exhibit symptoms such as leaf curling, wilting, and yellowing. The panicle number and yield decreased by 35 and 86%, respectively, per plant under 40/35 °C day/night temperature for 15 days versus plants grown under 28 °C. In the reproductive stage, events such as panicle initiation, gametogenesis, anthesis, pollination, and fertilization are sensitive to heat stress, and the impacts of heat stress are greater than those in the vegetative stage. Spikelets are reduced in number and degenerate before grain formation. Development and pollen viability are impaired as heat stress disintegrates tapetal cells, infringing on the nutrient supply to microspores and resulting in pollen abortion. Compared to pollen viability, ovule development is less prone to high temperatures. At the anthesis stage, episodes of high temperature alter anther shape, diminish anther and pollen viability, obstruct pollen from landing on the stigma and germination, inhibit pollen tube growth, and eventually restrict double fertilization. Seed formation becomes abnormal.

In the grain filling stage, the incidence of high temperature expedites the grain filling rate while curtailing the length of the grain filling period. The high temperature at the early stage of grain filling impairs embryo and endosperm development, which reduces caryopsis size and weight. Grain weight decreased by 24.6 and 39.1% for the N22 and IR64 rice cultivars, respectively, when exposed to 38/30 °C day/night temperature for 20 days compared to 31/23 °C day/night temperature (Shi et al. 2017). The grain quality decreases significantly under high temperatures, while the grains become deformed and exhibit increased chalkiness. The amylose content decreases, and the granules decrease in size as they pack loosely into the endosperm. These findings imply that late-stage tillers become more susceptible to heat stress than early-stage dominant tillers.

7.4.2 Physiological Symptoms

7.4.2.1 Membrane Injury

A biological membrane comprising a balanced proportion of proteins and phospholipids is considered the most heat-susceptible component of plant cells. At elevated temperatures, the strength of the hydrogen bonds and electrostatic interactions between proteins in the cytoplasmic fluid decrease. In addition, heat stress can modify the membrane composition by altering the proportion of saturated to unsaturated fatty acids and inducing protein denaturation. Disruption of the membrane structure and composition increases fluidity and permeability, which compromises membrane integrity, leading to leakage of organic and inorganic substances. Disruption of these proteins also inhibits photosynthesis and respiration, which rely on the activity of membrane-associated electron carriers and enzymes.

7.4.2.2 Reactive Oxygen Species (ROS) Accumulation

Elevated temperature triggers the synthesis of reactive oxygen species in cell organelles, including mitochondria, peroxisomes, and the endoplasmic reticulum. Free radicals can damage the membrane structure of organelles, ultimately disrupting internal cellular homeostasis. In tolerant plants, various heat shock genes, and other related genes encode an array of heat shock proteins, antioxidants, osmoprotectants, and osmolytes to maintain membrane integrity and cellular homeostasis (Zafar et al. 2018). Heat shock proteins protect cells from the injurious impact of high temperatures. Heat tolerance in rice is linked to the production of a high RNA content, strong antioxidative defence, and decreased malondialdehyde levels, preferably during meiosis (Cao et al. 2008). The damaging effects of short-term exposure to stress can be reduced by the synthesis of heat shock proteins (Maestri et al. 2002).

Compared to those under normal temperatures (28/22 °C day/night), the exposure of rice plants to high temperatures (38/32 °C day/night) at the meiosis stage increases the ROS content in anthers more than threefold (Zhao et al. 2018a). Stress injury upregulates a series of respiratory burst oxidase homologue genes in rice pistils to promote the accumulation of ROS (Fu et al. 2016; Zhang et al. 2018).

Under heat stress, the activities of the antioxidant enzymes superoxide dismutase (SOD) and catalase are impaired, leading to the accumulation of SOD in developing anthers of rice (Zhao et al. 2018b; Sailaja et al. 2015). The other antioxidant enzymes involved in heat tolerance are peroxidase, ascorbate peroxidase, dehydroascorbate reductase, glutathione-S-transferase, glutathione peroxidase, and guaiacol peroxidase (Zafar et al. 2018), and they are often used as biomarkers for screening rice genotypes for heat tolerance. Intracellular ROS accumulation leads to disturbed ROS homeostasis and oxidative damage, cell death, growth retardation, spikelet sterility, seedling death, and grain chalkiness and are symptoms of severe heat stress injuries in rice. Excessive accumulation of ROS impairs the structure and functions of biomembranes, accentuating lipid peroxidation and protein oxidation, which results in the accumulation of intracellular malondialdehyde to the detriment of normal protein and nucleic acid functions in rice plants (Bahuguna et al. 2015; Chakraborty and Bhattacharjee 2015). When heat stress disrupts cellular ROS homeostasis, resulting in a ROS burst and consequential oxidative damage, photosystems I and II in chloroplasts and the electron transport chain of mitochondria become major sites for the production of reactive oxygen species, and a thermotolerant plant survives oxidative stress through numerous antioxidant regulatory pathways, including enzymatic (SOD, CAT, APX, GR, etc.) and non-enzymatic (ASH, GSH, etc.) pathways (Kan and Lin 2021).

7.4.2.3 High-Temperature Damages Carbon Metabolism

High temperature is injurious to both respiration and photosynthesis. The temperature at which the amount of CO_2 fixed by photosynthesis equals the amount liberated by respiration in a particular time interval is called the compensation point. At temperatures above the compensation point, the amount of carbon fixation by photosynthesis cannot replenish the carbon assimilates used in respiration, amounting to a reduction in the carbon reserve in the plant and ultimately to starvation effects. Photosynthesis is susceptible to elevated temperature. Carbon metabolism in the stroma and photochemical activities in thylakoid lamellae are primary targets for high-temperature injury. Stress destroys the integrity of the thylakoid membrane, which triggers chlorophyll loss and alters photochemical reactions by reducing the ratio of variable fluorescence to maximum fluorescence (Fv/Fm) in the PSII complex and the photosynthetic rate. The D1 core protein of the PSII complex is susceptible to heat stress; the chloroplast is the organ responsible for activation of the heat stress response, and disruption of the PGL (pale green leaf encoding chlorophyllide oxygenase 1) increases thermo-sensitivity in rice (Kan and Lin 2021). These authors proposed a model showing heat stress perception via the RNA secondary structure and the phyB gene and the phase separation and response in plants through membrane fluidity and heat shock factor in rice.

Furthermore, elevated temperature-induced oxidative stress dissociates the oxygen-evolving complex (OEC), leading to the inhibition of electron transport from the oxidizing side of PSII to the reducing side and ultimately the transfer of electrons in the light reaction of photosynthesis for the synthesis of NADPH and ATP. Ribulose-1, 5-bisphosphate (RuBP) carboxylase-oxygenase, also known as

RuBisCo or Rubisco, catalyzes the formation of 3-phosphoglycerate from RuBP + CO_2, and it is a key enzyme in the Calvin cycle of the dark reaction of photosynthesis responsible for the synthesis of carbon assimilates. At high temperatures, Rubisco activity in rice is impaired because of the inactivation of Rubisco activase (Wang et al. 2010a, b). Like in carbon assimilation, elevated temperature upregulates carbohydrate metabolism and assimilate partitioning. The activities of two key enzymes of the glycolytic pathway, phosphoglucoisomerase and phosphofructokinase, are reduced when rice plant cells are exposed to extreme temperatures ranging from 12 °C to 20 °C or 36 °C to 44 °C (Gammulla et al. 2010). High temperature affects starch and sucrose synthesis through the reduced activity of sucrose phosphate synthase, ADP-glucose pyrophosphorylase, and invertase (Hasanuzzaman et al. 2013). Additionally, heat stress has a negative influence on leaf water potential through expedited evaporation. A reduction in leaf area and premature leaf senescence also depletes leaf photosynthesis and carbohydrate reserves. Compared to sensitive rice cultivars, heat-tolerant cultivars exhibit greater expression of the transporter gene *OsSUT1* (Miyazaki et al. 2013), ensuring adequate carbon assimilate supply to filling grains under heat stress. Stress also downregulates a series of genes encoding granule-bound starch synthase 1 and branching enzymes, disturbing starch synthesis and increasing grain chalkiness (Yamakawa et al. 2007).

7.4.2.4 Phytohormone Imbalance

An altered hormonal balance occurs in rice plants under high-temperature stress. Stress causes a reduction in the levels of GA, IAA, and cytokinin in spikelets during the grain-filling period, leading to impaired endosperm cell proliferation and the promotion of spikelet degeneration, pollen sterility, and loss of kernel weight. The incidence of high-temperature stress decreases pollen activity, pollen germination, and spikelet fertility in both tolerant and susceptible rice cultivars, but the effect is more pronounced in the latter. When subjected to stress, the contents of IAA, GA, free proline, and soluble proteins decrease in the anthers, but there is a concomitant increase in the level of ABA (Tang et al. 2008; Wu et al. 2016). A susceptible rice cultivar produces fewer spikelets per panicle under heat stress because heat-induced transportation of cytokinin is slower and degradation is faster in the organ (Wu et al. 2017).

7.5 Molecular Biology of the High-Temperature Response

7.5.1 Perception of Heat Stress

Identification of the molecular regulatory mechanism involved in the response to elevated temperature and development of high-throughput genetic manipulation via gene introgression or genome editing hold promise for breeding rice adapted to target temperatures without yield penalties (Kan and Lin 2021). Mitler et al. (2012) proposed four putative sensors for the heat stress response in plants: A plasma membrane channel that initiates an inwards calcium flux; a histone-type sensor in the

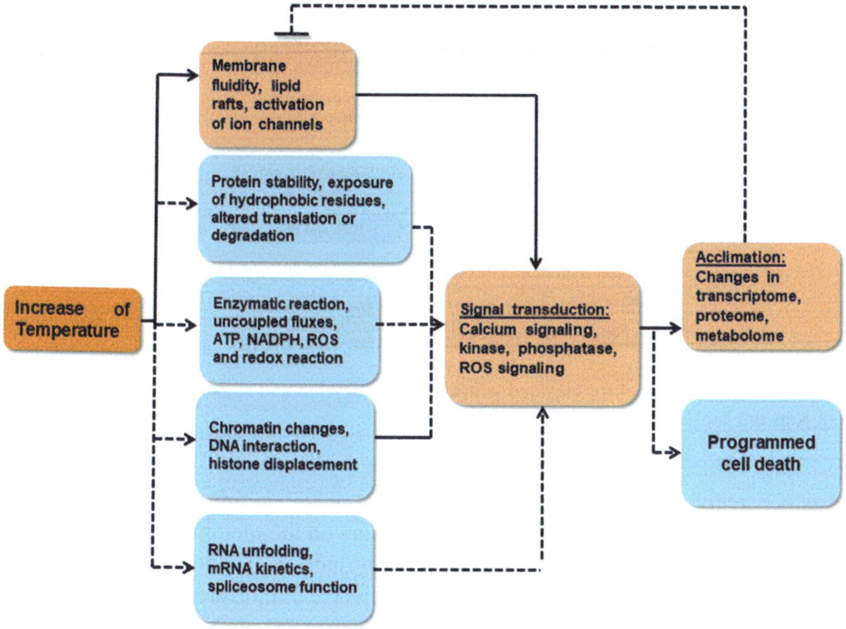

Fig. 7.5 Schematic model for temperature sensing in plants. (Adapted from Mitler et al. 2012)

nucleus and two unfolded protein sensors in the endoplasmic reticulum and cytosol. It is believed that each of the sensors activates a similar HSR gene set for thermotolerance, but the hierarchical order of response is unclear. A schematic model for temperature sensing in plants was proposed by these workers (Fig. 7.5).

In a more recent review (Kan and Lin 2021), the perception of temperature sensing in plants, with particular emphasis on rice plants at the reproductive stage, was illustrated by incorporating precise information on the effects of stress at the cellular component level, such as information about the membrane, endoplasmic reticulum, chloroplasts, mitochondria, and functional levels, such as protein denaturation, ROS scavenging, nucleic acid stabilization, translation, and metabolic flux changes. The perception of heat stress occurs via the RAN secondary structure, the phyB gene, and phase separation, which leads to a response in membrane fluidity, heat shock factor production, heat shock protein production, and reactive oxygen species generation. Accordingly, these authors (Kan and Lin 2021) revised the schematic model for the heat stress response as described in Fig. 7.6.

7.5.2 Heat Stress Signalling

The primary response to heat stress occurs in the membranes of chloroplasts and mitochondria, and these effects could result in increased fluidity or the induction of secondary oxidative stress. Both processes involve several intermediary factors. The

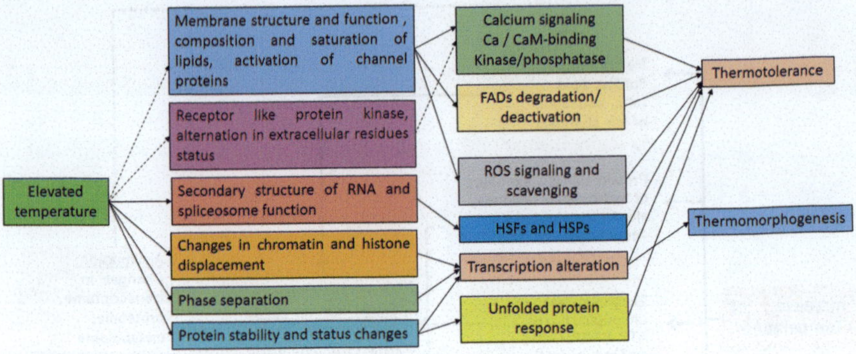

Fig. 7.6 Schematic model for elevated temperature perception and response in plants. (Adapted from Kan and Lin 2021)

stress is perceived by the membrane system first, which is followed by a change in fluidity and the activation of channels or receptors on the membrane. This response induces transient Ca^{2+} signalling and signal transducer molecules in the cytoplasm. DeFalco et al. (2009) reported the presence of diverse unique proteins that bind Ca^{2+} using the conserved EF-hand motif and these proteins undergo conformational changes to function as Ca^{2+} sensors after conjugation. This binding facilitates the regulation of target proteins and thereby coordinates downstream signalling pathways. The EF-hand Ca^{2+} sensors include CaMs (calmodulins), CDPKs (calcium-dependent protein kinases), and CBLs (calcineurin B-like proteins). Many Ca^{2+} and Ca^{2+} calmodulin-binding transcription factors that control stress signalling pathways in plants are involved in intracellular responses (Reddy et al. 2011).

In rice, calcium signalling is critical for both heat and chilling tolerance. Two closely related cyclic nucleotide-gated ion channel proteins, OsCNGC14 and OsCNGC16, are crucial for the regulation of thermotolerance by altering the heat-induced increase in cytosolic Ca^{2+}. These channels are non-specific cation channels that regulate calcium flux across the membrane. Under stress conditions, Ca^{2+} signal induction occurs within the first few minutes of exposure. The loss-of-function mutants *cngc14* and *cngc16* exhibit poor survival under both heat conditions because the gene mutation abolishes cytosolic calcium signalling (Cui et al. 2020). The leucine receptor-like kinases ERECTA and Hwi1 are located on the membrane and control thermotolerance in rice. Calcium channels are also directly or indirectly activated when elevated temperature alters membrane fluidity, resulting in changes in lipid saturation and composition and the generation of ROS. The cytoplasmic calcium concentration increases, which activates or represses the activity of Ca^{2+}/CaM-related kinases, phosphatases, and transcription factors. The heat stress-triggered decrease in membrane saturation is remedied by the downregulation of FAD. A regulatory network involved in heat stress response in plants has been proposed by Xu et al. 2021 (Fig. 7.7).

As shown in the figure, the life cycle of rice is divided into vegetative, reproductive, and ripening stages. Heat stress during the vegetative stage causes inhibited

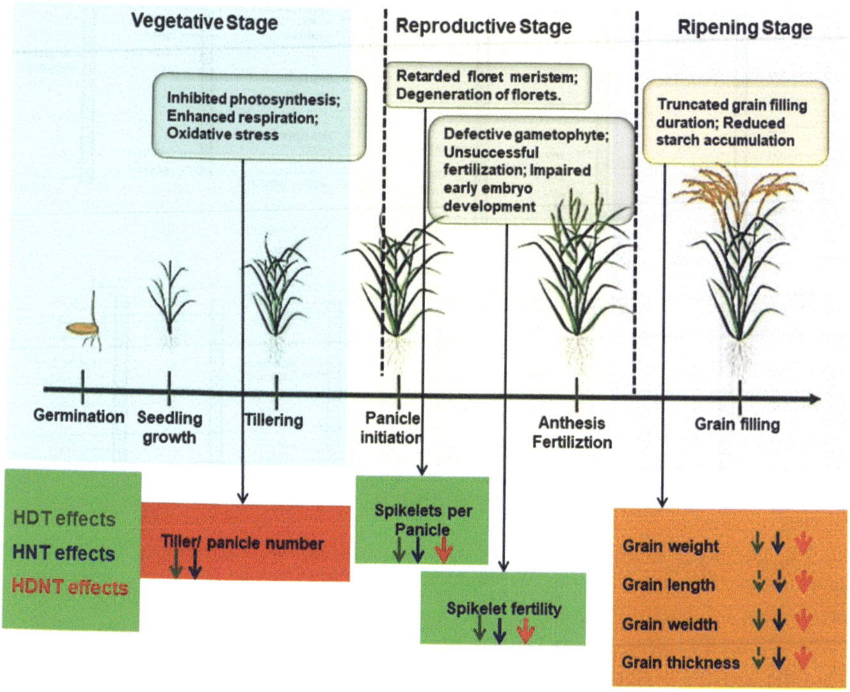

Fig. 7.7 Mapping the rice yield components to specific developmental stages and comparing their distinct responses to HDT (high day temperature), HNT (high night temperature), and high HDNT (high day and night temperature). (Adapted from Xu et al. 2021)

photosynthesis, enhanced respiration, and oxidative damage. These negative effects are reflected by a reduction in tiller/panicle number, which is differentially affected by HDT, HNT, and HDNT. In the early reproductive stage, floret meristem formation and development are affected by high temperatures, resulting in variable reductions in spikelet numbers per panicle. Later in the reproductive stage, heat stress can impair gametophyte development, fertilization, and early embryo development, thereby reducing spikelet fertility. During the grain ripening stage, a shortened grain filling duration and reduced grain starch accumulation are observed at high temperatures. As a result, the grain weight/size ratio is reduced by episodes of high temperature. High temperature also impacts tiller/panicle number in the vegetative stage until panicle initiation; spikelet fertility and number in the reproductive period extending from panicle initiation to anthesis, and grain weight in the maturation period.

According to Kan and Lin (2021), the RNA secondary structure is also considered a temperature sensor. The RNA secondary structure is influenced by elevated temperature via the spliceosome or another pathway. The transcription of heat shock transcripts is inhibited when the secondary structure traps the ribosomal-binding site, whereas elevated temperature releases the site for normal translation. PhyB is another temperature sensor that perceives temperature in the dark via reversion

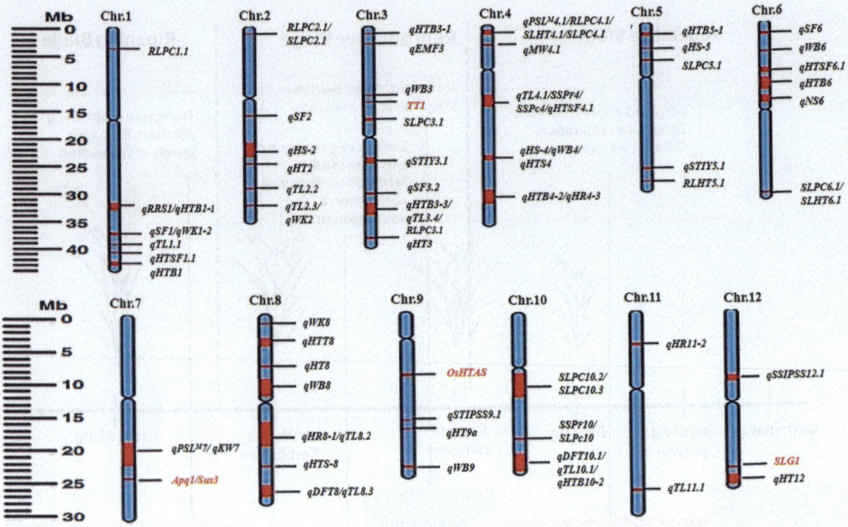

Fig. 7.8 Location of quantitative trait loci (QTLs) associated with heat tolerance in rice chromosomes (Adapted from Kan and Lin 2021 and Xu et al 2021)

from the active Pfr state to the inactive Pr state, and it binds to promoter regions of genes responsive to temperature. This proposition that high-temperature perception mediated by light receptors is supported by Lamers et al. (2020). A diagrammatic flow model showing the regulatory network involved in the heat stress response in rice is presented below concerning Kan and Lin (2021) and Xu et al. (2021) (Fig. 7.8). Xu et al. (2021) identified the QTLs associated with 12 rice chromosomes.

The QTLs shown in the figure were identified using heat stress-induced phenotypes, such as seedling root length or shoot length, survival rate, spikelet fertility or sterility, flowering time, yield per plant, seed-setting rate, frequency of white-back, basal-white, or milky-white kernels, or kernel weight. A single locus bearing different names indicates that the locus was identified in at least two studies. It suggests that the QTL may have large and stable effects. The red-coloured parts indicate functional validation of the QTLs using transgenic approaches (Adapted from Xu et al. 2021).

7.6 Cold Stress Response

A majority of the rice-growing areas in Asia are located in warm tropical climates. However, due to the increasing incidence of climate change, cold weather events are becoming more common, causing severe constraints on rice production in some countries (Ding et al. 2020; Zeng et al. 2017). Hence, it is important to unravel the underlying mechanisms involved and identify potential cold tolerance strategies for breeding rice. Cold stress causes various types of chilling injuries in rice, such as delayed germination, seedling death, reduced tillering, and spikelet sterility. Leaf chlorosis, stiff seedlings, reduced tillering, withering, and seedling death can be phenotypic indicators of chilling injury.

7.6.1 Physiological Responses

Normally rice plants are susceptible to chilling stress, which affects growth and grain yield. For a long time, there have been considerable efforts made to study the phenotypic and biochemical changes caused by cold stress and their physiological explanations underplaying these effects (Li et al. (2022). Physiological responses to cold stress in rice include alterations in the water balance, leaf photosynthesis, respiration rate, and hormone homeostasis. A decrease in groundwater temperature infringes water absorption capacity in the root system, although water loss by transpiration is not as much affected. In a temperate climate, *japonica* rice survives low-temperature stress by maintaining a water balance while maintaining resilient attributes such as small leaf size, thick cuticles, and low stomatal density. Because of their ability to adapt to warm climates, rice plants are adapted to a temperature range of 25–30 °C in terms of photosynthesis. A decrease in the ambient temperature below this range damages the photosynthetic apparatus and inhibits leaf photosynthesis. Low temperatures inhibit protochlorophyll formation and destroy existing chlorophylls, leading to leaf chlorosis. In addition to a reduction in the plastoquinone pool damaging the PSII reaction centre, the photochemical activity of the centre ultimately declines, causing a net decline in leaf photosynthesis. Chilling stress also decreases the rate of respiration.

Cold stress alters the hormonal balance in plant cells. Reports indicate that chilling-induced ABA synthesis is necessary for triggering a defence response to stress; however, the excess synthesis of this hormone is counterproductive for development (Li et al. 2022). ABA levels increase significantly in cold-tolerant rice cultivars but not in cold-sensitive cultivars in response to chilling (Lee et al. 1993). Pretreatment with ABA increases chilling tolerance in sensitive rice cultivars by increasing osmolyte concentrations, increasing antioxidant enzyme activity and regulating soluble carbohydrate and malondialdehyde contents (Hongtao et al. 2017).

7.6.2 Biochemical Responses

Chilling-injured leaves exhibit inhibited photosynthesis, a reduced respiration rate, decreased carbohydrate translocation, inhibited protein synthesis, and degradation of existing proteins in the cell. The common primary mechanism responsible for this process is the loss of membrane functions in various organelles. In chilling-sensitive plants, there is a high percentage of saturated fatty acid chains, and the membrane therefore tends to solidify into a semicrystalline state under a low-temperature regime. Changes in the physical properties of the membrane impact the activities of integral membrane proteins, which include carriers, ion channel proteins, and H^+-ATPase, ultimately impairing solute/ion transport and membrane-bound enzyme activities crucial for cell metabolism. In stress-tolerant plants, the presence of a high proportion of unsaturated fatty acids in the membrane reduces the phase transition temperature of lipids, permitting the membrane to remain relatively fluid at low temperatures and maintaining cellular functions.

Under cold stress, several biochemical reactions are activated in response to reactive oxygen species. Oxidative stress can negatively affect cellular metabolism by enhancing the amount of MDA generated through the peroxidation of polyunsaturated fatty acids (Nakashima et al. 2007). In cold-acclimated plants, the levels of antioxidant species increase to scavenge reactive oxygen species and protect cells from oxidative damage. As reviewed by Li et al. (2022), cold-acclimated plants under low-temperature stress exhibit significant induction of antioxidant enzymes, such as superoxide dismutase, catalase, and ascorbate peroxidase, and nonenzymatic antioxidants, such as ascorbic acid and glutathione, to detoxify surplus ROS.

Chilling stress also induces osmotic adjustment, which is a net increase in solute content irrespective of cell volume changes that result from dehydration. Most of the adjustments can be accounted for by increase in the concentration of a variety of common solutes. The solutes accumulate in the cytoplasm to maintain water potential equilibrium with the environment and protect cells against dehydration-induced injuries. Organic solutes are called compatible solutes (osmolytes) and do not interfere with normal cell metabolism. The commonly accumulated osmolytes are the amino acid proline; alcohol derivatives of sugars, such as sorbitol and mannitol, and a quaternary amine named glycine betaine. Proline is a natural osmoprotectant that plays multiple roles under stress conditions, including the storage of amino nitrogen, protection of the cell membrane from freezing injuries, and maintenance of cell functions by scavenging ROS. The accumulation of soluble sugars in the cytoplasm also protects plant cells by working as osmoprotectants, providing carbon skeletons and energy to other essential metabolites and protecting membrane integrity by interacting with the lipid bilayer (Liu et al. 2018). These sugars include glucose, fructose, sucrose, raffinose, and trehalose. The biochemical pathways involved in the synthesis of proline and trehalose during cold acclimation in rice are described below (Li et al. 2022) in Fig. 7.9.

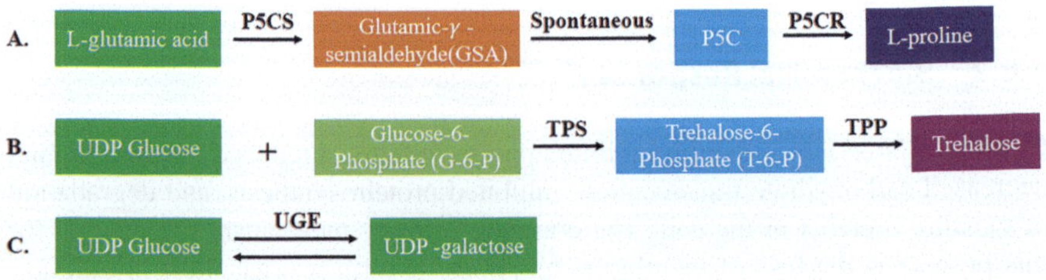

Fig. 7.9 Reactions related to osmolyte biosynthesis. (**A**) Proline biosynthesis in plants. *P5CS*, 1-pyrroline-5-carboxylic acid synthetase; *P5C*, 1-pyrroline-5-carboxylic acid; *P5CR*, P5C reductase (**B**) Biosynthesis of trehalose. *TPS*, trhalose-6-phosphate synthase; *TPP*, trehalose-6-phosphate phosphatase. (**C**) Interconversion of UDP-glucose and UDP-galactose by UGE (UDP-glucose-4-epimerase). (Adapted from Li et al. 2022)

7.7 Effect of the Water Depth

By nature, rice is a semi-aquatic water-loving crop that exhibits optimum growth in water-stress-free environments. Even if a crop is cultivated with adequate water, the surface water level significantly influences growth and yield. Moreover, the desired level of surface water varies with the stage of growth of the plant. Talpur (2013) recommended water depths of 5 cm and 10 cm for vegetative- to mid-stage crop growth and for the middle-stage of harvesting, respectively, as ideal for achieving maximum rice yield. A water depth of 5 cm at the transplanting stage was also found to be most effective at harnessing high rice yields (Saleem et al. 2015). However, partial submergence of the crop significantly reduces yield by decreasing the number of tillers.

Experiments conducted at the National Rice Research Institute, Cuttack, India, revealed that grain yield severely decreased with increasing standing water depth. Rice yield increased in the presence of 5–10 cm of standing water as was revealed by summarizing the grain yield production data from the experiment in which 20 genotypes were used. Grain yield decreased in the order of water depth: Control (5–10 cm) > shallow depth (35 cm) > medium-depth (50 cm) standing water (Fig. 7.10). Compared with those of the control, the varieties that are tolerant to shallow-depth water stagnation showed little or no reduction; however, they exhibited a reduction in the number of ear-bearing tillers or panicles. Compensation for yield mainly contributed to the increase in grain number per panicle. The effects of water depth on grain yield of rice per unit area (Fig. 7.10), varietal difference in grain yield (Fig. 7.11), panicle number per unit area (Fig. 7.12), and grain number per panicle (Fig. 7.13) have been shown.

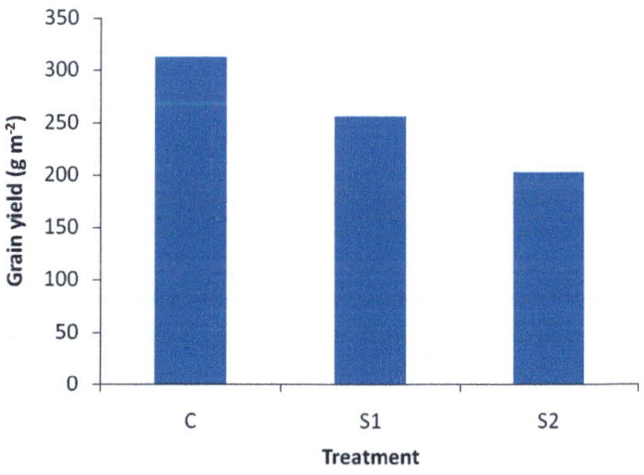

Fig. 7.10 Influence of water depth on average grain yield production. *C* control (5 cm standing water), *S1* shallow depth (35 cm standing water), *S2* medium depth (50 cm standing water). (Adapted from Sarkar 2016)

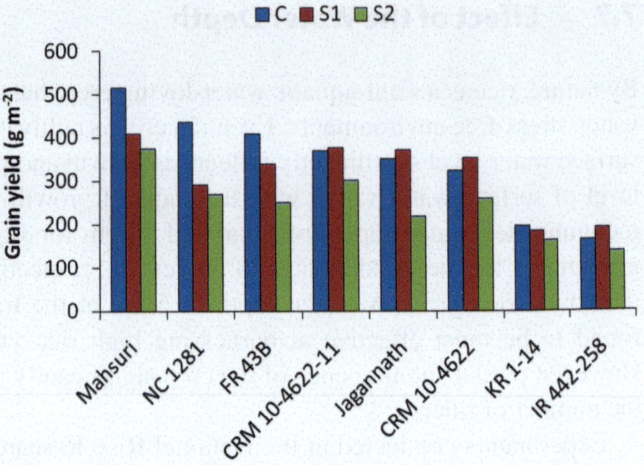

Fig. 7.11 Influence of water depth on the grain yield of different varieties of rice. *C* control (5 cm standing water), *S1* shallow depth (35 cm standing water), *S2* medium depth (50 cm standing water. (Adapted from Sarkar 2016)

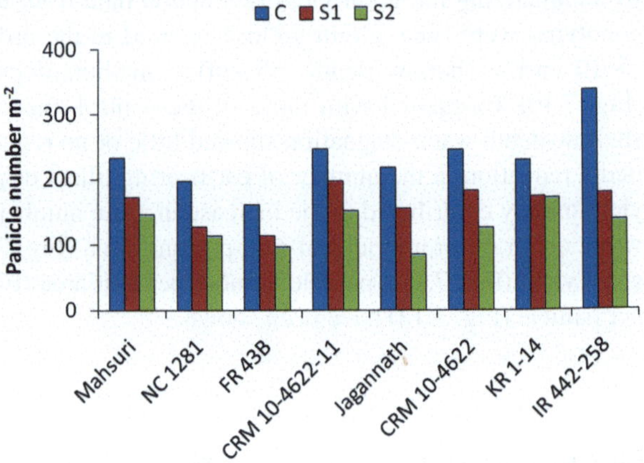

Fig. 7.12 Influence of water depth on the panicle number of different varieties of rice. *C* control (5 cm standing water), *S1* shallow depth (35 cm standing water), *S2* medium depth (50 cm standing water). (Adapted from Sarkar 2016)

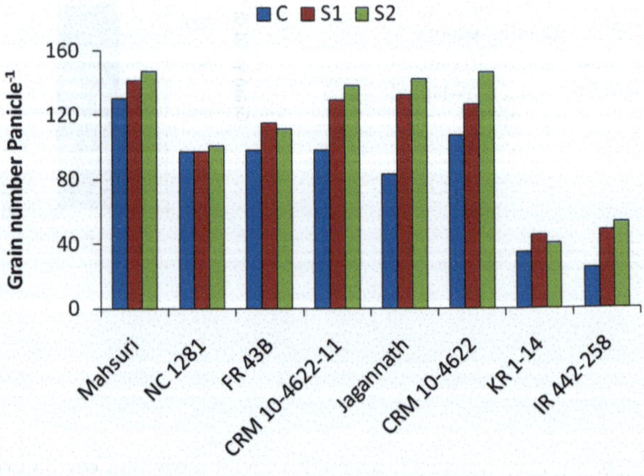

Fig. 7.13 Influence of water depth on the grain number per panicle of different varieties of rice. *C* control (5 cm standing water), *S1* shallow depth (35 cm standing water), *S2* medium depth (50 cm standing water). (Adapted from Sarkar 2016)

7.7.1 Plant Height, Tiller Number, and Yield Under Medium-Depth Water Stagnation

Under stagnant flooding, some parts of the plants remain inside the water. The impact on tiller numbers increases with increasing standing water. The greater the depth of standing water, the greater the reduction in tiller numbers (Sarkar 2016). Semi-dwarf rice cultivars with the submergence tolerance gene *SUB1A* perform poorly under medium-depth stagnant flooding conditions compared with their parents without *SUB1A*. A 10–15 cm difference in plant height has a great impact on the number of ear-bearing tillers/panicles number. Introgression of *Sub1* restricts elongation under waterlogging; hence, the elongation of plants with *Sub1* was shorter than that of the wild-type without *SUB1* at a water depth of 10–15 cm. However, the tiller numbers per unit area decreased by 20% and 35%, respectively, and the grain yields decreased by 27% and 65%, respectively, in Savitri-Sub1 and Swarna-Sub1, corresponding to genotypes without *Sub1* under stagnant flooding (Table 7.2). In general, with increasing stagnation of water in rice fields, tiller numbers decrease, as do other yield-related factors; however, some landrace species exhibit greater yield and yield stability under moderate-depth stagnant flooding than under normal-growth conditions.

The stability indices (SI) (values under waterlogged conditions/values under normal conditions) for different yield and yield attributes revealed that the impact of SF (stagnant flooding) differed among the different genotypes (Fig. 7.14). Correlation coefficient studies among different parameters and SI values revealed that the correlation was significant with the SI value. Based on these findings, it was concluded that maintenance of equivalent panicle weight and panicle number, plant length at maturity, leaf area, leaf and stem dry weight, root oxidase activity, and tiller numbers at flowering under SF compared to those of the control helps the plant counteract the adverse effects of SF (Sarkar 2016).

Table 7.2 Changes in yield and yield attributes of primary and secondary tillers (excluding the mother tiller) of rice plants with and without the submergence tolerance quantitative trait locus (QTL) *SUB1*

Genotype	Treatment	Plant height (cm)	Grain weight (g panicle^{-1})	Filled grain (%)	Panicle number (m^{-2})	Grain weight (g m^{-2})	Straw weight (g m^{-2})
Swarna	C	115 ± 0.8	2.56 ± 0.06	83 ± 2	246 ± 10	635 ± 11	666 ± 15
	SF	133 ± 1.5	2.40 ± 0.09	79 ± 11	178 ± 18	464 ± 15	601 ± 21
Swarna-Sub1	C	111 ± 1.0	2.61 ± 0.05	85 ± 3	240 ± 08	626 ± 12	671 ± 14
	SF	115 ± 1.4	1.21 ± 0.10	58 ± 14	123 ± 16	187 ± 16	495 ± 18
Savitri	C	116 ± 0.9	2.52 ± 0.07	75 ± 4	264 ± 09	654 ± 9	834 ± 14
	SF	127 ± 1.8	2.24 ± 0.11	68 ± 9	186 ± 17	443 ± 19	685 ± 24
Savitri-Sub1	C	114 ± 0.7	2.58 ± 0.06	74 ± 4	255 ± 09	657 ± 13	824 ± 13
	SF	118 ± 1.4	2.28 ± 0.12	64 ± 15	132 ± 15	325 ± 20	658 ± 20

C control, *SF* stagnant flooding with a medium depth (0–50 cm) of water. (Adopted from Kuanar et al. 2019)

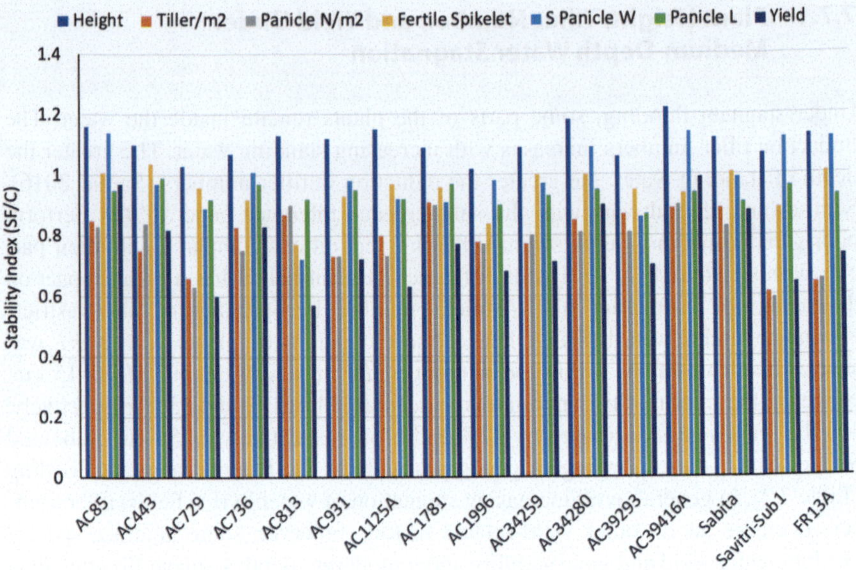

Fig. 7.14 Stability indices (values at SF/values at C) of yield and yield parameters of rice varieties under medium-depth stagnant flooding. (Adapted from Kuanar et al. 2017)

7.7.2 Tiller Mortality and Basis of Tolerance to Stagnant Flooding

Due to the stagnation of water on the soil surface, not only does the tiller number decrease but even the weak plants also die out to provide more space and nutrients to stronger plants (Das et al. 2002). The number of plants/tillers per hill decreases with time under stagnation of water beyond a depth of 20 cm. The decreases in the number of tillers after transplanting and after direct seeding were 34% and 50%, respectively, from the maximum tillering stage to maturity (Fig. 7.15) (Das et al. 2002).

7.7.3 Physiological Basis of Tolerance to Stagnant Flooding

Unlike other types of abiotic stresses, such as drought and salinity, the survival of plants plays a greater role in determining submergence tolerance. Submergence for 1–2 weeks completely kills the plants, while salinity and drought stunt plant growth, and with time, the plants start to experience symptoms of withering that lead to drying and death. However, under stagnant flooding, the event is somewhat different. Rice plants are water-loving. Therefore, rice can grow very well in standing water.

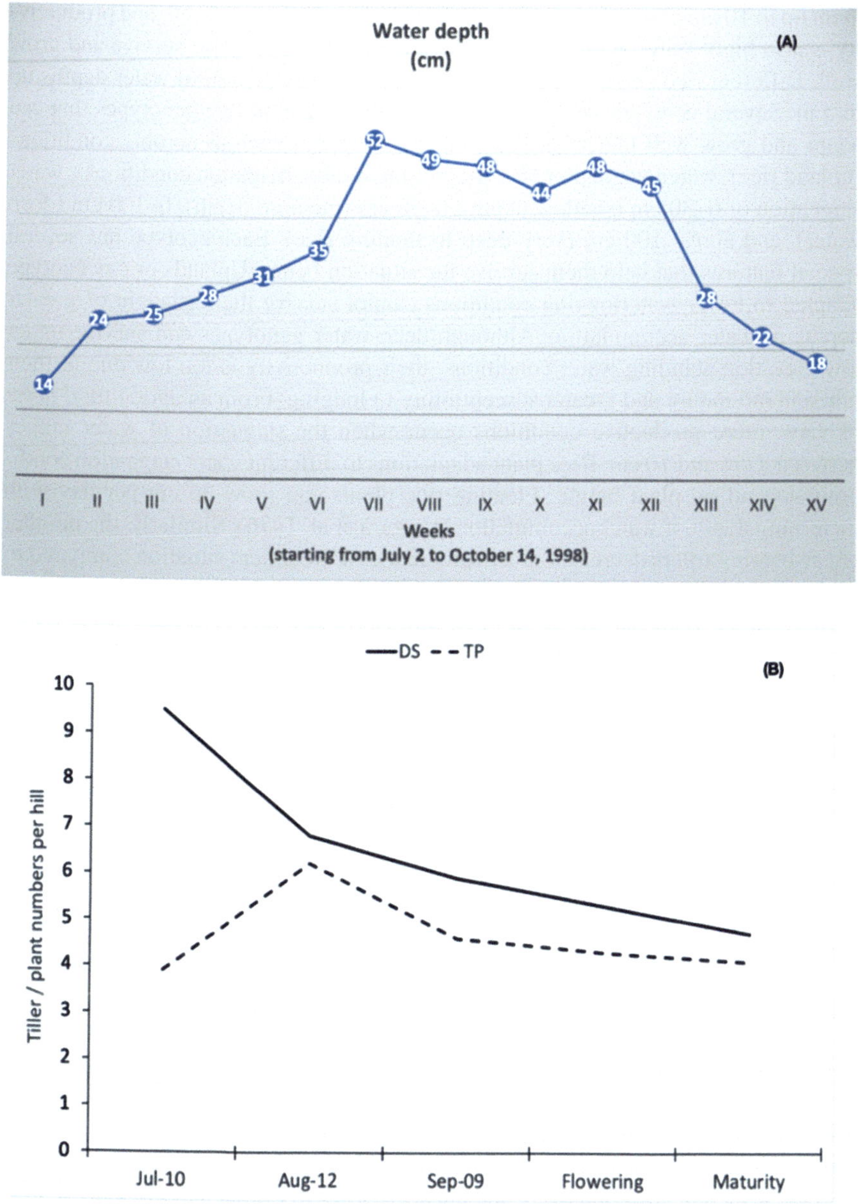

Fig. 7.15 Standing water and death of tillers. The water level at different times during the growth period (**A**) and decreases in plant/tiller numbers under direct seeding (DS) and transplanting (TP) conditions (**B**). (Adapted from Das et al. 2002)

With up to 10 cm of standing water in the field, rice plants grow well, and productivity is also high. With increasing water depth, rice plants can also survive and grow well. Different rice genotypes have evolved that can adapt well to water depths up to 5 m. Several ecotypes are available among the abundant rice genotypes that can adapt and grow well under various soil water regimes such as aerobic conditions (upland rice), water stagnation of 2–10 cm (favourable irrigation conditions), water stagnation of 0–30 cm (shallow depth), 0–50 cm (medium depth), 0–100 cm (deep water), and above 100 cm (very deep to floating rice). Each ecotype has several special features that help them survive the situation better. Uplands or any ecotype adapted to lower waterlogging conditions cannot survive the onslaught of greater depths of water accumulation. Although deep-water genotypes can survive under low-elevation standing water conditions, their productivity is too low due to their inherent tall nature and greater susceptibility to lodging. From an agricultural point of view, more productive conditions occur when the stagnation of water ranges between 2 cm and 10 cm. Rice plant adaptations to different water stagnation conditions depend on plant height. Floating rice plants can grow 35 cm per day with increasing depth of water accumulation (Vergara et al. 1976). Similarly, the number of ear-bearing tillers decreases drastically in the SF treatment situation compared to the control condition. At least >10 cm above the water surface at the onset of waterlogging, taller tillers survive better than submerged tillers. Plants with the capacity to produce sufficient numbers of stout and healthy tillers before water stagnation are useful under SF. This type of plant maintains a greater number of tillers under SF. As the water table increases, the supply of oxygen to the root zone decreases. The resilient cultivars can supply more oxygen to the root zones, survive better, and give fair yields.

7.7.4 Kneeing Ability

Root oxidase activity, an indirect method for measuring the oxygen-liberating power of the root system, is greater in tolerant cultivars than in susceptible cultivars under different depths of stagnant flooding. The tillers that are productive and survive the onslaught of stagnant flooding in both the tolerant and susceptible cultivars show greater activities of root oxidase than the non-performing weak tillers. Compared with secondary tillers, which accumulate greater quantities of dry matter before the onset of stagnant flooding, primary tillers survive better. Increased water depth also decreases leaf area and dry matter accumulation in secondary tillers. The chlorophyll a and chlorophyll b contents are greater in primary tillers than in weak secondary tillers (Sarkar 2016). To compensate for the contribution of tillers to grain yield, a greater weight of panicle is needed. The following characteristics contribute to panicle weight: high-density grain, low sterility (<10% per panicle), intermediate grain weight (3 g/100 grains), and greater grain number per panicle (250 grains per panicle). Very deep-water rice/floating rice plants possess certain peculiarities. In addition to extensive elongation, adventitious roots are produced from the nodes of water–air junctions. Nodal roots act like normal roots, but here,

Fig. 7.16 Photograph showing rice cultivars with kneeing ability (left panel) and without kneeing ability during the post-flooding phase. (Source: Photo gallery of R. K. Sarkar)

the medium is floodwater instead of soil. Stem elongation occurs due to both cell division and elongation of cells of intercalary meristems through the action of plant hormones, mainly ethylene and gibberellic acid (Kende et al. 1998). Before the initiation of flooding, these plants reserve non-structural carbohydrates in their stem that supply initial energy for elongation so that a portion of the shoot exits above the water surface (Sarkar 2016). When the water level recedes, the procumbent culm/the lower part of the culm rests on the ground, whereas the terminal part bends upwards, keeping the reproductive part above the water/soil surface. This ability is called the 'kneeing ability'. Kneeing ability helps the survival of reproductive parts from drowning so that rice plant could produce grain for consumption (Fig. 7.16). Floating rice plants in Bangladesh are the most elongators. The length of the plant can reach up to 6 m under a water stagnation of 4 m. However, the plant fails to survive if the water stagnation exceeds 4 m due to the over-expenditure of non-structural carbohydrates and the failure to replenish the stock through current photosynthesis. Depending on the nature of the flood, rice plants can be exposed to different microenvironments. Therefore, the basis of tolerance differs greatly and strongly and impacts tiller number and plant productivity considerably (Fig. 7.16).

7.7.5 Role of Ethylene in Tolerance to Submergence and Stagnant Flooding

The mechanisms of tolerating submergence and stagnant flooding differ, where ethylene plays a key role in enhancing CO_2 and decreasing O_2. Subjected to excess water stress, aerenchyma formation and elongation control determine survival and plant productivity. Zero to negligible elongation and minimum formation of aerenchyma are suitable for submergence and shallow water stagnation, whereas greater elongation and aerenchyma formation occur under floating as well as deep-water stagnation (Fig. 7.17).

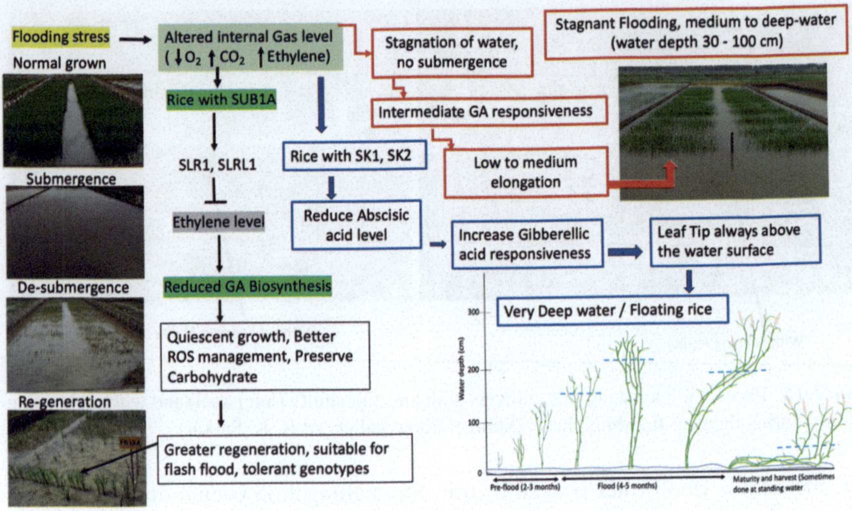

Fig. 7.17 Effect of excess water stress such as submergence and stagnant flooding in the induction of hormones and gene expression in response to alteration in internal gas exchange in rice

Mechanisms to tolerate submergence and stagnant flooding differ where ethylene plays a key role in the environment of enhancing CO_2 and decreasing O_2. Elongation due to flooding determines the survival as well as plant productivity under different levels of water depth. Elongation is greater in floating or very deep-water stagnation, followed by deep water (50–100 cm), medium deep water (30–50 cm), and shallow water (15-30 cm). *Snorkel 1* (*SK1*) and 2 (*SK2*) increase ethylene levels. Gibberellic acid (GA) response increases dramatically in plants with *SK1* and *SK2* resulting in greater internodal elongation in floating rice. Medium expression of *SK1* and *SK2* or lack of any one induces GA responsiveness comparatively less. It is suitable for deep water conditions. Lack of *SK1*, *SK2*, and *SUB1A* induces elongation both under submergence and water logging conditions in medium-deep water rice genotypes. Under stagnant flooding with less than 100 cm of water stagnation, low to medium elongation is important for better plant productivity. The presence of *SUB1A* quiescence growth decreases GA responsiveness and GA synthesis by the control of the expression of *SLENDER RICE 1* (*SLR1*) and *SLR Like1* (*SLRL1*) genes. The quiescence strategy preserves non-structural carbohydrates (NSC) and accelerates de-submergence growth and survival under submergence.

Ethylene in addition to GA biosynthesis induces (ROS) H_2O_2 generation leading to programmed cell death. RBOH, a ROS generator, triggers H_2O_2 formation in the cortex. Metallothionine (MT), ROS quencher if over-express dampens the activity of H_2O_2. Suppression of MT under stagnant flooding induces the burst of H_2O_2 and depending on the rate of burst initiates aerenchyma formation under different levels of waterlogging (Fig. 7.18).

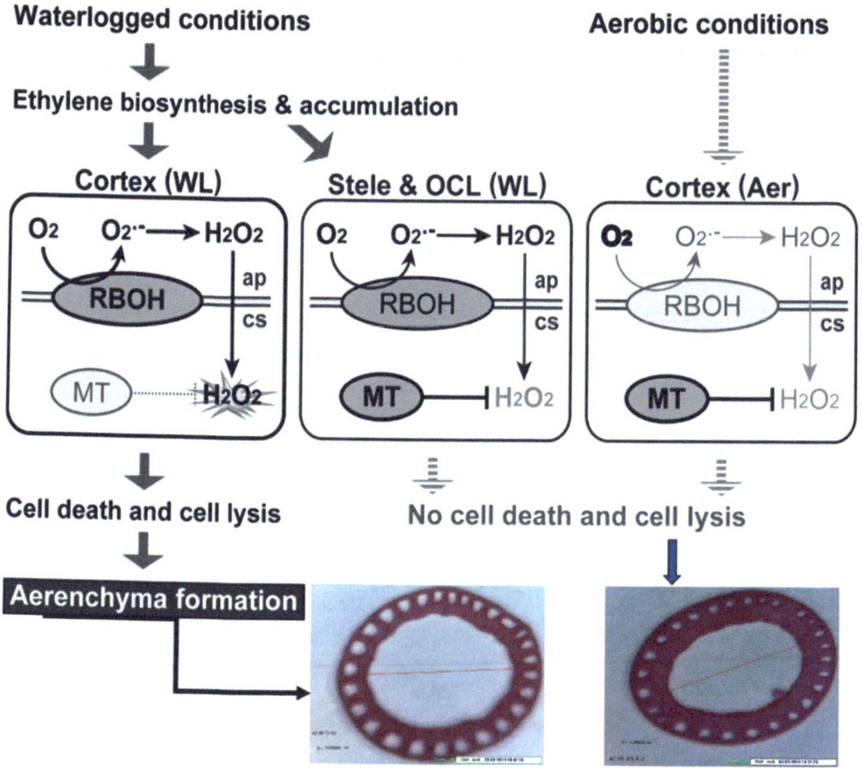

Fig. 7.18 Role of ethylene in the formation of aerenchyma. *RBOH* respiratory burst oxidase, *MT* metallothionine, *ap* apoplast, *cs* cytoplasm. (Adopted from Kende et al. 1998; Sarkar et al. 2014; Chakraborty et al. 2021)

References

Afza H, Andarini YN, Risliawati A (2020) Tiller profile diversity of upland rice germplasm in ICABIOGRAD gene bank. IOP Conf Ser Earth Environ Sci 482:012041

Bahuguna RN, Jha J, Pal M, Shah D, Lawas LMF, Khetarpal S, Jagadish KSV (2015) Physiological and biochemical characterization of NERICA-L-44: a novel source of heat tolerance at the vegetative and reproductive stages in rice. Physiol Plant 154:543–559

Blanc E, Strobl E (2016) Assessing the impact of typhoons on rice production in the Philippines. J Appl Meteorol Climatol 55:993–1007

Cao YY, Duan H, Yang LN, Wang ZQ, Zhou SC, Yang JC (2008) Effect of heat stress during meiosis on grain yield of rice cultivars differing in heat tolerance and its physiological mechanism. Acta Agron Sin 34:2134–2142

Chakraborty A, Bhattacharjee S (2015) Differential competence of redox-regulatory mechanism under extremes of temperature determines growth performances and cross-tolerance in two *Indica* rice cultivars. J Plant Physiol 176:65–77

Chakraborty K, Ray S, Vijayan J, Molla KA, Nagar R, Jena P, Mondal S, Panda BB, Shaw BP, Swain P, Chattopadhyay K, Sarkar RK (2021) Preformed aerenchyma determines the differential tolerance response under partial submergence imposed by fresh and saline water flooding in rice. Physiol Plant 173:1597–1615

Chauhan BS, Abugho SB (2013) Effects of water regime, nitrogen fertilization, and rice plant density on growth and reproduction of lowland weed *Echinochloa crus-galli*. Crop Prot 54:142–147

Clerget B, Bueno C, Domingo AJ, Layaoen HL, Vial L (2016) Leaf emergence, tillering, plant growth, and yield in response to plant density in a high-yielding aerobic rice crop. Field Crop Res 199:52–64

Cui Y, Lu S, Li Z, Cheng J, Hu P, Zhu T, Wang X, Jin M, Wang X, Li S, Huang S, Zou B, Hua J (2020) CYCLIC NUCLEOTIDE-GATED ION CHANNELs 14 and 16 promote tolerance to heat and chilling in rice. Plant Physiol 183:1794–1808

Das S, Ramakrishnayya G, Sarkar RK (2002) Tiller mortality and its mitigation through GA3 and IAA in unfavorable rainfed lowland rice. Plant Arch 2:299–304

DeFalco TA, Bender KW, Snedden WA (2009) Breaking the code: Ca^{2+} sensors in plant signaling. Biochem J 425:27–40

Ding YL, Shi YT, Yang SH (2020) Molecular regulation of plant responses to environmental temperatures. Mol Plant 13(4):544–564

Dingkuhn M, Luquet D, Kim H, Tambour L, Clement-Vidal A (2006) Eco meristem, a model of morphogenesis and competition among sinks in rice. 2. Simulating genotype responses to phosphorus deficiency. Funct Plant Biol 33:325–337

Dreccer MF, Chapman SC, Rattey AR, Neal J, Song Y, Christopher JT, Reynolds M (2013) Developmental and growth controls of tillering and water-soluble carbohydrate accumulation in contrasting wheat (*Triticum aestivum* L.) genotypes: can we dissect them? J Exp Bot 64:143–160

Dubey AN, Veram S, Goswami SR, Devedee AK (2018) Effect of temperature on different growth stages and physiological process of rice crop—a review. Bull Environ Pharmacol Life Sci 7:129–136

Ezin V, Ahanchede WW, Ayenan MAT, Ahanchede A (2022) Physiological and agronomical evaluation of elite rice varieties for adaptation to heat stress. BMC Plant Biol 22:236. https://doi.org/10.1186/s12870-022-03604-x

Fitzgerald MA, McCouch SR, Hall RD (2009) Not just a grain of rice: the quest for quality. Trends Plant Sci 14:133–139

Fu G, Feng B, Zhang C, Yang Y, Yang X, Chen T, Zhao X, Zhang X, Jin Q, Tao L (2016) Heat stress is more damaging to superior spikelets than inferiors of rice (*Oryza sativa* L.) due to their different organ temperatures. Front Plant Sci 7:1637

Fukushima A (2019) Varietal differences in tiller and panicle development determining the total number of spikelets per unit area in rice. Plant Prod Sci 22:192–201

Fukushima A, Shiratsuchi H, Yamaguchi H, Fukuda A (2011) Effects of nitrogen application and planting density on morphological traits, dry matter production and yield of large grain type rice variety *Bekoaoba* and strategies for super high-yielding rice in the Tohoku region of Japan. Plant Prod Sci 14:56–63. https://doi.org/10.1626/Pps.14.56

Gammulla CG, Pascovici D, Atwell BJ, Haynes PA (2010) Differential metabolic response of cultured rice (*Oryza sativa*) cells exposed to high- and low temperature stress. Proteomics 10:3001–3019

Gravois KA, Helms RS (1992) Path analysis of rice yield and yield components as affected by seeding rate. Agron J 84:1–4

Hanada K (1993) Tillers. In: Matsuo T, Hoshikawa K (eds) Science of the rice plant. Food and Agriculture Policy Center, Tokyo, pp 222–258

Hasanuzzaman M, Nahar K, Alam MM, Roychowdhury R, Fujita M (2013) Physiological, biochemical, and molecular mechanisms of heat stress tolerance in plants. Int J Mol Sci 14:9643–9684. https://doi.org/10.3390/ijms14059643

Hayashi S, Kamoshita A, Yamagishi J (2006) Effect of planting density on grain yield and water productivity of rice (*Oryza sativa* L.) grown in flooded and non-flooded fields in Japan. Plant Prod Sci 9:298–311

Honda T (1977) A study of tillering in paddy rice with special reference to dry matter production and its translocation. Bull Inst Agric Res Tohoku Univ 28:171–312. (in Japanese with English summary)

Hongtao X, Tongtong W, Dianfeng Z, Lizhi W, Yanjiang F, Yu L, Rui L, Zhongjie L, Ying M, Wan L, Lianmin W, Chunjie Y (2017) ABA pre-treatment enhances the chilling tolerance of a chilling-sensitive rice cultivar. Braz J Bot 40(4):853–860

IPCC (2014) Contribution of working groups I, II and III to the fifth assessment report of the intergovernmental panel on climate change, Synthesis report. IPCC, Geneva

Jones DB, Snyder GH (1987) Seeding rate and row spacing effects on yield and yield components of drill-seeded rice. Agron J 79:623–626

Kakizaki Y (1987) Effect of light and temperature on tillering of rice plant. Agric Hortic 62:1323–1330. (in Japanese)

Kan Y, Lin HX (2021) Molecular regulation and genetic control of rice thermal response. Crop J 9:497–505

Kariali E, Kuanar SR, Mohapatra PK (2008) Individual tiller dynamics of two wild *Oryza* species in contrasting habitats. Plant Prod Sci 11:355–360

Kende H, van der Knaap E, Cho H-T (1998) Deepwater Rice: a model plant to study stem elongation. Plant Physiol 118:1105–1110

Khush GS (2000) New plant type of rice for increasing the genetic yield potential. In: Nanda JS (ed) Rice breeding and genetics. Science Publishers, Enfield, NH, pp 99–108

Krishnan P, Ramakrishnan B, Reddy K, Reddy VR (2011) High-temperature effects on rice growth, yield, and grain yield. Adv Agron 111:87–206

Kuanar SR, Ray A, Sethi SK, Chattopadhyay K, Sarkar RK (2017) Physiological basis of stagnant flooding tolerance in rice. Ric Sci 24:73–84

Kuanar SR, Molla AM, Chattopadhyay K, Sarkar RK, Mohapatra PK (2019) Introgression of Sub1 (SUB1) QTL in mega rice cultivars increases ethylene production to the detriment of grain-filling under stagnant flooding. Sci Rep 9:18567

Kumar N, Jeena N, Kumar A, Khwairakpam R, Singh H (2021) Comparative response of rice cultivars to elevated air temperature in Bhabar region of Indian Himalaya: status on yield attributes. Heliyon 7:e07474. ISSN 2405-8440. https://doi.org/10.1016/j.heliyon.2021.e074

Kwak Y, Arifuzzanman B, Iwami Y (2015) Prompt proxy mapping of flood damaged rice fields using modis-derived indices. Remote Sens (Basel) 7:15969–15988

Lafarge T, Tubana B, Pasuquin EM (2004) Yield advantage of hybrid rice induced by its higher control in tiller emergence. In: Turner N, Angus J, McIntyre L, Robertson M, Borrell A, Lloyd D (eds) New directions for a diverse planet: proceedings for the 4th International Crop Science Congress, Brisbane, Australia, 26 Sep–1 Oct 2004. Regional Institute, Gosford. ISBN 1-920842-20. http://www.cropscience.org.au/icsc2004/poster/2/7/1/862 lafargeta.htm

Lamers J, van der Meer T, Testerink C (2020) How plants sense and respond to stressful environments? Plant Physiol 182:1624–1635

Lee TM, Lur HS, Chu C (1993) Role of abscisic acid in chilling tolerance of rice (*Oryza sativa* L.) seedlings. I. Endogenous abscisic acid levels. Plant Cell Environ 16(5):481–490

Li J, Zhang Z, Chong K, Xu Y (2022) Chilling tolerance in rice: past and present. J Plant Physiol 268:153576

Liu CT, Wang W, Mao BG, Chu CC (2018) Cold stress tolerance in rice: physiological changes, molecular mechanism, and future prospects. Yi Chuan 40(3):171–185

Liu L, Lu H, Li J, Cao Z (2020) High-throughput rice density estimation from transplantation to tillering stages using deep networks. Plant Phenomics 2020:1375957. https://doi.org/10.34133/2020/1375957

Maestri E, Klueva N, Perrotta C, Gulli M, Nguyen HT, Marmiroli N (2002) Molecular genetics of heat tolerance and heat shock proteins in cereals. Plant Mol Biol 48:667–681

Miller BC, Hill JE, Roberts SR (1991) Plant population effects on growth and yield in water-seeded rice. Agron J 83:291–297

Mitler R, Finka A, Goloubinoff P (2012) How do plants feel the heat? Trends Biochem Sci 37(3):118–125

Mittler R, Blumwald E (2010) Genetic engineering for modern agriculture: challenges and perspectives. Annu Rev Plant Biol 61:443–462

Miyazaki M, Araki M, Okamura K, Ishibashi Y, Yuasa T, Iwaya-Inoue M (2013) Assimilate translocation and expression of sucrose transporter, OsSUT1, contribute to high-performance ripening under heat stress in the heat tolerant rice cultivar Genkitsukushi. J Plant Physiol 170:1579–1584

Mohanan KV, Mini CB (2008) Relative contribution of rice tillers of different status towards yield. Int J Plant Breed Genet 2:9–12

Mohapatra PK, Kariali E (2008) Time of emergence determines pattern of dominance of rice tillers. Aust J Crop Sci 1:53–62

Mohapatra PK, Panda BB, Kariali E (2011) Plasticity of tiller dynamics in wild rice *Oryza rufipogon* Griff.: a strategy for resilience in suboptimal environments. Int J Agron 2011:543237. https://doi.org/10.1155/2011/543237

Nakano H, Morita S, Kitagawa H, Wada H, Takahashi M (2012) Grain yield response to planting density in forage rice with a large number of spikelets. Crop Sci 52:345–350

Nakashima K, Tran L-SP, Nguyen DV, Fujita M, Maruyama K, Todaka D, Ito Y, Hayashi N, Shinozaki K, Yamaguchi-Shinozaki K (2007) Functional analysis of a NAC-type transcription factor OsNAC6 involved in abiotic and biotic stress-responsive gene expression in rice. Plant J 51:617–630

Nemato K, Morita S, Baba T (1995) Shoot and root development in rice related to the phyllochron. Crop Sci 35:24–29

Nuruzzaman M, Yamamoto Y, Nitta Y, Yoshida T, Miyazaki A (2000) Varietal differences in tillering ability of fourteen *japonica* and *indica* rice varieties. Soil Sci Plant Nutr 46(2):381–391

Ottis BV, Talbert RE (2005) Rice yield components as affected by cultivar and seeding rate. Agron J 97:1622–1625

Peng S, Cassman K, Virmani S, Sheehy J, Khush G (1999) Yield potential trends of tropical rice since the release of IR8 and the challenge of increasing rice yield potential. Crop Sci 39:1552–1559. https://doi.org/10.2135/cropsci1999.3961552x

Peng S, Khush GS, Virk P, Tang Q, Zou Y (2008) Progress in ideotype breeding to increase rice yield potential. Field Crop Res 108:32–38. https://doi.org/10.1016/j.fcr.2008.04.001

Reddy ASN, Ali GS, Celesnik H, Day IS (2011) Coping with stresses: roles of calcium- and calcium/calmodulin-regulated gene expression. Plant Cell 23:2010–2032

Sailaja B, Subrahmanyam D, Neelamraju S, Vishnukiran T, Rao YV, Vijayalakshmi P, Voleti SR, Bhadana VP, Mangrauthia SK (2015) Integrated physiological, biochemical, and molecular analysis identifies important traits and mechanisms associated with differential response of rice genotypes to elevated temperature. Front Plant Sci 6:1044

Saleem MS, Akhtar M, Farooq U, Iqbal N, Haider Z (2015) Effect of different water depths on plant population and rice (*Oryza sativa*) yield at time of transplanting using mechanized transplanter. Int J Sci Eng Res 6:76–80

San-Oh Y, Kondo M, Ookawa T, Hirasawa T (2008) Eco-physiological analysis on effect of planting pattern on biomass production and grain yield in rice. JQRA 42:79–89

Sarkar RK (2016) Stagnant flooding tolerance in rice: endeavours and achievement. Research Bulletin 11. National Rice Research Institute, Cuttack

Sarkar RK, Das KK, Panda D, Reddy JN, Patnaik SSC, Patra BC, Singh DP (2014) Submergence tolerance in rice: biophysical constraints, physiological basis and identification of donors. Research Bulletin 7. Central Rice Research Institute, Cuttack

Shi W, Yin X, Struik PC, Solis C, Xie F, Schmidt RC, Huang M, Zou Y, Ye C, Jagadish SKV (2017) High day- and night-time temperatures affect grain growth dynamics in contrasting rice genotypes. J Exp Bot 68:5233–5245

Stoop WA, Uphoff N, Kassam A (2002) A review of agricultural research issue raised by the System of Rice Intensification (SRI) from Madagascar: opportunities for improving system for resource poor farmers. Agric Syst 71:249–274

Talpur MA (2013) Effect of different water depths on growth and yield of rice crop. Afr J Agric Res 8:4654–4659

Tanaka A, Garcia CV (1965) Studies on the relationship between tillering and nitrogen uptake of the rice plant. 2. Relation between tillering and nitrogen metabolism in the plant. Soil Sci Plant Nutr 11:31–37
Tang RS, Zheng JC, Jin ZQ, Zhang DD, Huang YH, Chen LG (2008) Possible correlation between high temperature-induced floret sterility and endogenous levels of IAA, GAs and ABA in rice (*Oryza sativa* L.). Plant Growth Regul 54:37–43. https://doi.org/10.1007/s10725-007-9225-8
Tollefson J (2020) How hot will earth get by 2100? Nature 580(7804):443–445
Vergara B, Jackson B, De Datta SK (1976) Deep water rice and its response to deep water stress. In: Climate and rice. International Rice Research Institute, Los Banos, pp 301–319
Wang F, Cheng F, Zhang G (2007) Difference in grain yield and quality among tillers in rice genotypes differing in tillering capacity. Ric Sci 14:135–140
Wang D, Li XF, Zhou ZJ, Feng XP, Yang WJ, Jiang DA (2010a) Two Rubisco activase isoforms may play different roles in photosynthetic heat acclimation in the rice plant. Physiol Plant 139:55–67
Wang M, Liang JW, Wang ZC, Huang L, Ma H, Liu M, Gu XY (2010b) Effect of irrigation water depth on rice growth and yield in a saline-sodic soil in Songnen plain, China. J Food Agric Environ 8:530–534
Wang Y, Lu J, Ren T, Hussain S, Guo C, Wang S, Cong R, Li X (2017) Effects of nitrogen and tiller type on grain yield and physiological responses in rice. AoB Plant 9:plx012. https://doi.org/10.1093/aobpla/plx012
Wu GW, Wilson LT, McClung AM (1998) Contribution of rice tillers to dry matter accumulation and yield. Agron J 90:317–323
Wu C, Cui K, Wang W, Li Q, Fahad S, Hu Q, Huang J, Nie L, Peng S (2016) Heat induced phytohormone changes are associated with disrupted early reproductive development and reduced yield in rice. Sci Rep 6:34978
Wu C, Cui K, Wang W, Li Q, Fahad S, Hu Q, Huang J, Nie L, Mohapatra PK, Peng S (2017) Heat-induced cytokinin transportation and degradation are associated with reduced panicle cytokinin expression and fewer spikelets per panicle in rice. Front Plant Sci 8:371
Xu Y, Ramanathan V, Victor DG (2018) Global warming will happen faster than we think. Nature 564:30–32
Xu J, Henry A, Sreenivasulu N (2020) Rice yield formation under high day and night temperatures—a prerequisite to ensure future food security. Plant Cell Environ 43:1595–1608
Xu Y, Chu C, Yao S (2021) The impact of high-temperature stress on rice: challenges and solutions. Crop J 9:963–976
Yamakawa H, Hirose T, Kuroda M, Yamaguchi T (2007) Comprehensive expression profiling of rice grain filling-related genes under high temperature using DNA microarray. Plant Physiol 144:258–277
Yamamoto Y, Kurokawa H, Nitta Y, Yoshida T (1995) Varietal difference of tillering response to shading and nitrogen levels in rice plant. Jpn J Crop Sci 64:227–234. (in Japanese with English abstract)
Yang JC, Zhang H, Zhang J (2012) Root morphology and physiology in relation to the yield formation of rice. J Integr Agric 11:920–926
Yoshida S (1973) Effects of temperature on growth of the rice plant (*Oryza sativa* L.) in a controlled environment. Soil Sci Plant Nutr 19:299–310
Yoshida S (1981) Fundamental of rice crop science. Scientific Research Publishing, International Rice Research Institute, Los Banos, Laguna
Yoshida S, Hayakawa Y (1970) Effects of mineral nutrition on tillering of rice. Soil Sci Plant Nutr 16:186–191
Zafar SA, Hameed A, Nawaz MA, Wei MA, Noor MA, Hussain M, Rahman M (2018) Mechanisms and molecular approaches for heat tolerance in rice (*Oryza sativa* L.) under climate change scenario. J Integr Agric 17:726–738
Zeng Y, Zhang Y, Xiang J, Uphoff NT, Pan X, Zhu D (2017) Effects of low temperature stress on spikelet-related parameters during anthesis in *Indica-Japonica* hybrid rice. Front Plant Sci 8:1350

Zhang C, Li G, Chen T, Feng B, Fu W, Yan J, Islam MR, Jin Q, Tao L, Fu G (2018) Heat stress induces spikelet sterility in rice at anthesis through inhibition of pollen tube elongation interfering with auxin homeostasis in pollinated pistils. Rice 11:14

Zhao Q, Zhou L, Liu J, Cao Z, Du X, Huang F, Pan G, Cheng F (2018a) Involvement of CAT in the detoxification of HT-induced ROS burst in rice anther and its relation to pollen fertility. Plant Cell Rep 37(5):741–757

Zhao Q, Zhou L, Liu J, Du X, Asad MAU, Huang F, Pan G, Cheng F (2018b) Relationship of ROS accumulation and superoxide dismutase isozymes in developing anther with floret fertility of rice under heat stress. Plant Physiol Biochem 122:90–101

Zheng H, Chen Y, Chen Q, Li B, Zhang Y, Jia W, Mo W, Tang Q (2020) High-density planting with lower nitrogen application increased early rice production in a double-season rice system. Agron J 112:205–214

Physiology of Tiller Production and Development

8

Abstract

Tiller growth and development depend on several nutritional and hormonal factors. A nodal tiller derives these biochemicals from the mother shoot before gaining independence for survival. The new dispensation of super/NPT rice has favoured the exclusion of unproductive tillers bereft of a thought given to making them productive by exploring the intrinsic factors regulating their occurrence. Auxin, cytokinins, gibberellins, ethylene, strigolactone, and brassinosteroids are the growth regulators that affect tillering. Of these hormones, the roles of auxin, cytokinin, and strigolactone are primary and central to the control of tiller bud activation. These three hormones move throughout the plant, forming a systemic signalling network, and interactions between them determine the final plant architecture and tiller number. Other hormones, such as GA, ABA, and ethylene, also influence the tiller production of rice in combination with the growth regulators of the primary group. The interplay of these hormones determines the tiller outgrowth of a plant. The paucity of information on this topic has hindered the identification of designer rice plants.

Keywords

Tiller production · Hormone signalling · Ethylene · Auxin · Gibberellin · Strigolactone · cytokinin

8.1 Introduction

The number of panicles is a yield component of rice and depends largely on the number of tillers produced during the vegetative stage of the plant. Tiller development initiates at the early growth stage, during which tillers grow vigorously, bear panicles on stem tips, and ultimately contribute to grain yield as productive tillers.

© The Author(s), under exclusive license to Springer Nature Singapore Pte Ltd. 2025

P. K. Mohapatra et al., *Tillering Behavior of Rice Plant*,
https://doi.org/10.1007/978-981-97-5235-5_8

The effect of productive tillers on grain yield depends on various environmental conditions during growth. The amount of variation caused by the environmental conditions during the vegetative period is most noticeable in terms of the number of tillers. Subject to the congenial environment, a rice genotype may produce more than 50 tillers at the maximum tiller number stage, but in practice, this number is restricted to a few because of limitations in cultural management and environmental conditions. Additionally, it has been discussed in previous chapters of this book that tiller development in rice follows a hierarchical pattern, where early-initiated tillers on basal stem nodes grow vigorously and contribute more to overall grain yield than younger tillers located on higher nodes. On most occasions, late-stage tillers become less productive or unproductive. The physiological mechanisms dictating tiller asynchrony, therefore, assume immense importance for the achievement of synchronized growth and development of tillers and enable optimum use of biomass production by plants for grain production for the rational design of plant architecture by rice breeders.

8.2 Variation in Tiller Growth and Survival

As soon as the rice seedling becomes self-supporting through the development of five leaves, tillering starts on the stem. Tillering finishes at the panicle initiation stage. The first tiller emerges from the space between the main stem and the second leaf from the base of the plant. Similarly, a second tiller develops between the third leaf and main stem when the sixth leaf emerges, and the process of tillering continues in the same order thereafter. As discussed before, tillers are secondary shoots that appear on the main shoot, and their formation is completed within 2–3 weeks when the plant transitions from the vegetative to the reproductive stage of growth (Table 8.1, Nawaz and Farooq 2017). The tillering stage is highly important for determining grain yield because it determines panicle density in the field. In rice cultivation, this stage is one of the most critical stages and can be influenced by agronomical management practices. Under good management, the number of tillers initiated on the plant at the maximum tillering stage could be two times greater than what is needed for high yield. However, many of the tillers die before flowering (Table 8.2). Additionally, the seeding rate strongly affects the tiller number of rice plants. At a low seeding rate, all tillers survive and contribute to grain yield, whereas at a high seeding rate, tiller numbers rise high at the maximum tillering stage, but approximately half of them die before maturity (CRPW 2018).

Tillers produced on the main stem are primary tillers, which in turn produce secondary tillers, and the secondary tillers generate tertiary tillers in the same fashion. Plant density, light, nutrient supply, water depth, culture practices, and temperature affect tillering. A tiller immediately after emergence from the leaf sheath enclosure remains connected by vascular bundles to the stem at the point of origin for the supply of nutrients and assimilates until it develops root structure

Table 8.1 Growth stages of rice plants at different maturity durations (Adapted from Nawaz and Farooq 2017)

Growth duration			
120 days after sowing	150 days after sowing	Growth stages	Description
4–6	4–6	Germination	Coleoptile and radicle development
5–7	5–7	Seedling emergence	Shoot comes out from the soil surface and seedlings are visible
14–22	14–22	Leaf 2nd, 3rd, 4th	Development and initiation of early leaf structure
19–24	19–24	5th leaf and initiation of the first tiller	The first tiller visible
43–46	Variable time may be beyond 45 days	Active tillering	The specific stage is when the rate of tiller formation is at its peak
60–66	90–96	Panicle initiation stage	Panicles start to emerge from the stem tip, plant enters the vegetative to reproductive stage
79–85	110–117	Booting stage	During this stage panicle is inside the swollen uppermost leaf sheath
90–95	120–127	Flowering stage	During this stage, flowers are visible on the panicle
120	150	Physiological maturity	Ripe grain is ready to harvest

Table 8.2 Tiller number hill^{-1} with growth. Ear-bearing tiller number hill^{-1} was noted at maturity

	July 10		August 12		September 9		Flowering		Maturity	
Name	DS	TP	DS	TP	DS	TP	DS	TP	DS	TP
Tulasi	9.7a	4.0a	6.6a	6.7a	5.8a	4.8a	5.4ab	4.3a	5.0a	4.1ab
FR13A	7.9b	4.0a	6.2a	6.0ab	4.8b	4.7a	4.6b	4.5a	4.3bb	4.5a
T1471	10.0a	4.5a	7.2a	6.3a	6.8a	4.9a	5.9a	4.4a	4.9ab	4.2ab
Sabita	10.0a	3.8a	7.3a	6.8a	6.5a	4.8a	5.4ab	4.2a	4.9ab	3.9ab
Kolasali	9.7a	3.9a	7.1a	5.1b	6.5a	4.1a	5.6a	3.9a	4.8ab	3.8b
CH19	9.6a	4.1a	6.6a	6.3a	4.9b	4.4a	4.7b	4.3a	4.4b	4.2ab
Mean	9.5	4.0	6.8	6.2	5.9	4.6	5.3	4.3	4.7	4.1

DS, TP: Direct seeding and transplanting respectively. In a column, means followed by a common letter are not significantly different at the 5% level by the Duncan Multiple Range Test. (Adapted from Das et al. 2002)

and gains independence. Not all the tillers generated on plants survive to bear panicles and become productive. In particular, tillers that initiate later have short life spans and decimate prematurely; however, they recycle the essential nutrients embedded in their cellular structure to healthy tillers before death. Eventually, the growth and survival of individual tillers of a plant are not identical, and unravelling the physiological mechanisms regulating heterogeneity is expedient for improving rice yield.

8.3 Tillering Concerning Plant Growth

Before analyzing the physiological mechanism regulating tiller formation and survival in rice, it is imperative to understand the relationship between tillering and plant growth (Hanada et al. 1993). In the early stage of seedling growth, the tiller number increases with the increase in leaf number. The temporal increase in the number of tillers follows an S-shaped pattern, revealing that the increase in the number of tillers, which is very slow, initially accelerates with time. The increment is small at the early stage because of the initiation of primary tillers on the main stem; tillering becomes faster later as the primary tillers give rise to secondary and tertiary tillers. The pattern of change in tiller number over time is mostly similar to that of whole-plant growth. However, unlike dry mass, growth continues to increase until the maturity stage, and the tiller number declines after a certain stage because of the decimation of unproductive tillers. Primary tillers, which arise from basal stem nodes early in the vegetative period, are productive and synchronized in terms of growth and development, but tillers produced later are nonproductive. The growth stage during which tillering is vigorous is called the most active tillering stage (Fig. 8.1). The stage when the tiller number reaches a maximum and tends to decrease is called the maximum tiller number stage. A tiller bearing a panicle at the stem tip is productive but devoid of this trait, it is unproductive. The event at which the number of tillers equals the number of productive tillers is called the time of the last productive tiller; tillers emerging before this time are productive, and those emerging later are nonproductive.

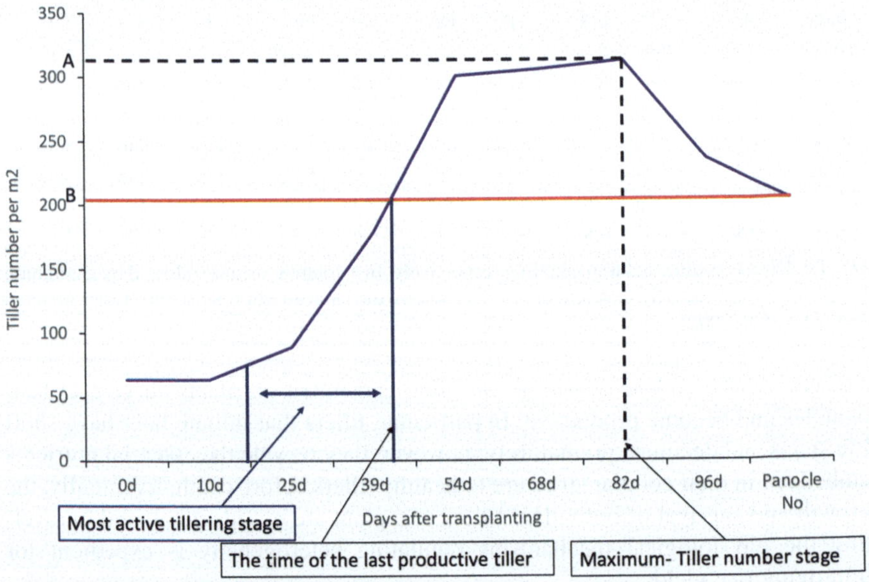

Fig. 8.1 Changes in the number of tillers during the growth of rice plants. (A) Maximum number of tillers, (B) Number of productive tillers. (B/A) Percentage of productive tillers (Sarkar 1995)

8.4 Relationships Between Maximum Tiller Number and Panicle Differentiation Stage

In many rice cultivars, the maximum tiller number coincides with the panicle differentiation stage. Coincidence logically surmises an assumption for competition between these two types of organs for assimilates available within the same source, i.e. the stem. Young tillers that cannot compete successfully tend to grow on their resources, and die prematurely when panicle differentiation occurs at the stem tip. A deficiency of mineral nutrients in the soil and a shortage of photosynthetic assimilates caused by mutual competition and/or shading inhibit increases in tiller number and productivity. Hence, these tillers remain entirely dependent on the main stem for resources, and unsuccessful competition for nutrients and photosynthetic assimilates curtails their growth. Mutual competition is accentuated by an increase in plant density, which reduces tiller survival. The coincidence between the maximum tiller number and panicle differentiation stage is often associated with variation in the maturity duration of the cultivar. In early-duration cultivars, panicle differentiation occurs before the maximum tiller number stage, whereas this stage is delayed to posterity in late-duration cultivars. In contrast, both events occur simultaneously in the medium-duration cultivars.

8.5 Distinction Between Productive and Non-productive Tillers

According to Matsushima (1957), for a tiller that has two or fewer green leaves, a tiller length ratio (the ratio of the length of the tiller concerned to the length of the longest tiller) less than 2/3 or a tiller in which the increase in the number of leaves significantly decreases at the maximum tiller number stage is considered non-productive for the sake of accurate forecasting and enumerating the number of panicles. It is suggested that a tiller must accumulate a certain amount of biomass to become productive, and this margin is varietal-specific (Kariali et al. 2008).

8.6 Varietal Distinction in Tiller Development

Tiller growth is partly based on the characteristics of the genotypes regarding yield-related attributes. The yield components are broadly divided into two types, i.e. the number of panicles per area and the weight of a panicle. The number of panicles plays a more important role in varieties bearing a large number of panicles, whereas the weight of panicles is more desirable for yield in rice cultivars bearing a small number of panicles. The cultivars belonging to the first group are called the panicle number type, while those in the second group are called the panicle weight type.

8.7 Physiology of Rice Tillering

The ability of rice plants to produce more than one stalk (tiller) has been recognized by scientists since the beginning of rice cultivation. This ability is of great agronomic importance in enabling plants to take advantage of surplus space, soil water, and mineral fertilizers available in the growth media. Subject to favourable growth conditions, the main stem of rice produces branches. The *MONOCULM1* (*MOC1*) gene controls the process of tillering. The *MOC1* gene encodes the GRAS nuclear family of proteins that are expressed in leaf axils on the stem nodes of rice, thus initiating the formation of axillary buds and promoting their outgrowth (Li et al. 2003). Therefore, tillering in rice occurs in a two-stage process: first, an axillary bud initiates at each leaf axil of the stem, followed by subsequent growth. According to Hanada et al. (1993), the differentiation of tiller buds in leaf axils is rarely influenced by environmental conditions, while tiller growth is susceptible to an environment that cues intrinsic physiological changes during development. This proposition is further corroborated by Ongaro and Leyser (2008); the authors suggested that the process of activating a dormant branch and its transition into a growing branch is complex and fine-tuned through interaction between environmental signals and endogenous factors. Such physiological changes occurring in the nascent tiller bud throughout development could be mediated by either hormonal dynamics or nutritional factors (Xu et al. 2015). Although both factors are interactive, hormones are likely the primary regulators of tiller bud outgrowth, and the role of nutrients is supplementary.

8.8 Phytohormones and Shoot Branching

The plant hormones known to have major effects on rice tillering are auxin, cytokinin, and strigolactone, although some influences of ABA and ethylene have been reported. According to Domagalska and Leyser (2011), auxin, cytokinin, and strigolactone are central to the control of tiller bud activation. These three hormones move throughout the plant, forming a network of systemic signals, and interactions between them determine the final plant architecture and tiller number. Several influences of GA, ABA, and ethylene on the tiller production of rice have also been reported. The interplay of these hormones determines the tiller outgrowth of a plant. While auxin in the nodes reduces tiller bud outgrowth, cytokinin movement to the buds promotes growth. GA3 and NAA inhibit tiller bud outgrowth by exerting their effects through the control of auxin and cytokinin contents in the tissue (Liu et al. 2011a). In contrast to auxin and cytokinin contents, panicle removal does not influence ABA contents in the stem node or tiller buds. Furthermore, Arite et al. (2007) showed that *D10* is a rice orthologue of *MAX4/RMS1/DAD1* and encodes carotenoid cleavage dioxygenase 8 (CCD8), which is involved in the synthesis of an inhibitor of shoot branching. CCD8 is involved in the synthesis of strigolactones, and RMS is involved in the inhibition of bud release by strigolactone. Furthermore, these authors showed that exogenous auxin application induces the accumulation of

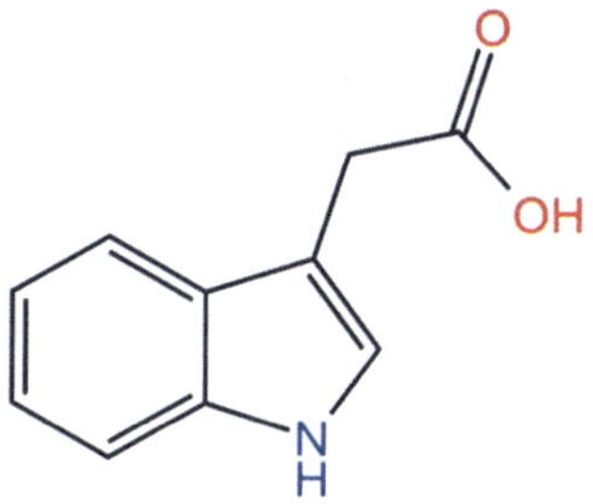

Fig. 8.2 Chemical structure of Indole-3-acetic acid

D10 mRNA. These lines of evidence are significant for supporting the interaction between auxin and strigolactone in rice tiller production. The chemical structure of auxin like IAA is depicted in Fig. 8.2.

8.8.1 Indole-3-Acetic Acid (IAA or 3-IAA)

Auxin was the first plant hormone discovered in the plant kingdom. The hormone is prominently involved in the mechanism of cell expansion and apical dominance. Several auxins have been discovered in higher plants, but indole-3-acetic acid (IAA) is by far the most abundant and physiologically relevant. It is the most common naturally occurring plant hormone of the auxin class. The shoot apical meristem, young leaves, and developing fruits and seeds are the primary sites for IAA biosynthesis. IAA is structurally related to the amino acid tryptophan and is believed to be the probable precursor of its biosynthesis. However, plants can synthesize the hormone via multiple pathways. Nevertheless, among these products, indole-3-pyruvic acid (IPA) is the most common tryptophan-dependent pathway product leading to IAA synthesis. Here, tryptophan is deaminated to form IPA, which is decarboxylated to indole-3-acetaldehyde. In the next step, indole-3-acetaldehyde is oxidized to IAA by a specific dehydrogenase (Taiz and Zeiger 2006). IAA is a colourless solid, soluble in organic solvents. The hormones promote growth in stems and coleoptiles but it is inhibitory for root growth. IAA synthesized in the shoot apex is transported down the stem and it actively promotes growth by cell elongation in the tissue below the apex.

8.8.2 Auxin and Shoot Branching

The action of auxin in shoot branching is modulated by the interplay of two secondary hormones, cytokinin and strigolactones. These three hormones move throughout the plant, forming a network of systemic signals, and integration of the signal controls shoot branching (Domagalska and Leyser 2011). Auxin synthesized in the shoot apex is actively translocated basipetally downwards in the stem vasculature, preferably in the xylem parenchyma, and inhibits axillary bud outgrowth. Axillary bud outgrowth is activated when the apical bud, the primary source of auxin synthesis, is removed. The role of auxin in axillary bud inhibition is indirect. The hormone does not enter the bud to regulate its growth. The indirect action of auxin in regulating bud outgrowth has been explained in two hypothetical models (Domagalska and Leyser 2011), namely, the auxin transport canalization-based model and the second messenger model, which provide mechanistic explanations for apical dominance. The models explain the participation of strigolactone and cytokinin hormones.

8.8.3 *RICE LAX PANICLE1* Functions in Rice Tillering

In addition to hormones and nutrients, another factor involved in rice tillering is the function of the *RICE LAX PANICLE1* gene. It has been shown that the function of *RICE LAX PANICLE1* (*LAX1*) is needed for the generation of all types of axillary meristems in both the vegetative and reproductive phases of rice plant growth (Oikawa and Kyozuka 2009). The authors showed a close relationship between axillary meristem initiation and leaf development on the shoot apex. The plastochron4 (P4) stage of leaf development coincides with axillary meristem formation because *LAX1* expression begins in the leaf axil at this stage. The LAX1 protein is formed at the boundary between the subtending leaf and future axillary meristem, and it moves towards the growing meristem to establish a new axillary meristem. LAX1 protein accumulation in the axillary meristem is strictly regulated by mRNA expression and protein trafficking. The protein controls the proliferation of meristematic cells, which is crucial for the complete development of axillary buds.

8.8.4 Auxin Transport Canalization-Based Model

A trove of evidence gathered since the last several decades of research on the role of auxin in apical dominance has led to the proposition of two models (previously stated), namely, auxin transport canalization and second messenger. The first model proposes that effective bud outgrowth is based on auxin efflux from the cell. Auxin synthesized in young expanding leaves and shoot apices moves rootwards in the stem through the polar auxin transport stream (PATS), and the direction of movement is determined by the polar localization of plasma membrane anion efflux carrier proteins (PINs). In the stem, efficient rootwards polar auxin flow requires the presence of PIN1 at the basal end of xylem parenchyma cells (Fig. 8.3). The presence of auxin in PAT inhibits axillary bud outgrowth. Biochemical inhibition of PAT or removal of the main stem apex reduces auxin flow in the stem to promote axillary bud outgrowth. Hence, downwards auxin transport in the stem from the apex can inhibit bud outgrowth by compromising the ability of axillary buds to establish their own polar auxin transport stream to export auxin into the main stem (Li and Bangerth 1999). Auxin export from a branch that develops earlier inhibits auxin transport from a bud that develops later; auxin saturation in the former inhibits auxin transport from the latter, thereby inhibiting its outgrowth (Brewer et al. 2009). Thus,

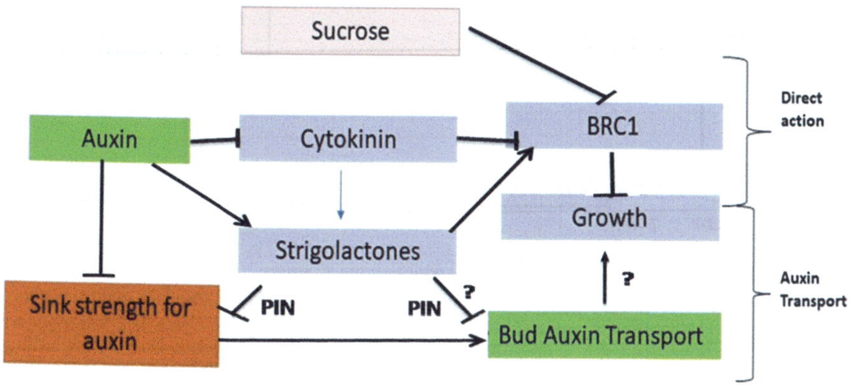

Fig. 8.3 Relationships between key branching signals and bud outgrowth. The model support the direct action or auxin transport/canalization hypothesis of bud outgrowth. Arrows represent promotion, while flat-ended lines represent inhibition. (Adapted from Brewer et al. 2015)

axillary buds are also sources of auxin, and the main stem acts as an auxin sink. Auxin transport canalization can connect the bud to the stem by removing auxin from the bud apex.

However, an abundance of auxin in the main stem can obstruct the canalization of auxin transport out of the bud by reducing the strength of the stem sink for auxin and limiting auxin efflux from the bud, thereby preventing positive feedback during the canalization process. Because the transport of auxin through the axillary bud is essential for growth, the effect of auxin must be indirect, bereft of any role for secondary hormones. Hence, both the axillary bud and main stem shoot apex compete for a common auxin transport pathway downstream to the root (Shinohara et al. 2013). It was observed that bud outgrowth increases in some genetic mutants owing to the loss of a long-distance mobile signal termed SMS (Beveridge 2006), which interferes with auxin transport. It is conceptualized that auxin from apical and lateral sources competes for primary channels of auxin transport in the stem to regulate axillary bud outgrowth. While auxin export from buds could be critical for ongoing bud growth, it is not the initial trigger (Brewer et al. 2009). In more recent work, Brewer et al. (2015) proposed a model demonstrating relationships among key branching signals and bud outgrowth to support the direct action or auxin transport/canalization hypothesis (Fig. 8.3).

According to the direct action or auxin transport/canalization hypotheses of bud outgrowth, auxin moving basipetal in the stem inhibits cytokinin levels and promotes the expression of genes biosynthesizing strigolactone. Strigolactones, cytokinins, and sucrose can move acropetally into buds and regulate branching. The pathway of action involves the regulation of *BRANCHED 1* (*BRC1*), which encodes a transcription factor necessary for branching inhibition. Reduced auxin transport/canalization from axillary buds suppresses their growth. Strigolactones can reduce auxin transport in the stem and enhance PIN cellular internalization. *N*-1-Naphthylphthalamic acid (NPA) inhibits the function of PIN and ABCB auxin transport proteins (Brewer et al. 2015).

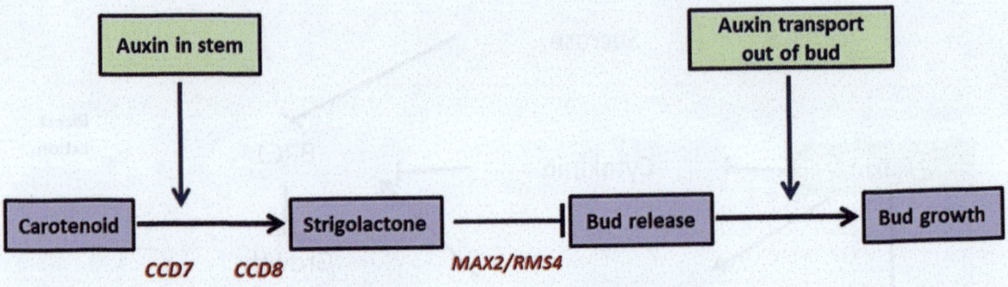

Fig. 8.4 Pathway through which auxin and strigolactone regulate axillary bud growth. Arrows represent promotion, while flat-ended lines represent inhibition. Auxin promotes the expression of the strigolactone biosynthesis genes *CCD7* (*CAROTENOID CLEAVAGE DIOXYGENASE 7*) and *CCD8*. The MAX2/RMS4 F-box protein is needed for strigolactone-mediated inhibition of bud release. Auxin transport out of an axillary bud, which can be inhibited by NPA, is then needed for a bud to be released to proceed to sustained bud growth. (Adapted from Brewer et al. 2009)

8.8.5 Strigolactone-Mediated Auxin Action

The auxin transport canalization model explains the association of high branching with high PIN1, auxin transport, and auxin levels in loss-of-function strigolactone mutants. It is postulated that strigolactones reduce the accumulation of PIN1 on the plasma membrane, thereby increasing the difficulty of transporting auxin to the axillary bud to achieve high efflux of auxin to drive bud release. The interaction between strigolactone and auxin is shown in Fig. 8.4.

In addition, the regulatory effect of the PIN1 protein on bud activity by strigolactone is illustrated in Fig. 8.5.

8.8.6 Second Messenger Hypothesis and the Role of Phytohormones

The second messenger hypothesis implicates auxin regulation by the production of a second messenger, which can enter the bud and directly regulate its activity (Sachs and Thimann 1967; Bangerth 1994). Apical dominance is a prominent control mechanism in the development of branching patterns and determines plant architecture. The physiological mechanisms controlling shoot branching and apical bud dominance via cognitive interactions between the three hormones auxin, cytokinin, and strigolactone were primarily illustrated by contemplating results obtained in investigations conducted by plant physiologists using mostly dicot plants, as has been discussed previously. Very few investigations have been conducted on monocots to elucidate the hormonal control of apical dominance. Unlike those of dicots, the axillary buds of monocot plants remain embedded within the precinct of the leaf sheath until emergence, which obstructs calibration for growth. However, because of the immense contribution of tillers to the grain yield of cereals, the physiology of apical dominance has been extended to monocots, and rice has been used as a model system. Like in dicots, the removal of the shoot apex (panicle) stimulates the growth of dormant tiller buds located at the elongated upper internode (Liu et al. 2011a). In

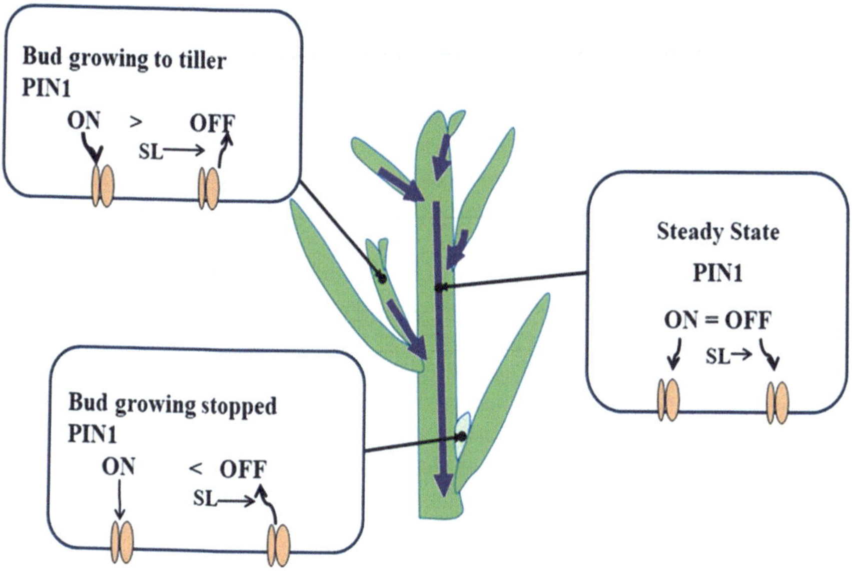

Fig. 8.5 Schematic diagram of the mechanism by which the PIN1 protein is regulated by strigolactone and its effect on axillary bud activity. Strigolactone (SL) signalling via *MORE AXILLARY GROWTH 2* (*MAX 2*) depletes PIN1 from the plasma membrane of cells in the shoot. The hormone systemically depletes PIN1 throughout the shoot. In the main stem, PIN1 on the plasma membrane is at a steady state. In an activating bud, canalization is underway, with rapid PIN1 insertion outstripping PIN1 removal. In an inhibited bud, PIN1 insertion is slower than PIN1 depletion, resulting in no PIN1 accumulation on the plasma membrane. Systemically higher strigolactone levels reduce the number of active buds and steady-state levels of PIN1 on the plasma membrane. In contrast, lower strigolactone concentrations exhibit the opposite effect. (Adapted from Shinohara et al. 2013)

intact plants, auxin synthesized in the shoot apex moves basipetal and inhibits axillary bud release. The removal of the apical bud for the inhibition of polar auxin transport leads to bud outgrowth, while the exogenous application of auxin reverses this inhibition. Apical bud removal promotes auxin transport out of the bud through the low auxin concentration in the main stem by enhancing the strength of the auxin sink, resulting in the outgrowth of axillary buds. Downstream messengers, also called secondary messengers, such as strigolactones and cytokinins, relay auxin signal transduction during axillary bud release by directly entering the bud.

8.8.7 Auxin Downregulates Cytokinin Biosynthesis

In apical dominance, the shoot apex grows predominantly and represses axillary bud growth. Auxin synthesized in the apical bud moves downwards in the stem in a basipetal direction, where the pattern of polar flow is determined by the expression of PIN1 proteins on the basal end of the cells. Auxin downregulates cytokinin biosynthesis in stem nodes to the detriment of axillary bud growth, while cytokinin promotes growth. Cytokinin is the only hormone that triggers axillary bud release.

The hormone promotes auxin production and basipetal auxin transport out of axillary buds, eventually inhibiting the production of cytokinin in the stem and limiting its availability for other buds (Bangerth et al. 2000; Tanaka et al. 2006). According to Shimizu-Sato et al. (2009), auxin derived from the main shoot apex inhibits local biosynthesis of cytokinin in the stem node through regulation of *IPT* gene expression. It is also possible that auxin reduces cytokinin levels in the stem by inducing overexpression of the *cytokinin oxidase* (*CKX*) gene, which synthesizes enzymes for cytokinin breakdown (Fig. 8.6).

Exogenous application of cytokinin to plants enhances axillary bud release irrespective of the presence or absence of apical buds. The inhibitory effect of auxin on cytokinin biosynthesis has been proposed to be controlled by the downregulation of the gene *adenosine phosphate-isopentenyl transferase* (*IPT*) (Tanaka et al. 2006), which encodes a key enzyme in the cytokinin biosynthesis pathway. In an intact plant, the transcript level of this gene was not detectable in the nodal stem, but the level increased significantly following decapitation of the apical bud. The significance of this study is that cytokinin accumulation in the stem nodes of decapitated plants results in the promotion of axillary bud release via stimulation of local biosynthesis and is not sourced from roots. Under these observations, the transcriptome dynamics underlying the interactions among strigolactones, auxin, and cytokinin controlling shoot branching in rice were further evaluated by Zha et al. (2019). In this study, axillary bud outgrowth at the second node from the panicle at the full heading stage was induced by panicle removal, but the exogenous application of strigolactone inhibited this process. These responses are induced by several genes associated with hormone signal transduction. Strigolactone application minimized the expression of auxin early response genes induced by the excision of panicles independent of ARF. Panicle removal induced cytokinin signal transduction, but *type-A ARRs* were the only cytokinin signal-associated genes responsive to strigolactone treatment.

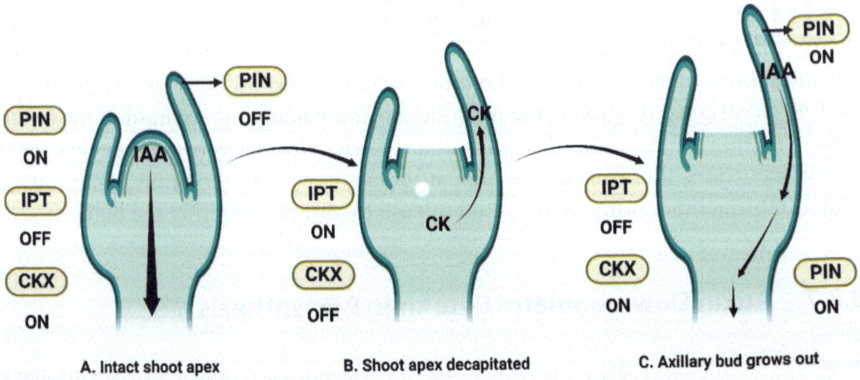

Fig. 8.6 A model of the interactions between auxin and CK (cytokinin) in shoot branching through the functions of the *IPT* (*isopentenyl transferase*), *CKX* (*cytokinin oxidase/dehydrogenase*), and *PIN1* (*pin formed 1*) genes shown in intact shoot apex (**A**), shoot apex decapitated (**B**) and when axillary bud grows out (**C**). (Adapted from Shimizu-Sato et al. 2009)

8.8.8 Antagonism Between Strigolactone and Cytokinin

Cytokinin and strigolactones play specific roles in plant development by regulating separate genetic processes. These genes function independently in adventitious rooting and additively in lateral root formation but antagonistically regulate dormant axillary bud release (Dun et al. 2012; Hu et al. 2014; Faizan et al. 2020). Both cytokinin and strigolactone act directly on the axillary bud, one in contrast to the other. An increase in the endogenous strigolactone concentration inhibits rice tiller growth by reducing the cytokinin content of rice tiller buds and vice versa. An exogenous supply of strigolactones inhibited cytokinin-promoted axillary bud growth at node two at the full heading stage in rice plants (Zha et al. 2022, Fig. 8.7).

The contrary effects of strigolactone and cytokinin were explored further by the exogenous application of synthetic strigolactone and benzyl adenine (BA) in rice (Zha et al. 2022). The type-A RR (Response regulator) genes are defined as the cytokinin-mediated genes that are needed for axillary bud release, and the transcript levels of ten *OsRR* genes increased upon treatment with BA and decreased to different levels following application of the synthetic strigolactone GR24 (Fig. 8.8).

8.8.9 Interaction Between Strigolactones and ABA

ABA and strigolactones both participate in shoot branching. Both hormones are derived from a common source, i.e. from the biosynthesis of β–carotene, and inhibit plant growth and development in challenging environments (Cheng et al. 2013, Fig. 8.9). Although strigolactone extensively inhibits shoot branching in rice, the effect of ABA on the tillering process is not significant. It was reported that introgression of the *HDT1* allele into the rice cultivar *Huazhang* greatly improved the tiller number compared to that in the *japonica* cultivar *Nekken 2*, but there was no

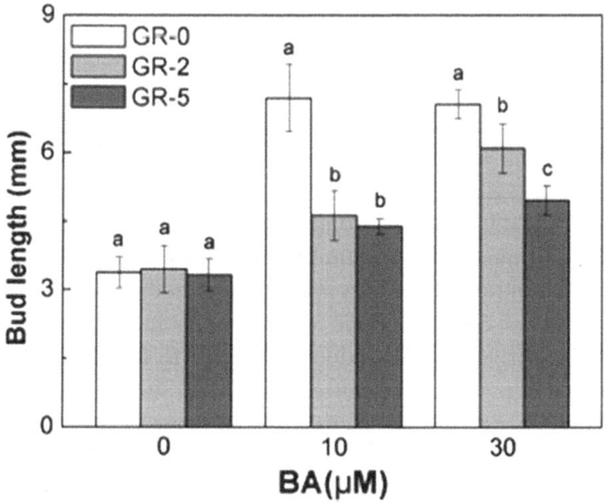

Fig. 8.7 Strigolactone reduced the stimulatory effects of benzyl adenine (BA)-induced bud elongation in rice. Synthetic strigolactone (GR) at three levels and BA were supplied to axillary rice buds on node 2 at the full heading stage. The decrease in bud elongation was enhanced with increasing strigolactone concentration. (Adapted from Zha et al. 2022 with permission)

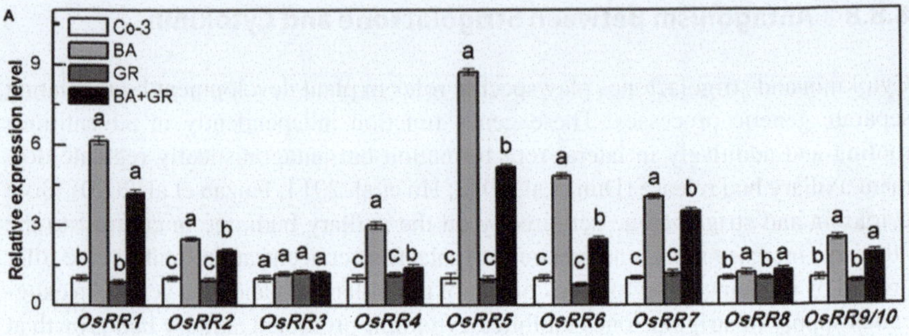

Fig. 8.8 Effects of strigolactone and/or BA treatment on cytokinin-type-A RR gene (*OsRR*) expression in axillary buds at node 2 in the full heading stage of rice plants. (Adapted from Zha et al. 2022 with permission)

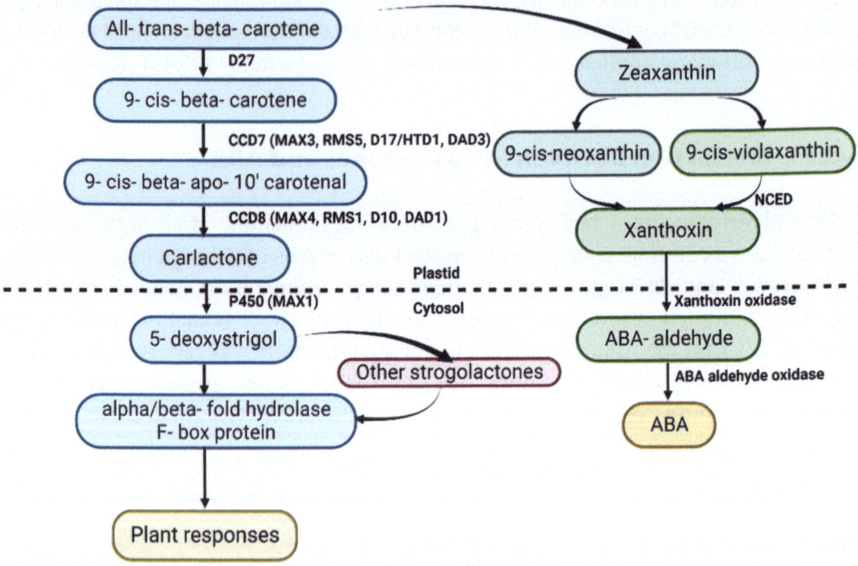

Fig. 8.9 Strigolactone and ABA biosynthetic pathways share a common origin for β-carotene. (Adapted from Cheng et al. 2013)

difference in the ABA content of the plants (Wang et al. 2020). Similarly, the excision of panicles in rice stimulates axillary tiller bud outgrowth by altering the auxin/cytokinin ratio, but there is no change in the ABA content in the stem node or bud. However, there is some evidence of interactions between strigolactones and ABA. Strigolactones induce tolerance to drought and salinity by activating ABA signalling. The adverse effect of salinity on rice plants is alleviated by pretreatment with exogenous strigolactone GR24 (Ling et al. 2020). Another study showed that the strigolactone biosynthesis-deficient mutants *D10* and *D17* and the strigolactone signalling-deficient mutant *D13* of rice possessed greater ABA contents than did the corresponding wild-type plants, resulting in induced drought tolerance, and the low ABA-producing line *D27* was susceptible to stress (Haider et al. 2018).

In the signal transduction pathway, ABA acts downstream of strigolactone and the TCP transcription factor BRANCHED1 (BRC1), which suppresses bud activity in *Arabidopsis*. *OsTB1/PC1* is the orthologue of *BRC1* in rice (Wu et al. 2022). The interaction between strigolactones and ABA was further elucidated by the discovery of the high-tillering rice mutant *tillering 20* (*tc20*) (Liu et al. 2020). The dominant allele *T20* encodes a chloroplast ζ-carotene isomerase (Z-ISO), which is involved in the biosynthesis of carotenoids and their metabolites strigolactones and ABA. The *t20* mutant has a low strigolactone and ABA pool, suggesting that these two genes are involved in rice tillering. Strigolactone application stimulated *T20* expression, increased β-carotene biosynthesis, and promoted the expression of *Oryza sativa 9-CIS-EPOXYCAROTENOID DIOXYEGASE1* (*OsNCED1*) through the induction of *Oryza sativa HOMEOBOX12* (*OsHOX12*) to increase ABA biosynthesis in the shoot bases of rice. In contrast, ABA application depresses strigolactone biosynthesis, and the content of strigolactone increases significantly in ABA biosynthesis mutants. Taken together, these findings suggest that ABA inhibits tiller bud growth and suppresses the formation of unproductive tillers on the upper nodes of rice plants.

8.8.10 Interaction Between Strigolactones and Ethylene

In plants, strigolactones are synthesized in roots and transported to the shoot. In roots, they determine root growth, while in shoots, they modulate plant architecture by regulating axillary bud outgrowth. In addition to having an independent role, strigolactones cross-regulate with phytohormones such as auxin, cytokinin, and ethylene for root development (Sun et al. 2022); however, comprehensive information on the mechanism underlying the cross-talk of strigolactones with ethylene is lacking. Strigolactones interact with ethylene signalling to control root hair elongation (Wu et al. 2022), but there are no reports on the associations between shoot branching and axillary bud outgrowth. It has been shown that a synthetic analogue of strigolactones GR-24 promotes the upregulation of the ethylene biosynthesis genes *ACC synthase* and *ACC oxidase,* and the accumulation of ethylene promotes root hair elongation in *Arabidopsis* plants similar to that of auxin. However, the GR24 application is ineffective in the ethylene signalling mutants *ein2-1* and *etr1-1*. It is implied that strigolactone signalling is not necessary for the ethylene response but that the effects of strigolactones on root hair elongation require ethylene biosynthesis. The application of strigolactones stimulates the expression of ethylene biosynthesis genes, and ethylene forms a cross-talk junction between the strigolactone and auxin pathways (Kapulnik et al. 2011).

8.8.11 Auxin Signal Transduction Pathway

The auxin signal transduction pathway extends from the receptor-binding site on the membrane to physiological responses through auxin-regulated gene expression. Auxin influx occurs in the cell through two processes. In the first mechanism, auxin enters the cell by passive diffusion of the protonated form IAAH across the phospholipid bilayer of the membrane. When protonated, the carboxyl group of IAA becomes lipophilic and easily permeable to cross the membrane barrier. The passive permeability depends on the apoplastic pH. The dissociated form of IAA is negatively charged and therefore needs the help of carriers to cross the membrane. Because plasma membrane H^+-ATPase maintains the cell wall pH at approximately 5, half of the IAA in the apoplast remains undissociated and crosses the membrane by passive diffusion in the apical end of the cell. However, the cytoplasm inside the cell has a neutral pH, where an anionic form of IAA^- predominates and these anions exit through anion efflux carriers (PIN1 proteins), which are abundant at the basal end of the cell during polar transport of auxin (Fig. 8.10). The PIN carriers are assisted by AUX/LAX auxin influx carriers, which are also located at the basal end.

The change in pH, as envisaged in the model, is regulated by the activity of the plasma membrane-located H^+-ATPase. The activity of the enzyme is also necessary for cell expansion. Auxin entering the cell induces proton extrusion, which acidifies the cell wall and increases cell expansion. The source of H^+ is the plasma membrane H^+-ATPase. The activity of H^+-ATPase increases in response to auxin. The auxin-induced proton extrusion rate involves both activation and synthesis through two mechanisms: (i) activation of preexisting H^+-ATPase and (ii) synthesis of new H^+-ATPase. (Adopted from Taiz and Zeiger 2006, Fig. 8.11).

8.8.12 Auxin-Induced Gene Regulation

IAA serves as a signalling molecule necessary for gene regulation, inducing cell expansion and cell division. IAA begins when a hormone binds to its receptor, activating a select group of transcription factors. The activated transcription factors enter the nucleus and promote the expression of IAA-inducible genes. Genes whose activity is stimulated by the activation of pre-existing transcription factors are referred to as primary response genes or early response genes. Early response genes encode transcription factors that stimulate secondary response genes or late response genes. They are also involved in cell-to-cell communication and stress resistance.

The expression of early auxin genes is regulated by the auxin response factor (ARF). The auxin-responsive genes possess a conserved early response element (Aux-RE) in the promoter region, and ARF in combination with Aux-RE stimulates gene expression. It is also proposed that proteins encoded by the *AUX/IAA* gene family can inhibit the transcription of early response genes by forming inactive heterodimers with ARF. The inactive heterodimers block the expression of the ARF-Aux-RE combination-mediated gene downregulation. Thus, AUX/IAA proteins work as ARF inhibitors. A schematic model of auxin regulation of the

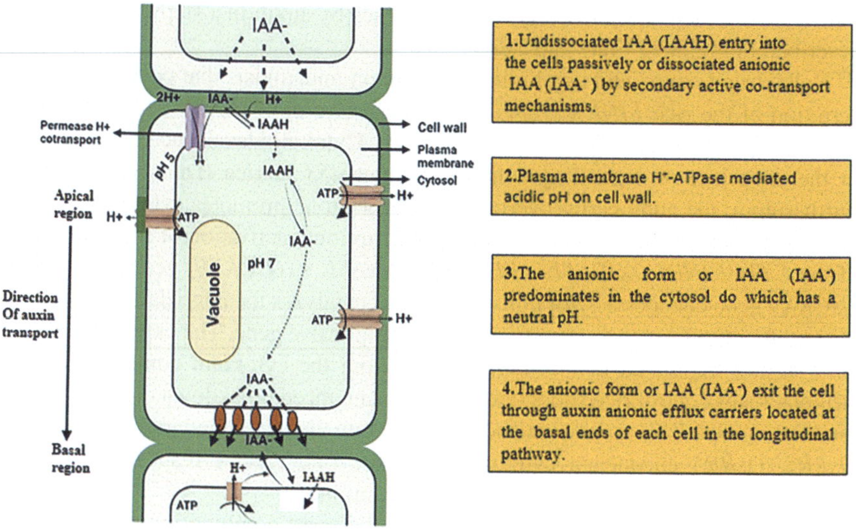

Fig. 8.10 Basipetal transport of auxin in stem cells. Auxin is transported by influx and efflux carriers in the plasma membrane. The polar transport of auxin occurs by either passive diffusion of protonated IAAH or secondary active transport of auxin using the IAA influx carrier proteins of the AUX/LAX family. IAA enters the cytoplasm and dissociates to form IAA^{-1} because of the neutral pH of 7.0, and IAA is not permeable through the plasma membrane. The anion efflux carriers PIN1 and ABC family proteins are needed for escape. (Adapted from Taiz and Zeiger 2006)

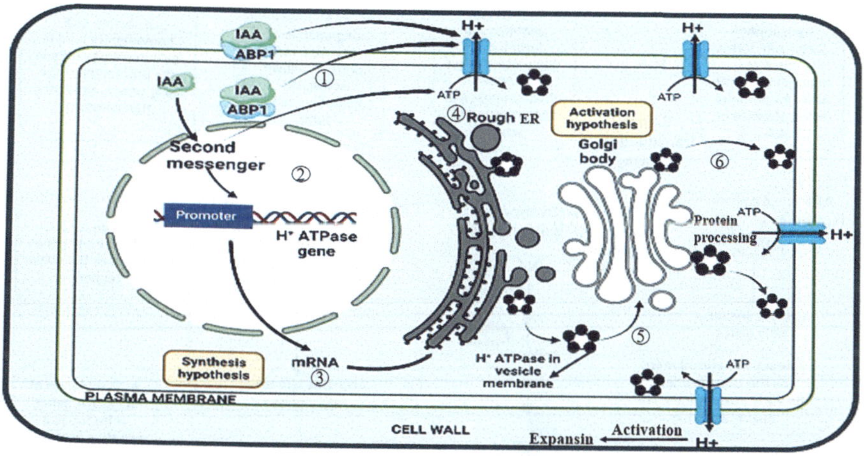

Fig. 8.11 Activation hypothesis: Auxin binds to an auxin-binding protein (ABP1) located either on the cell surface or in the cytosol. ABP1-IAA then interacts directly with plasma membrane H^+-ATPase to stimulate proton pumping (step 1). Second messengers, such as calcium or intracellular pH, could also be involved. Synthesis hypothesis: IAA-induced second messengers activate the expression of genes (step 2) that encode the plasma membrane H^+-ATPase (step 3). The protein is synthesized on the rough endoplasmic reticulum (step 4) and targeted via the secretory pathway to the plasma membrane (steps 5 and 6). An increase in proton extrusion results from an increase in the number of proton pumps on the membrane (Adapted from Taiz and Zeiger 2006)

transcriptional activation of early response genes by auxin in a stepwise fashion is described in Fig. 8.12 (Taiz and Zeiger 2006).

Both strigolactone and cytokinin are common antagonists that control the transcription of the rice *FINE CULM1* (*FC1*) gene (Xu et al. 2015). *FC1* expression modulates the bud activation potential of plants. Cytokinin levels increase markedly in the strigolactone signalling-deficient mutant *d53* in rice (Duan et al. 2019). Furthermore, the authors discovered that strigolactone promotes cytokinin catabolism to alter shoot architecture through transcriptional activation of the expression of the *CYTOKININ OXIDASE/DEHYDROGENASE 9* (*OsCKX9*) gene in rice. The *OsCKX9* gene encodes a cytokinin oxidase that catalyzes the degradation of cytokinin and functions as a primary strigolactone-responsive gene. The expression of the gene is downregulated in the mutant, increasing the cytokinin content. Notably, strigolactones control rice shoot branching via enhanced breakdown of cytokinin by modulating *OsCKX9* expression and gene function via the secondary response gene *OsRR5*. *OsRR5* encodes a cytokinin-inducible rice type-A response regulator. Cytokinins may be degraded by a feedback reaction.

Zha et al. (2022) analyzed the impact of different combinations of strigolactone and cytokinin treatments, which influence tiller and axillary bud development on node 2 from the panicle base at the tillering and full heading stages, respectively. At the tillering stage, strigolactone promotes cytokinin degradation in rice buds. The genes encoding CKX target endogenous cytokinin and are

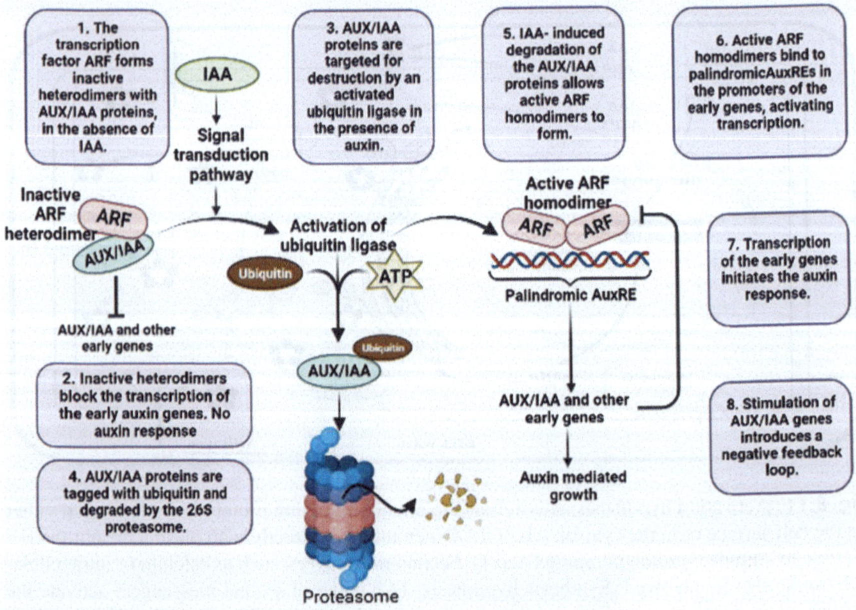

Fig. 8.12 A model representing auxin regulation of transcriptional activation of early response genes leading to auxin-mediated growth. (Adapted from Taiz and Zeiger 2006)

involved in the irreversible degradation of the hormone. Exogenously or endogenously synthesized strigolactone inhibitors affected the expression level of *OsCKX* (*cytokinin oxidase*) genes but had little impact on *OsIPT* genes. The expression of the cytokinin catabolism genes *CKX1*, *CKX2*, *CKX3*, and *CKX5* was found to be downregulated in the strigolactone signalling-deficient mutant *MAX2* compared to that in wild-type plants (Ha et al. 2014). Thus, the induction of *OsCKX* expression in rice is believed to occur via a CK-independent pathway. Furthermore, strigolactones may affect cytokinin content through their interaction with auxin. A composite model for the interaction of various phytohormones and their influence on various aspects of yield formation in rice was constructed according to Deveswar et al. (2020) (Fig. 8.13).

The nature of the proteins encoded by the QTLs is shown as follows: *PIN (PIN-FORMED)*: auxin efflux transporter; *LAX1 (LAX PANICLE 1), a* bHLH transcription factor for the generation of axillary meristem; *PAY1 (PLANT ARCHITECTURE AND YIELD 1),* a nuclear-localized peptidase that affects polar auxin transport by altering the levels of endogenous IAA; *BG1 (BIG GRAIN 1),* a plasma membrane-localized protein and a positive regulator of grain size; *NAL1 (NARROW LEAF 1),* a novel protein regulating polar auxin transport; and *ASP1 (ABERRANT SPIKELET*

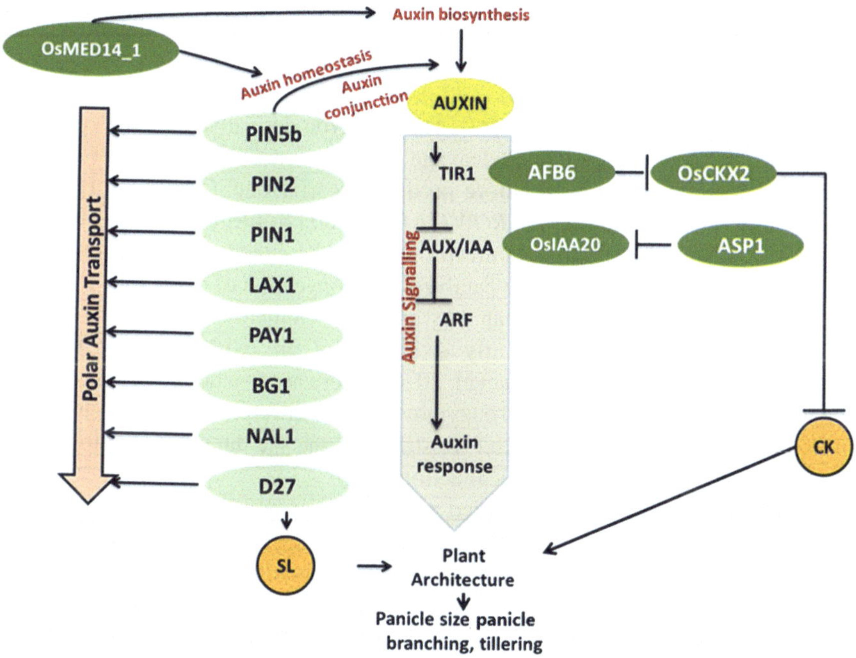

Fig. 8.13 Regulators of auxin signalling controlling plant architecture and affecting yield attributes. Factors affecting auxin biosynthesis, homeostasis, signalling, and transport were identified in rice, and the functions known to affect these traits are shown. Interconnections of auxin signalling with CK and SL also contribute to plant architecture. Arrows represent positive regulation; bar-headed lines represent negative regulation. (Adapted from Deveswar et al. 2020)

AND PANICLE, a TPL-like transcriptional corepressor that regulates panicle morphology, *D27* β-carotene isomerase, panicle size, and tillering.

8.9 Properties of Strigolactones

Strigolactones are plant hormones synthesized in roots. The structure of strigolactones is based on a tricyclic lactone (A, B, and C parts of the molecule), which is linked to a hydroxymethyl butanolide (D part of the molecule). While the D part remains common, strigolactones exhibit variation in the A, B, and C parts. Strigolactones lose activity when the C-D section is modified. Strigolactones are soluble in nonpolar solvents, such as benzene and acetone, but insoluble in hexane. These hormones are the root-to-shoot phytohormones and suppress shoot branching by inhibiting axillary bud release. Chemically strigolactones are a group of secondary metabolites, carotenoid-derived terpenoid lactones. Carotenoids (β-carotenes) are converted to strigolactone precursor carlactone by the sequential action of three biosynthetic enzymes, namely, DWARF27 (D27), HIGH TILLERING AND DWARF 1/DWARF 17 (HDT1/D17), and DWARF10 (D10) (Wang et al. 2020). The first step is the isomerization of the ninth chemical bond of the β-carotene molecule, which changes from the *trans*- to the *cis*-configuration, carried out by the enzyme β-carotene isomerase, also named DWARF27 or D27, with iron serving as a cofactor. The second step involves the chemical separation of 9-*cis* β-carotene into two compounds, namely, 9-*cis* aldehyde and β-ionone. This step is catalyzed by carotenoid cleavage dioxygenase 7 (CCD7). The aldehyde derivative of the second step is catalyzed by another carotenoid cleavage oxygenase, CCD8, which converts and rearranges it to 9-*cis* β-apo-10 before producing carlactone. Carlactone is further modified by *MORE AXILLARY GROWTH 1* (*MAX 1*) to form the active strigolactone in rice (Zhang et al. 2014). In this process, *MAX1* encodes a cytochrome P450 monooxygenase (CYP711A1) that catalyzes the conversion of carlactone to carlactonoic acid (CLA). CLA is methylated by an unknown methyltransferase to methyl carlactonate, which is subsequently oxidized by an oxidoreductase to form strigolactone-like compounds (Wu et al. 2022). There are five homologues of *MAX1* in rice, and these cytochrome P450 enzymes catalyze different steps in strigolactone biosynthesis in rice. Plant cells perceive strigolactone via the dual receptor/hydrolase protein DWARF14 for biological action.

Strigolactones (Fig. 8.14) identified so long show three distinct physiological responses: (A) prevention of excess bud growth in the stem apex and the stopping mechanism of branching; (B) promotion of germination of parasitic organisms growing in the host plant roots in some cases; and (C) recognition of plants by symbiotic fungi, especially arbuscular mycorrhizal fungi. These fungi establish a mutualistic association with the host plant by providing phosphate and other soil nutrients.

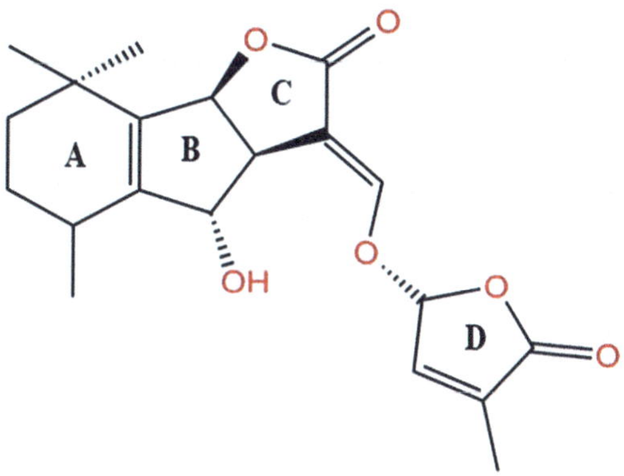

Fig. 8.14 Chemical structure of strigolactone showing the A, B, C, and D rings

8.9.1 Strigolactones and the Green Revolution in Rice

Strigolactones are a new class of phytohormones. These small signalling chemicals are carotene-derived and critical for designing plant architecture and under-adapting to challenging environments. In addition to the two previously known hormones cytokinin and auxin, this hormone has a profound influence on tiller production in rice plants. Instead of operating independently, strigolactones interact prominently with cytokinin and auxin and, to some extent, also with ABA and ethylene, forming an elaborate signalling network that controls shoot branching in many plant species and is responsible for the organization of shoot architecture. Strigolactones play systemic and local roles in directly inhibiting axillary bud release. Wang et al. (2020) demonstrated that a partial loss of function allele of the strigolactone biosynthesis gene *HIGH TILLERING AND DWARF 1/DWARF17* (*HTD1/D17*) (Arite et al. 2007), which encodes the CAROTENOID CLEAVAGE DIOXYGENASE 7 (CCD7) enzyme, increases tiller number and improves grain yield in rice. The *HTD1* allele has been extensively utilized and co-selected with the *semi-dwarf 1* (*SD1*) allele. These methods have made additive contributions to the improvement of plant phenotypes of modern rice varieties since the advent of the first Green Revolution of rice in the 1960s. The beneficial traits of both *HDT1s* have been pyramided together into modern semi-dwarf rice during the introgression process (Fig. 8.15) (Wang et al. 2020). Similarly, Ishikawa et al. (2005) reported that the strigolactone biosynthesis mutant *d10* and the strigolactone-signalling mutant *d3* increase the tiller number in rice because they interfere with strigolactone accumulation and signal transduction, respectively. The other strigolactone biosynthesis mutant, *d27*, also exhibits no adverse effects of strigolactone on plant morphology (Wu et al. 2022). Kebrom et al. (2013) also reported the presence of several mutants that control rice branching in both the shoot and panicle; these mutants are short-statured and exhibit high tillering. The genes involved in these features were *d3, d10, d14, d17 (or hdt1)*, and *d27,* and they negatively regulate tiller bud activity.

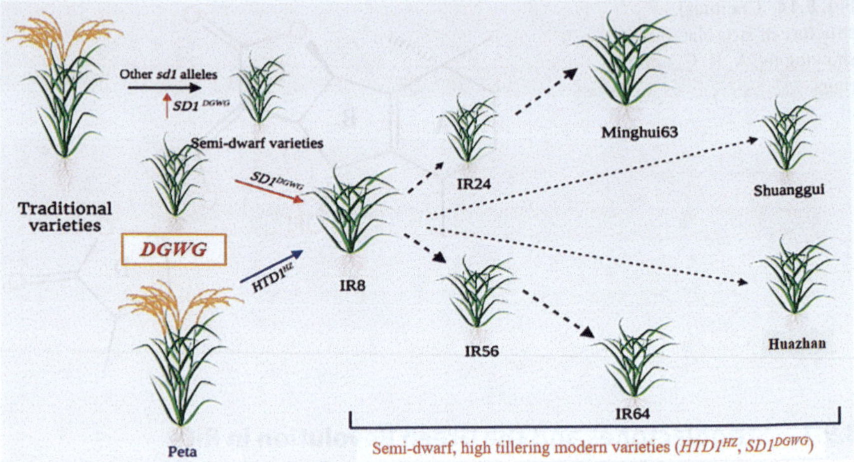

Fig. 8.15 The beneficial traits of *HTD1* and *SD1* contribute to rice grain yield. A schematic diagram showing the introgression of beneficial alleles of *SD1* and *HTD1* during the breeding of modern semi-dwarf rice. The different alleles of *SD1* have been used by breeders on the one hand to make traditional cultivars for obtaining semi-dwarf traits and high grain yield. On the other hand, the simultaneous introduction of *HTD1* from tall rice *Peta* and the *SD1* allele from short-stature rice *DGWG* (*Dee-geo-woo-gen*) into miracle rice IR8 has been performed. Therefore, pyramiding $SD1^{DGWG}$ and $HTD1^{HZ}$ have great potential for promoting the grain yield of *japonica* rice varieties. *HZ* Huazhan (*indica* rice). (Adapted from Wang et al. 2020)

8.9.2 Strigolactone Signalling

Strigolactones are a class of phytohormones responsible for the inhibition of shoot branching in rice. The tiller number in rice has been a critical target in rice breeding (Wang et al. 2018), and genetic or chemical modification of the strigolactone pathway can significantly alter the tiller number (Waters et al. 2017). It is a medium used to control shoot branching and root development. In rice, strigolactones have exhibited the potential to improve cultivar genetic makeup through enhanced stress resilience and grain yield. In principle, the signalling mechanisms involved are not different from those of other plant hormones. Signal transduction involves hormone-activated targeting of transcriptional regulators for the degradation of the α/β-fold hydrolase DWARF14 in rice. The pathway of strigolactone signalling is far from simple. Jiang et al. (2013) reported that the involvement of strigolactone signalling in the shoot branching of rice requires the hormone-dependent interaction of DWARF14 (D14). D14 was considered to be a strigolactone receptor together with DWARF 3 (D3), an F-box component of the Skp–Cullin–F-box (SCF) E3 ubiquitin ligase complex. Characterization of SL-insensitive rice mutant *d53* and cloning of *D53* elucidated that strigolactone signalling is mediated by proteasome system-induced D53 protein degradation in cells via the participation of D14 and the SCF^{D3} ubiquitin ligase for target selection. *D53* encodes a substrate of the SCF^{D3} ubiquitination complex and functions as a repressor of SL signalling. The application of synthetic analogues of strigolactones follows this pathway for the degradation of D53, but the dominant form of D53 is resistant. Furthermore, D53 can interact with transcriptional corepressors such as TOPLESS-RELATED (TPL) and TPL-related proteins (TPRs). Strigolactones work as a second message signal downstream of auxin to repress axillary bud release in rice (Domagalska and Leyser 2011). This is evident because the strigolactone biosynthesis genes *MAX3/D17* and *MAX4/D10* are positively regulated by auxin (Waters et al. 2012; Hayward et al. 2009). A model for strigolactone signal transduction through D53 action in rice was proposed by Jiang et al. (2013) as given below (Fig. 8.16). This model of SL signalling involves SL-dependent degradation of the D53 repressor mediated by the D14–D3 complex. D3 and D14 are the other two signalling components in addition to D53.

DWARF 53 (D53) is a key repressor of the strigolactone signalling pathway. In the absence of SLs, D53 is stable and may recruit TPL/TPR proteins and repress downstream responses. In the presence of SLs, the perception of SL leads to SCF^{D3}-mediated ubiquitination of D53 and its subsequent degradation by the proteasome system, which in turn releases the repression of downstream responses.

Chinese researchers subsequently modified the model of D53 action by elucidating the downstream transcriptional network of the strigolactone response (Song et al. 2017). These authors reported that *Ideal Plant Architecture* 1 (*IPA1*) could be a transcription factor in rice that functions as a direct downstream component of D53 in regulating tiller number and strigolactone-induced gene expression. Strigolactone interacts with *IPA1* to suppress its transcriptional activity. In this

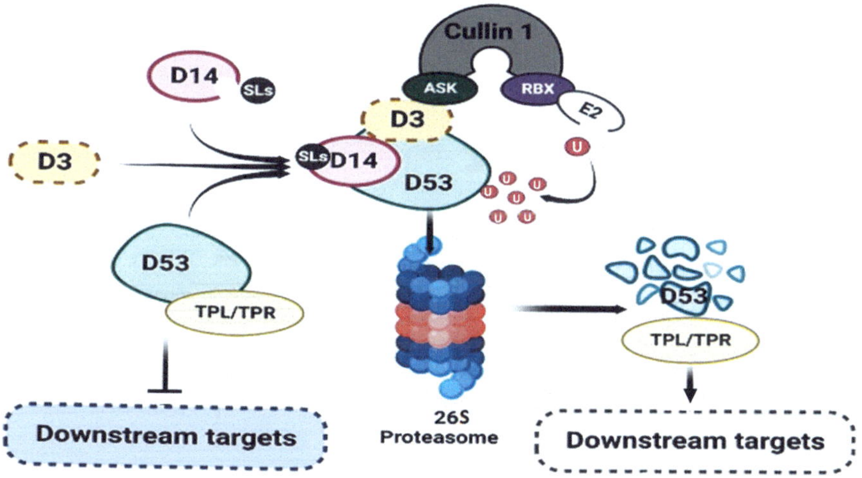

Fig. 8.16 Strigolactone signalling pathway. Model showing the action of D53 in the absence and presence of Strigolactones (SLs). D53 is stable in the absence of SLs and recruits TPL/TPR proteins leading to repression of downstream targets. On the other hand, perception of SL in presence of SLs causes SCF^{D3}-mediated ubiquitination of D53 and its subsequent degradation by 26S proteasome system leads to release of repression of downstream responses. (Adapted from Jiang et al. 2013)

process, *IPA1* directly binds to the D53 promoter and plays a crucial role in the feedback regulation of strigolactone-induced D53 expression. Hence, *IPA1* is concluded to be the transcriptional activator that works with D53 to mediate strigolactone-controlled tiller development in rice (Fig. 8.17).

8.9.3 Strigolactones Controlling Axillary Bud Release

Strigolactone signalling, which controls bud activity, is also proposed to be regulated by the ratio of strigolactone to cytokinin, where the former is an inhibitor and the latter is a promoter. Both hormones can modulate each other's levels. This antagonistic action between strigolactones and cytokinin was elucidated by Dun et al. (2012). A low concentration of cytokinin applied locally stimulates bud outgrowth in strigolactone-deficient mutant plants, whereas in wild-type plants, buds are recalcitrant to stimulation by cytokinin application. These hormones act antagonistically in bud outgrowth control, but the delivery of one hormone does not obstruct the other. However, during signal transduction, these two genes converge at a common point in the regulatory pathway related to bud outgrowth. Similarly, Dun et al. (2012) suggested that the point of convergence could be BRC1 (BRANCHED1), a TCP transcription factor expressed downstream of strigolactone in shoot

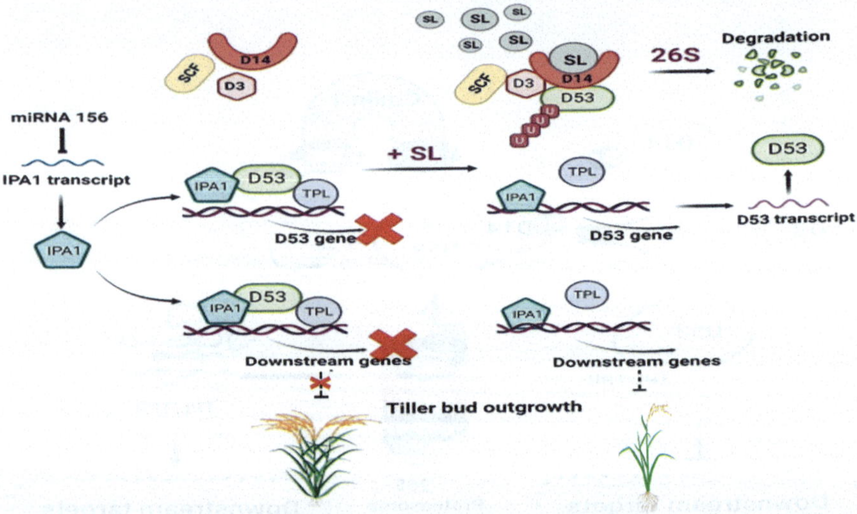

Fig. 8.17 A proposed model of the *IPA1*-mediated SL signalling pathway. In the absence of SLs, the D53 protein binds to *IPA1*, and together with the TPL/TPR proteins, the D53 protein represses the transcriptional activity of *IPA1*, which leads to tiller bud outgrowth (left panel). In the presence of SLs, the perception of SL leads to the degradation of D53 by the proteasome system, which in turn releases the repression of *IPA1*-regulated gene expression and leads to an SL response for the inhibition of tiller bud outgrowth (right side panel). (Adapted from Song et al. 2017)

branching. The transcription factor is regulated by both strigolactone and cytokinin without requiring protein synthesis. Thus, the hormones act antagonistically to each other to control bud release by regulating the expression of *BRC1* in axillary buds. *FC1* is an orthologous gene of *BRC1* in rice.

8.10 Cytokinin Action on Tillering

8.10.1 Role of Cytokinin in Shoot Branching

As discussed before, auxin is actively transported in the basipetal direction to inhibit shoot branching. In contrast, cytokinin travels acropetally from roots to shoots and promotes shoot branching. To date, cytokinin is the only hormone known to promote axillary bud outgrowth. The removal of the apical bud strongly increases cytokinin levels in the axillary bud within 24 h. External application of auxin, however, depresses the cytokinin level. The hormone is synthesized in the roots and transported upwards to the shoot in the xylem (Hopkins and Huner 2004). It is also possible that the cytokinin content increases locally within the axillary bud following the decapitation of the apical bud. Additionally, cytokinin triggers axillary bud release because of its capacity to promote auxin production and basipetal auxin transport in developing buds, thereby inhibiting cytokinin production in the stem and reducing its availability for other buds (Bangerth et al. 2000; Tanaka et al. 2006; Shimizu-Sato et al. 2009).

8.10.2 Cytokinin Properties

Plant cells are formed as a result of cell division in primary and secondary meristems. Way below the meristems, the cells elongate in size and become functionally differentiated into various types, such as leaf photosynthetic, vascular, and epidermal tissue cells. These cells do not divide again, leading to the belief that they are terminally differentiated, similar to animal cells. However, some of these cells retain the nucleus and can rejuvenate cell division long after a period of mitotic inactivity; for example, wound-induced mitotic activity can lead to the generation of new cells/tissues. The cell division activity at the wound site is stimulated by infection with the soil-living bacterium *Agrobacterium tumefaciens,* which causes crown gall disease. The re-initiation of cell division occurs through the release of a water-soluble substance named cytokinin. Coconut milk contains the cytokinin *zeatin*, chemically called *trans*-6-(4-hydroxy-3-methyl but-2-enylamino) aminopurine, which induces mature cells to enter the cell division cycle. Another cytokinin was identified from autoclaved herring sperm DNA and exhibited a strong cell division-promoting effect. It was named kinetin, an amino purine derivative, 6-furfurylaminopurine (Fig. 8.18). Kinetin is a heat-degraded product of DNA.

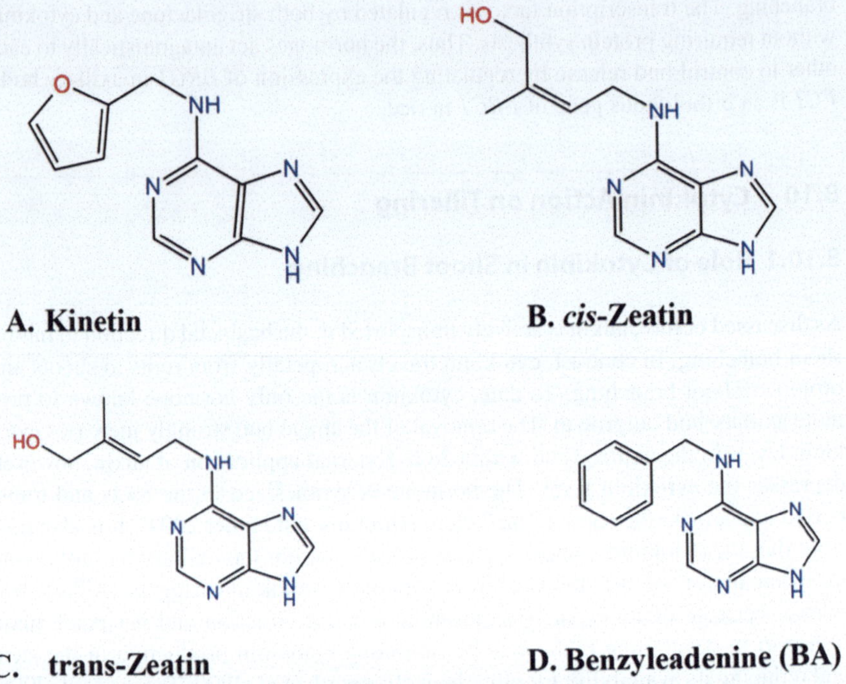

Fig. 8.18 Chemical structure of different cytokinins, such as (**A**) kinetin, (**B**) cis-zeatin, (**C**) trans-zeatin and (**D**) benzyladenine

8.10.3 Biosynthesis of Cytokinin

The side chains of naturally occurring cytokinins are related to carotenoids, rubber, phytoalexins, gibberellins, and abscisic acid. Cytokinin side chains are synthesized from an isoprene derivative. Isoprene units, in turn, are derived from mevalonic acid or pyruvate-3-phosphoglycerate pathways. These products are converted to the biological isoprene unit dimethylallyl diphosphate (DMAPP). Cytokinin biosynthesis begins with the transfer of the isopentenyl group of DMAPP to an adenosine moiety of ATP or ADP by the enzyme isopentenyl transferase (IPT). The end products of this reaction are iPDP or iPTP, which are converted to zeatin by a hydrolase (Fig. 8.19).

8.10.4 A Model for Cytokinin Signalling

The receptors that perceive cytokinin signals on the cytoplasmic membrane are CKI1 and CRE. Both the *CKI1* and *CRE* genes encode proteins similar in sequence to the bacterial two-component sensor protein histidine kinase. The system has two functional components: a sensor histidine kinase to which the cytokinin signal binds

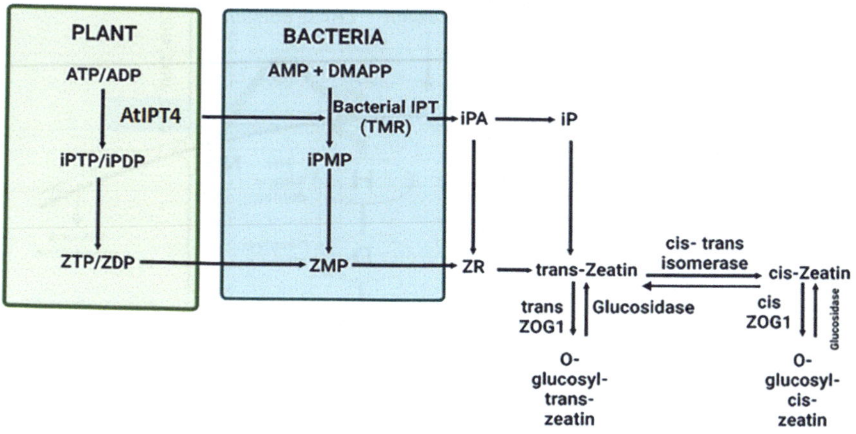

Fig. 8.19 Biochemical pathway for cytokinin biosynthesis. The first committed step in cytokinin biosynthesis is the addition of the isopentenyl side chain from DMAPP (Dimethylallyl pyrophosphate) to an adenosine moiety. The plant and bacterial IPT (Isopentenyl transferase) enzymes differ in the adenosine substrate used; plant enzymes use both ADP and ATP, whereas bacterial enzymes utilize AMP (Adenosine monophosphate). The products of these reactions (iPMP, iPDP, or iPTP) are converted to zeatin by an unidentified hydroxylase. The various phosphorylated forms can be interconverted, and free *trans*-Zeatin can be formed from riboside by enzymes involved in general purine metabolism. *Trans*-zeatin can be converted to *cis*-zeatin by an isomerase. Both *cis*- and *trans*-zeatin can be glucosylated. (Adapted from Taiz and Zeiger 2006)

and a downstream response regulator. The activity of the response regulator is regulated through phosphorylation by histidine kinase. Cytokinin binds to a cytokinin receptor sitting on the cytoplasmic membrane in the input component. Reception of cytokinin signal by the input component changes the activity of the dimeric histidine kinase domain. It transphosphorylates a conserved histidine residue, and this energized phosphate is subsequently transferred to a conserved aspartate residue in the receiver part of the response regulator. Phosphorylation changes the activity of the kinase enzyme and enables the output domain to work as a transcription factor. There are two other genes in the *Arabidopsis* genome, *AHK2* and *AHK3*, which are closely related to *CRE* and function as cytokinin receptors. The first set of genes is activated by the cytokinin *ARR* (*Arabidopsis Response Regulator*). There are two clades of *ARR* genes, type A and type B. Type A *ARR*s comprise solely a receiver domain, whereas type B *ARRs* consist of a transcription factor in addition to the receiver domain.

Cytokinin binds to an extracellular portion of the dimeric form of CRE1 called the CHASE domain. AHK2 and AHK3, which contain the CHASE domain, can also function as cytokinin receptors. A stepwise model describing cytokinin signalling via CRE is described in Fig. 8.20.

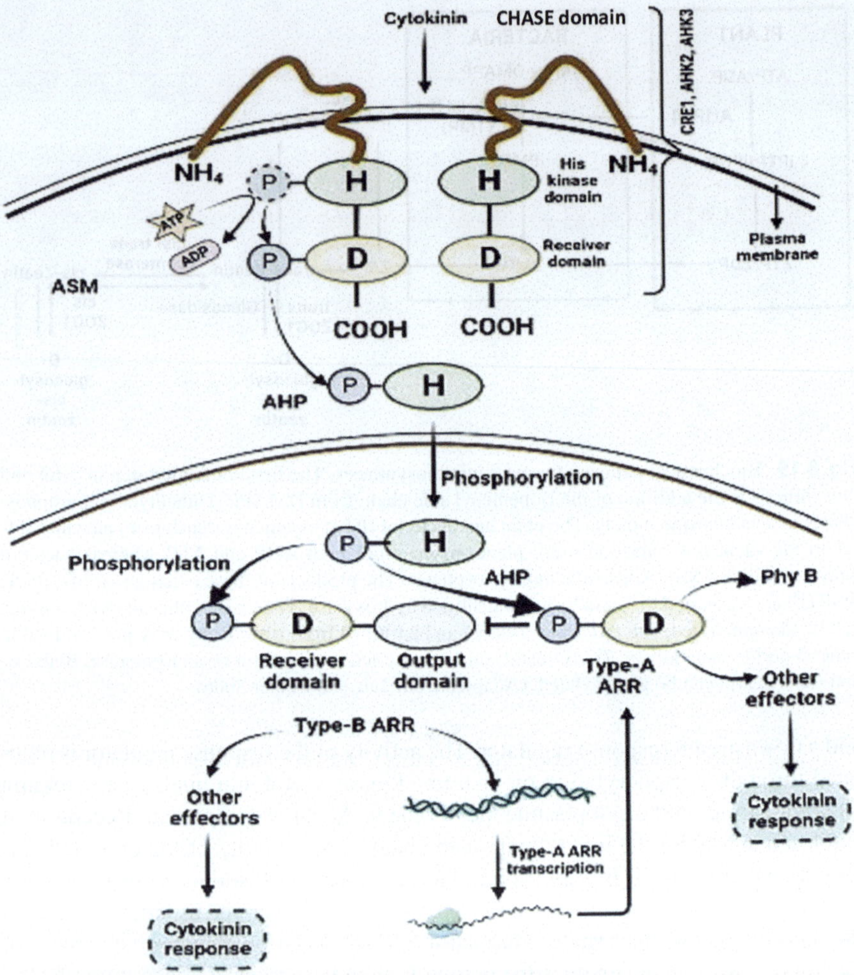

Fig. 8.20 Cytokinin signalling in plants. In the membrane of endoplasmic reticulum, cytokinin binds to the dimerised receptor CRE1 leading to the initiation of a phosphorylation cascade and resulting into cytokinin response. (Adapted from Taiz and Zeiger 2006)

1. This binding stimulates histidine kinase activity. The phosphate is transferred to an aspartate residue (D) on the fused receiver domain.
2. The phosphate is transferred to the conserved histidine of an AHP (*Arabidopsis* histidine phosphotransfer) protein. AHPs are set phosphotransfers in addition to CRE1, which is likely involved in this signalling cascade.
3. This phosphorylation causes the AHP protein to move from the cytoplasm to the nucleus. It transfers the phosphate to an aspartate residue present inside the receiver domain of a type-B ARR.
4. Phosphorylated type-B ARRs activate the output domain to induce transcription of genes encoding type-A ARRs. In turn, type A ARR proteins can regulate the

expression of many other cytokinin-responsive genes as well as the activities of various target proteins.

5. Type A ARRs can also be phosphorylated by AHP proteins.
6. Phosphorylated type-A ARRs interact with various intermediaries to mediate changes in cellular activities appropriate for cytokinin (cytokinin responses) (Taiz and Zeiger 2006).

8.11 Nutrient Deficiency and Tillering

8.11.1 Nutrient Deficiency Affects Rice Tillering

In addition to phytohormones, the activation of dormant buds on leaf axils in rice and their transformation into actively growing tillers are also regulated by mineral nutrients. A deficiency of nitrate, as well as phosphate, has been shown to strongly inhibit tillering in rice (Sun et al. 2014; Brewer et al. 2013; Zha et al. 2022). According to Sakakibara et al. (2006), nitrogen is the most essential element for growth and promotes tiller development in rice. The availability of nitrogen fluctuates significantly in the soil media both temporally and spatially. Plants can sense and uptake varying levels of nitrogen in the soil. The incoming nitrogen is assimilated and distributed to different parts of the plant. The functional mechanisms of nitrogen uptake, transportation, assimilation, and remobilization are tightly regulated by proteins and microRNAs in response to both external and internal nitrogen levels, the carbon status of the plant, and phytohormones (Reddy and Ulaganathan 2015). Shoot branching in rice is strongly influenced by nitrogen and three phytohormones, namely, auxin, strigolactone, and cytokinin, where nitrogen and cytokinins are promoters and strigolactones and auxins are inhibitors of bud outgrowth (Xu et al. 2015).

The positive effect of cytokinin-mediated tiller bud outgrowth is not sustainable long without nitrogen. Xu et al. (2015) showed that the effects of nitrogen on elevating all forms of endogenous cytokinin in both tiller buds and nodes of rice plants become visible within 12 h from the beginning of exogenous treatment with either nitrogen or cytokinin. Exogenous nitrogen and cytokinin treatments improved tiller bud release and increased IAA and zeatin plus zeatin riboside levels in tiller buds but significantly decreased ABA levels.

8.11.2 Nitrogen and Phosphorus Deficiencies Induce Strigolactone Action

As signalling mediators, strigolactones are crucial for plant adaptation to challenging environments. The hormone provides the balance of essential metabolite distributions through precise strategic modifications during plant development to adapt to nutrient deficiency. There are several reports on the increase in strigolactone contents in roots and root exudates under nitrogen- and phosphorus-deficient conditions in the literature (Foo et al. 2013; Xie et al. 2010; Brewer et al. 2013; Sun et al. 2014). Roots are the primary organs that experience deficiencies of major nutrients, such as nitrogen and phosphorus, in the growth environment of rice, and they coordinate the development of roots and shoots via strigolactone biosynthesis. An underground change in root morphogenesis involves the production of adventitious roots and root hairs, while an influence on aboveground architecture includes tiller production, leaf senescence, and reproductive development. In rice plants, strigolactones inhibit tillering in response to nitrogen and phosphorus deficiency (Umehara et al. 2010; Xu et al. 2015). However, the suppression of axillary tiller bud outgrowth under phosphorus deficiency does not occur in strigolactone-deficient or -signal mutants. In addition, strigolactones confer tolerance to suboptimal conditions such as drought and salinity (Saeed et al. 2017; Wu et al. 2022). Deficiencies in nitrogen and phosphorus increase strigolactone levels in roots, which move upwards into the shoot and ultimately inhibit tillering in rice plants (Umehara et al. 2010; Xu et al. 2015). It has been reported that the translocation of nitrogen in rice plants is inhibited by strigolactones (Luo et al. 2018), and bud outgrowth is thereby precluded. Under nitrogen deficiency, strigolactone inhibits cell division-related elongation but not the initiation of rice tiller bud formation (Zha et al. 2022; Luo et al. 2017). In addition, there is abysmal information available on the interaction of nitrogen with the inhibitors regulating tiller development, but the literature is rich in the cognitive effects of nitrogen and cytokinin on the system. In the present study, the interaction of nitrogen with cytokinin during rice tiller development is discussed.

8.11.3 Nitrogen and Phosphorus Deficiency Modulates Cytokinin Biosynthesis

Liu et al. (2011b) showed that nitrogen can regulate rice bud release by two mechanisms: controlling nitrogen metabolism and manipulating endogenous hormones. A nitrogen deficiency inhibits cell division-oriented elongation but not the initiation of rice tillers (Luo et al. 2017). According to Sakakibara et al. (2006), nitrogen functions as a signal triggering widespread changes in gene expression, and the signal is integrated at the whole-plant level through the use of multiple signalling routes. Nitrogen itself can be a route itself, but another route uses cytokinin as a messenger. In rice, nitrogen treatment becomes favourable for tiller bud outgrowth because it promotes local cytokinin synthesis. In this context, Zha et al. (2022) identified five *OsIPT* and *OsCKX* genes each and elucidated their expression in tiller buds and tiller nodes located at the fifth axil of rice plants in response to nitrogen and phosphorus deficiency and synthetic strigolactone treatment within the first 6 h. Nitrogen and phosphorus deficiency increased the expression of *OsIPT* in the tiller buds but decreased it in the tiller nodes, revealing that no cytokinin activity was responsible for tiller initiation. Strigolactone application did not change the expression of *OsIPT* genes (Fig. 8.21).

The characterization of the expression of various *OsCKX* genes in these organs corroborates the inference that stem nodes play an abysmal role in maintaining

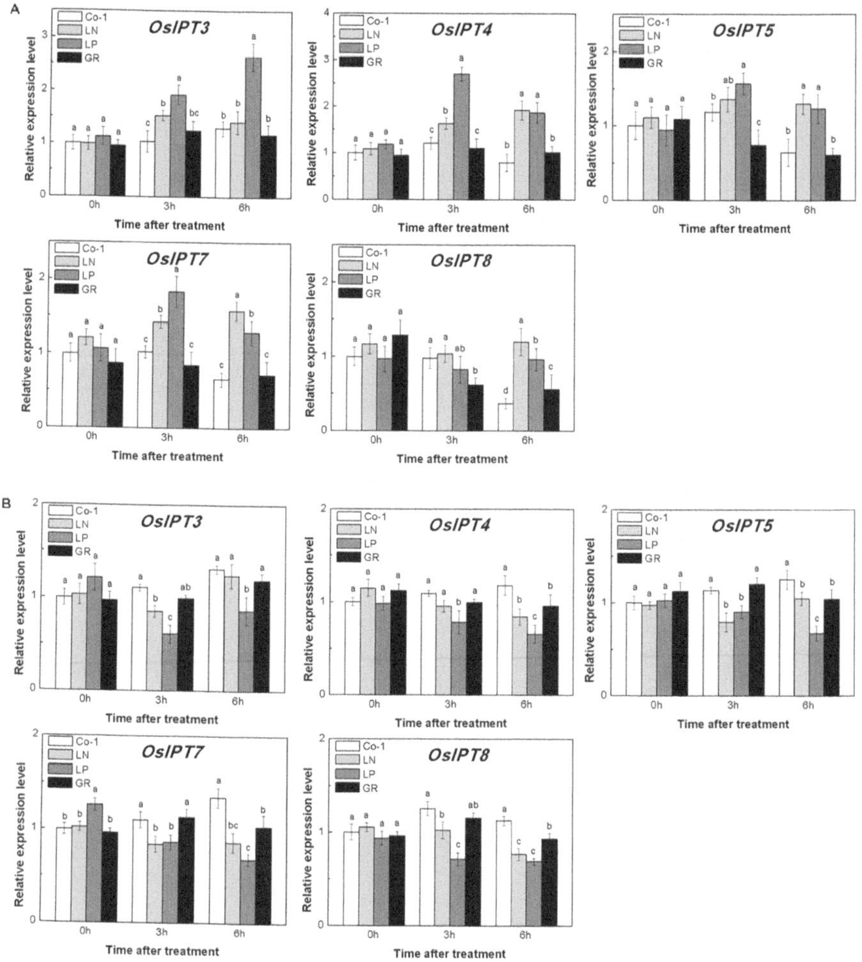

Fig. 8.21 Expression patterns of the *OsIPT3, OsIPT4, OsIPT5, OsIPT7*, and *OsIPT8* cytokinin biosynthesis genes in rice tiller buds (**A**) and tiller nodes (**B**) located at the fifth leaf axils expressed in response to nitrogen and phosphorus deficiency and strigolactone treatment. (Adapted from Zha et al. 2022 with permission)

cytokinin homeostasis during tiller growth in rice. In the tiller buds, nutrient deficiency promoted the expression of *OsCKX* genes, but there was no effect on the stem nodes. However, strigolactone treatment induces the expression of genes in the stem node (Zha et al. 2022). It is implied that strigolactones are capable of depleting cytokinin content in rice tillers by promoting cytokinin degradation (Fig. 8.22).

Xu et al. (2015) showed that the promotion of cytokinin biosynthesis was sustained by a corresponding increase in the expression of various forms of *OsIPT* genes in rice stem nodes. High nitrogen application significantly increased the expression patterns of the *OsIPT* genes, whereas the expression levels of the

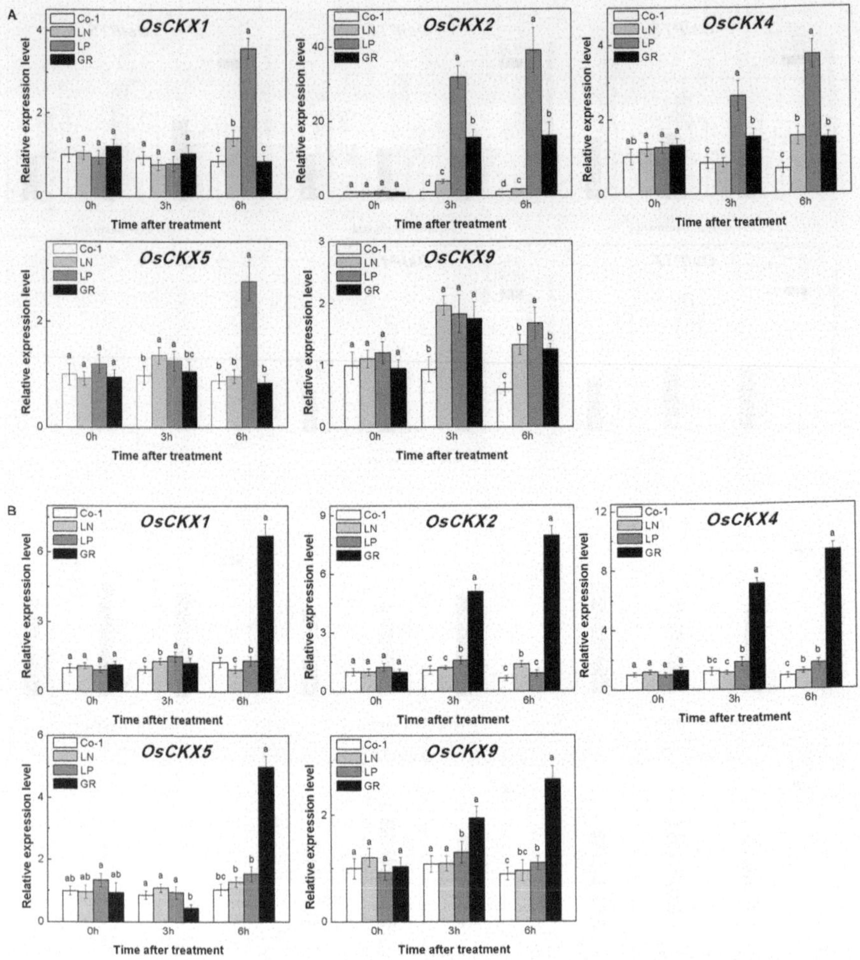

Fig. 8.22 Expression patterns of the *OsCKX1, OsCKX2, OsCKX4, OsCKX5*, and *OsCKX9* cytokinin degradation genes in rice tiller buds (**A**) and tiller nodes (**B**) located at the fifth leaf axils expressed in response to nitrogen and phosphorus deficiency and strigolactone treatment. (Adapted from Zha et al. 2022 with permission)

genes in cytokinin-treated plants were not as encouraging. Concerning a schematic model, these authors constructed a composite schematic model of the interaction between nitrogen availability and phytohormones such as auxin, cytokinin, and strigolactone in rice tiller bud release, as detailed in Fig. 8.23 (Xu et al. 2015).

Nitrogen availability enhanced the cytokinin pool by promoting the expression of *OsIPTs* in the axillary nodes. Nitrogen and cytokinin suppressed the expression of genes (*OsD10, OsD17, OsD27*) related to strigolactone synthesis

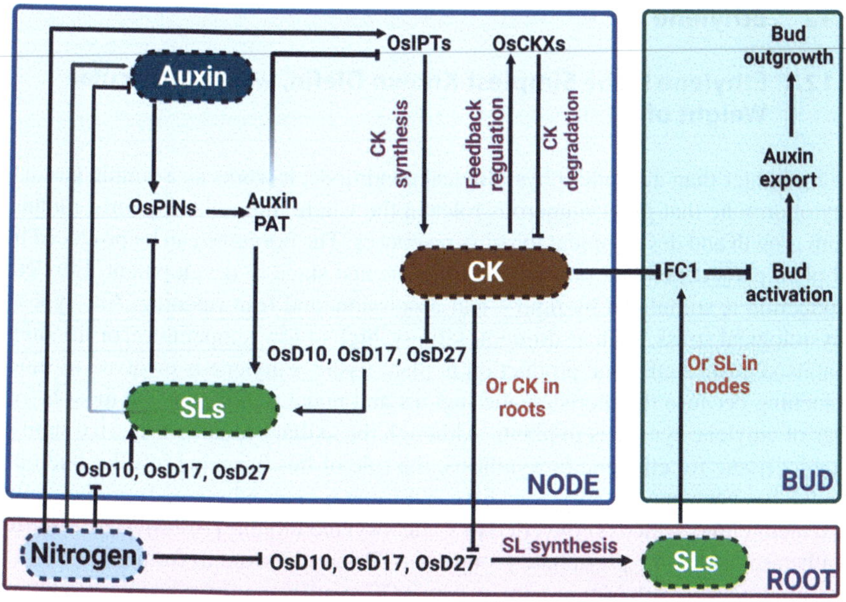

Fig. 8.23 A model explaining the interaction between nitrogen availability and auxin, cytokinin, and strigolactone in the control of shoot branching in rice. OsPINs, *Oryza sativa* PIN-FORMED1; SLs, Strigolactone; CK, Citokinin; OsIPT, *Oryza sativa* Isopentenyl transferase; OsCKX, *Oryza sativa* Cytokinin oxidase/dehydrogenase; FC1, FINE-CULM1. (Adapted from Xu et al. 2015)

in roots and nodes. It is implied that the inhibition of strigolactone synthesis by nitrogen could partly occur via the cytokinin pathway. Strigolactones do not reduce the amount of cytokinin or the expression levels of *OsIPT* genes, but they reduce the auxin pool and auxin transport capacity in the nodes. The polar auxin transporter (PAT) directly inhibits cytokinin biosynthesis but promotes strigolactone biosynthesis in the nodes rather than in the roots. Cytokinin reduces the expression of *FC1 (FINE CULM1)* in buds, but strigolactone promotes this expression. *FC1* is a common target for both cytokinin and strigolactones. It is surmised that OsFC1 (rice transcription factor 1) could be a common target for the cytokinin and strigolactone pathways determining the shoot branching of rice. Nitrogen and auxin delay the expression change patterns of *FC1* by altering the downstream signals for cytokinin and strigolactone. Bud activation promotes auxin export, resulting in bud outgrowth. Nitrogen and auxin delayed the expression change patterns of *FC1*, potentially by changing the downstream signals for cytokinin and strigolactone.

8.12 Ethylene

8.12.1 Ethylene Is the Simplest Known Olefin, with a Molecular Weight of 28

Being lighter than air under physiological conditions, it works as a simple gaseous phytohormone that plays numerous roles in the whole life cycle of plants, ranging from growth and development to stress responses. The hormone can be produced by all plant parts depending on the type of tissue and stage of development. Ethylene production is stimulated by injury, leaf abscission, and fruit ripening. Any type of physiological stress, such as drought, salinity, high or low temperature, or flooding, can also enhance ethylene production in plant tissue. Ethylene is involved in shoot branching because the meristematic regions and nodal regions are the most active sites of ethylene synthesis in plants. Although the axillary region of the rice stem is a target tissue for ethylene biosynthesis, the role of this hormone in tiller bud outgrowth has been unclear for a long time. Ethylene is biosynthesized from the amino acid methionine, which is converted to *S*-adenosyl methionine (AdoMet) by AdoMet synthetase, where the phosphate groups of ATPs are removed to form $PP_i + P_i$. In a rate-limiting step, AdoMet is catalyzed by ACC synthase to form 1-aminocyclopropane-1-carboxylic acid (ACC). In the last step of the pathway, ACC is converted to ethylene by the ACC oxidase enzyme, which requires oxygen. CO_2 and HCN are the by-products of this reaction (Fig. 8.24).

8.12.2 Ethylene Signalling

For the first time, among terrestrial plants, a linear ethylene signalling pathway was established in the dicot model plant *Arabidopsis*. The expression of homologous genes of the *Arabidopsis* ethylene signalling components in rice was subsequently

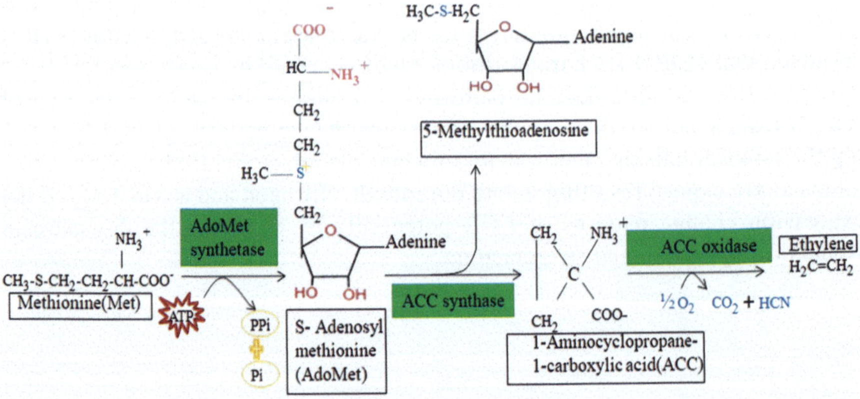

Fig. 8.24 Biosynthesis pathway of the hormone ethylene in plants

elucidated by characterizing ethylene-responsive mutants and their corresponding genes, as reviewed by Yang et al. (2015). These Chinese authors have pursued a linear ethylene signalling pathway from ethylene perception in the endoplasmic reticulum to transcriptional regulation in the nucleus. A complete picture of the ethylene signalling mechanism is provided in the schematic model (Fig. 8.25).

Ethylene is perceived in the endoplasmic reticulum by a family of receptors that are similar to bacterial two-component histidine kinase sensors. When ethylene levels are low in open-air conditions, active receptors interact with the N-terminal region of CTR1 to form a complex. Active CTR1 directly phosphorylates the C-terminal domain of EIN2 and leads to the degradation of EIN2 by the F-box protein ETP1/2 to repress downstream signal transduction. In the nucleus, EIN3/EIL1 are degraded through the ubiquitin proteasome-mediated F-box proteins EBF1/2. The ERFs remain inactive, revealing the response of ethylene-responsive genes.

An abundance of ethylene leads to the binding of ethylene to receptors; five receptors have been identified (Sub-family 1: ETR1 and ERS1; and Sub-family 2: ETR2, ERS2, and EIN4). This binding inactivates the receptors and CTR1, leading to the dephosphorylation of EIN2 and subsequent cleavage. The split EIN2 C-terminal fragment (CEND) enters the nucleus, participates in the stabilization and accumulation of EIN3/EIL1 and ultimately induces ERF1 and other ethylene-responsive genes. The mechanism of ethylene action controlling various agronomic features of rice plant are described in Fig. 8.26 (Yin et al. 2017).

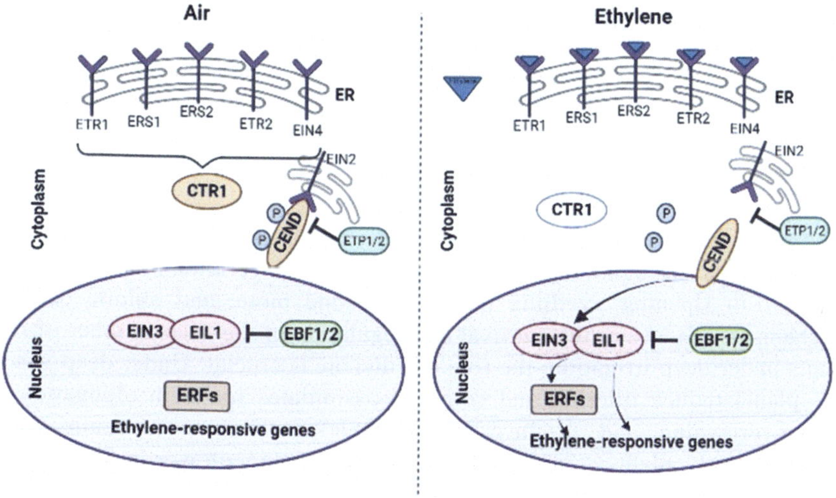

Fig. 8.25 Ethylene signalling pathway in *Arabidopsis*. Absence of ethylene (left side panel) and presence of ethylene (right side panel). (Adapted from Yang et al. 2015)

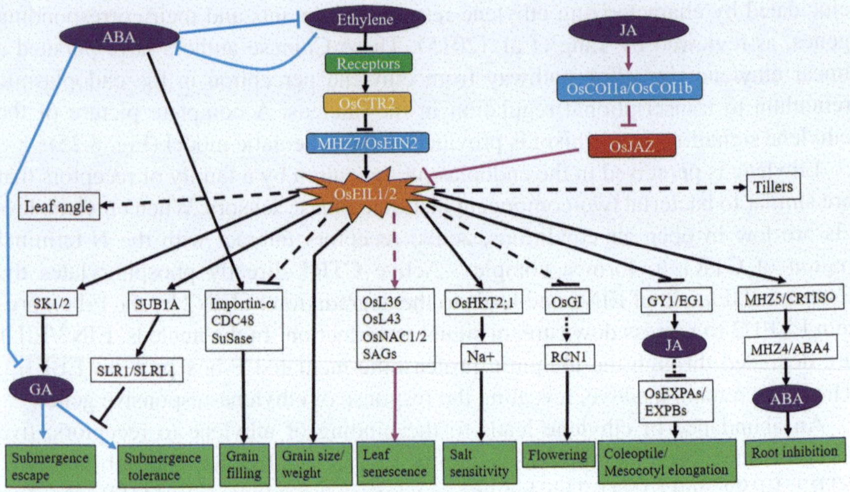

Fig. 8.26 Ethylene-mediated regulation of different agronomic traits in rice. Ethylene promotes growth of coleoptiles/mesocotyls by partially inhibiting JA biosynthesis whereas JA promotes rice leaf senescence through a cascade of OsCOI1b-OsJAZ-OsEIN3-OsORE1. JA and ethylene synergistically accelerate rice leaf senescence by activating common and differential SAGs. Ethylene positively regulates rice grain size/weight, flowering, tillering and leaf angle but negatively affects rice grain filling and salt tolerance. The lower ratio of ethylene to ABA is beneficial for rice grain filling. Upon submergence, SUB1A negatively regulates GA response by inhibiting the degradation of SLR1/SLRL1 to restrict plant growth in submergence-tolerance species. On the contrary, ethylene promotes internode elongation through increasing transcription of SK1/2 and GA production to escape flooding in deepwater rice. Arrows and T-bars indicate direct or indirect activation and suppression, respectively. Dotted lines indicate several or unknown steps involved in the pathway

8.12.3 Ethylene and Deep-Water Rice

Ethylene signalling has been studied extensively for its ability to regulate various functions throughout plant life, including seed germination, leaf senescence, fruit ripening, seedling morphology, and biotic and abiotic defence mechanisms. To ensure the survival and regulation of the tiller number of rice plants under deep irrigation, the role of ethylene is crucial. Under deep water, rice plants reduce tillering and conserve assimilates for stem elongation to escape drowning and produce larger panicles to increase grain yield. Simultaneously, plants gain strength to better compete with weeds. The genetic basis of internodal elongation in rice is largely obscure, and deep-water rice has become an excellent model for the study of basic aspects of plant growth and development, where the kinetics of ethylene production have been mostly studied in the stems of submerged plants.

8.12.4 Deep Water Alters/Suppresses Tillering

Water availability in rice production is a major problem for farmers in rice-growing nations. This issue can be resolved through either growing water-efficient rice varieties or efficient water management practices when resources are available (Mubeen et al. 2022). However, when excessive water results in prolonged inundation, the impacts on rice yield can be devastating. Under these circumstances, the identification of genetic variants that confer tolerance to calamity bears practical potential (Sasaki 2006). Hence, by the selective use of water-efficient rice varieties for success in deep-water situations, a most experienced farmer can achieve extremely high grain yields, nearly 800 g m^{-2}. Deepwater rice varieties are grown in flooded conditions with the level of water remaining more than 50 cm in the field for more than a month. For sustenance, more than 100 million people in Southeast Asia rely on deep-water rice. Two adaptations permit water-efficient rice plants to thrive in deep water, i.e. floating rice and traditional tall rice plants. In deep water situations, the benefits accrued by rice plants are significant. During submergence, plants escape night-time chilling injuries, reduced tillering increases panicle size, and stagnant water prevents weed growth (Myung et al. 2004).

8.12.5 Deepwater Induces Ethylene Biosynthesis

The adaptation of floating rice to submergence is influenced by ethylene. Rice plants growing under flood conditions accelerate the growth of stem internodes, which allows the plants to avoid drowning and maintain foliage above the water level for the exchange of gases. The presence of aerenchyma extends the percolation of essential gases to submerged organs of the plant. The elongation of the internodes of the main shoot of rice is triggered by a mechanism involving ethylene biosynthesis. Generally, ethylene is produced by plants and released into the air by diffusion. When floating rice is submerged in water, the diffusion process is disrupted. According to Jackson (1985), the diffusion coefficient of ethylene in water is 10,000 times lower than that in air, and the hormone is physically trapped in the submerged tissue. In the second method, ethylene accumulates in submerged plant parts when deep-water rice plants respond to submergence by sensing an altered concentration of the internal gas atmosphere. Under submergence, lowered O_2 and increased CO_2 tension elevate ethylene synthesis and simultaneously induce growth-promoting effects of ethylene (Raskin and Kende 1984).

Internodal elongation in deep-water rice is based on increased cell division activity and cell elongation. A longitudinal section of a rice stem containing the uppermost internode reveals the structural variation of cells in three different parts of the internode. Cell division intercalary meristem at the base of the internode generates new internodal cells, and these cells are displaced into the elongating zone where they reach maximum size. In the differentiating zone, growth ceases, and secondary wall and xylem formation occur (Kende et al. 1998, Fig. 8.27).

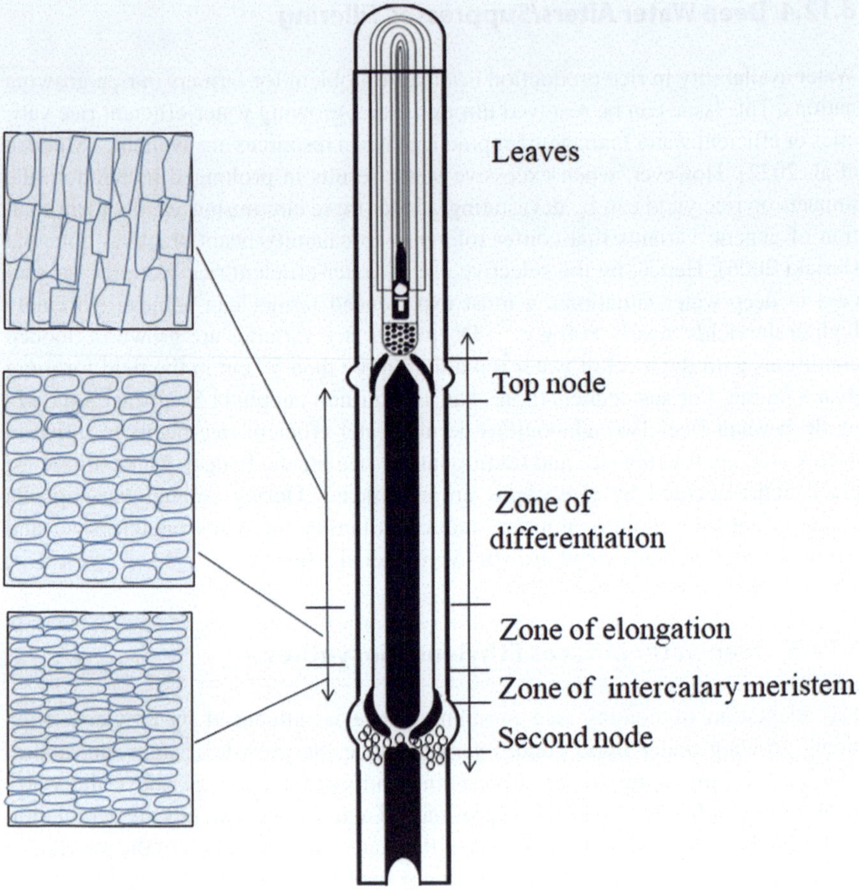

Fig. 8.27 Schematic diagram of a longitudinal section of a rice stem containing the uppermost internode. (Adapted from Kende et al. 1998)

According to Kende et al. (1998), GA activates the cell cycle in the intercalary meristem by inducing the expression of several growth-related gene products that are involved in different phases of the cell cycle and DNA replication, such as cyclin, expansin, and histone kinase. The rapid internodal elongation of deep-water rice occurs from an ethylene-mediated increase in the ratio of GA to ABA in the sequence given in Fig. 8.28. This proposition has been tested by Myung et al. (2004), who reported that deep water induced ethylene production in the leaf sheath within 24 h of exposure, during which the concentration of ethylene continued to increase 35-fold for 35 days. Rice cultivars in deep water exhibit potential for rapid internode elongation because of the interaction between ethylene and GA (Fukao and Bailey-Serres 2008), where an increase in ethylene concentration triggers increased responsiveness of the stem to GA action (Hattori et al. 2009) through the promotion of the ABA pool.

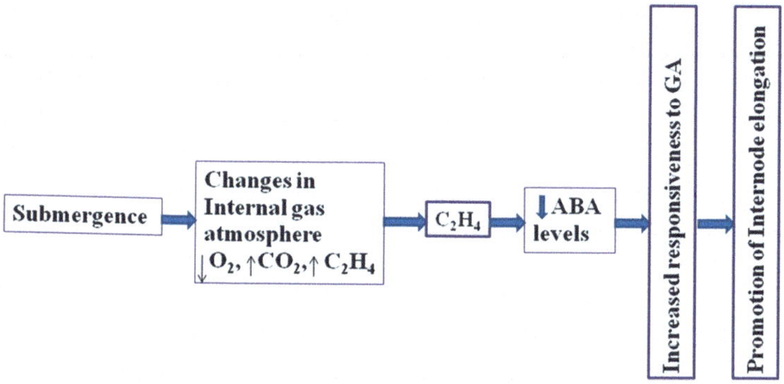

Fig. 8.28 Hormonal dynamics in a sequence of events connecting submergence and enhanced internode elongation in deep-water rice. (Adapted from Kende et al. 1998)

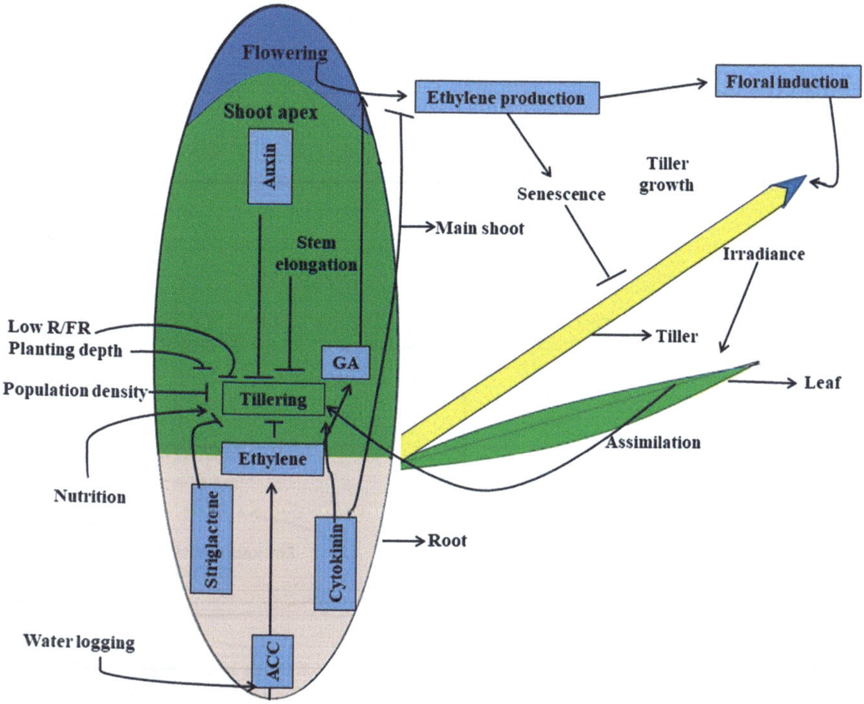

Fig. 8.29 Schematic of factors influencing tiller production and development in *Oryza species* plants. Arrowhead lines indicate promotion and flat lines indicate inhibition. (Adapted from Mohapatra et al. 2011)

Ethylene-induced stem elongation comes at the cost of diverting resources for tiller bud outgrowth in deep-water rice. A holistic view of the interaction of various physiological factors, including hormonal cues controlling tillering and stem elongation in rice, was proposed in a review article by Mohapatra et al. (2011) (Fig. 8.29).

8.13 Crosstalk Between Ethylene and Gibberellic Acid (GA)

The cross-talk between ethylene and GA is proposed to be mediated by the APETALA 2 (AP2)/ethylene-responsive element-binding factor (ERF) family of transcription factors. In *indica* rice, there are 161 *AP2/ERF* genes, of which Qi et al. (2011) isolated one gene, *OsEATB,* for the ERF protein associated with tillering and branching. These findings revealed that the rice ethylene response *AP2/ERF* gene *OsEATB* restricts internode elongation to promote tillering by downregulating the expression of GA biosynthetic genes. The *OsEATB* transgenic lines exhibited a dwarf phenotype. In the GA biosynthesis pathway, *ent*-kaurene is an early intermediate; *ent*-kaurene synthase A catalyzes the cyclization of geranyl geranyl diphosphate (GGDP) to *ent*-copalyl diphosphate (CDP), which in turn is converted to *ent*-kaurene by *ent*-kaurene synthase B. In rice seedling transgenic *OsETAB* lines, the expression of *OsCPS2* was sharply downregulated tenfold, while the expression of *GIBBERELLIN 20-OXIDASE* (*GA20ox2*) increased by 2.77-fold in comparison to that in the wild type. Similarly, in the elongating internodes of the transgenic plants, *OsCPS2* expression was downregulated by 11-fold, while *GA20ox2* expression was increased by twofold compared with the respective expression in the wild-type plants. The authors reiterated that *OsEATB* positively regulates rice yield components by promoting tillering and branching and that this effect occurs because the gene negatively regulates the ethylene-induced enhancement of GA responsiveness by reducing GA biosynthesis. A schematic model for OsETAB-dependent ethylene regulation of internode elongation and tillering is proposed in Fig. 8.30.

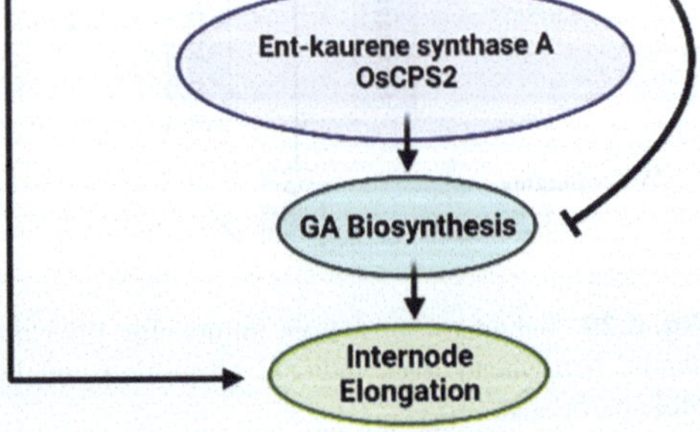

Fig. 8.30 Model of *OsEATB*-dependent hormonal regulation of internode elongation. *OsEATB* acts as a negative balance regulator of the ethylene-induced enhancement of GA responsiveness during the internode elongation. (Adapted from Qi et al. 2011)

References

Arite T, Iwata H, Ohshima K, Maekawa M, Nakajima M, Kojima M, Sakakibara H, Kyozuka J (2007) DWARF10, an RMS1/MAX4/DAD1 orthologue, controls lateral bud outgrowth in rice. Plant J 51:1019–1029

Bangerth F (1994) Response of cytokinin concentration in the xylem exudates of bean (*Phaseolus vulgaris* L.) plants to decapitation and auxin treatment, and relationship to apical dominance. Planta 194:439–442

Bangerth F, Li C-J, Gruber J (2000) Mutual interaction of auxin and cytokinins in regulating correlative dominance. Plant Growth Regul 32:205–217

Brewer PB, Dun EA, Ferguson BJ, Rameau C, Beveridge CA (2009) Strigolactone acts downstream of auxin to regulate bud outgrowth in peas and *Arabidopsis*. Plant Physiol 150(1):482–493

Brewer PB, Koltai H, Beveridge CA (2013) Diverse roles of strigolactones in plant development. Mol Plant 6:18–28

Brewer PB, Dun EA, Gui R, Mason MG, Beveridge CA (2015) Strigolactone inhibition of branching independent of polar auxin transport. Plant Physiol 168:1820–1829

Cheng X, Ruyter-Spira C, Bouwmeester H (2013) The interaction between strigolactones and other plant hormones in the regulation of plant development. Front Plant Sci 4:199. https://doi.org/10.3389/fpls.2013.00199

Das S, Ramakrishnayya G, Sarkar RK (2002) Tiller mortality and its mitigation through GA3 and IAA in unfavourable rainfed lowland rice. Plant Arch 2:299–304

Deveswar P, Prusty A, Sharma S, Tyagi AK (2020) Phytohormone mediated molecular mechanisms involving multiple genes and QTL govern grain number in rice. Front Genet 11:1360

Domagalska MA, Leyser O (2011) Signal integration in the control of shoot branching. Nat Rev Mol Cell Biol 12:211–221

Duan J, Yu H, Yuan K, Liao Z, Meng X, Jing Y, Liu G, Chu J, Li J (2019) Strigolactone promotes cytokinin degradation through transcriptional activation of CYTOKININ OXIDASE/DEHYDROGENASE 9 in rice. Proc Natl Acad Sci U S A 166:14319–14324. https://doi.org/10.1073/pnas.1810980116

Dun EA, de Saint Germain A, Rameau C, Beveridge CA (2012) Antagonistic action of strigolactone and cytokinin in bud outgrowth control. Plant Physiol 158:487–498. https://doi.org/10.1104/pp.111.186783

Faizan M, Faraz A, Sami F, Siddiqui H, Yusuf M, Gruszka D, Hayat S (2020) Role of strigolactones: signalling and crosstalk with other phytohormones. Open Life Sci 15:217–228. https://doi.org/10.1515/biol-2020-0022

Foo E, Yoneyama K, Hugill CJ, Quittenden LJ, Reid JB (2013) Strigolactones and the regulation of pea symbioses in response to nitrate and phosphate deficiency. Mol Plant 6:76–87

Fukao T, Bailey-Serres J (2008) Ethylene—a key regulator of submergence responses in rice. J Plant Sci 175:43–51

Ha CV, Leyva-González MA, Osakabe Y, Tran UT, Nishiyama R, Watanabe Y, Tanaka M, Seki M, Yamaguchi S, Van Dong N, Yamaguchi-Shinozali K, Shinozaki K, Herrera-Estrella L, Tran L-SP (2014) Positive regulatory role of strigolactone in plant responses to drought and salt stress. Proc Natl Acad Sci U S A 111:851–856

Haider I, Andreo-jimenez B, Bruno M, Bimbo A, Flokova K, Abuauf H, Otang-Ntui V, Guo X, Charnikhova T, Al-Babili S, Bouwmeester HJ, Ruyter-Spira C (2018) The interaction of strigolactones with abscisic acid during the drought response in rice. J Exp Bot 69:2403–2414

Hanada K, Kagawa K, Ohuchi A (1993) Effect of temporary short period of leveling off under deep water condition on growth of tillers and leaves of floating rice. Jpn J Trop Agr 37(2):84–92

Hattori Y, Nagai K, Furukawa S (2009) The ethylene response factors *SNORKEL1* and *SNORKEL2* allow rice to adapt to deep water. Nature 460:1026–1030. https://doi.org/10.1038/nature08258

Hayward A, Sternberg P, Beveridge C, Leyser O (2009) Interactions between auxin and strigolactone in shoot branching control. Plant Physiol 151:400–412

Hopkins WG, Huner NPA (2004) Cytokinins are synthesized primarily in the root. In: Hopkins WG, Huner NP (eds) Introduction to plant physiology. Wiley, New York, pp 325–326

Ishikawa S, Maekawa M, Arite T, Onishi K, Nakamura I, Kyozuka J (2005) Suppression of tiller bud activity in tillering dwarf mutants of rice. Plant Cell Physiol 46(1):79–86

Jackson MB (1985) Ethylene and responses of plants to soil waterlogging and submergence. Annu Rev Plant Physiol 36:145–174

Jiang L, Liu X, Xiong G, Liu H, Chen F, Wang L, Meng X, Liu G, Yu H, Yuan Y, Yi W, Zhao L, Ma H, He Y, Wu Z, Melcher K, Qian Q, Xu HE, Wang Y, Li J (2013) DWARF 53 acts as a repressor of strigolactone signalling in rice. Nature 504:401–405

Kapulnik Y, Resnick N, Mayzlish-Gati E, Kaplan Y, Wininger S, Hershenhorn J, Koltai H (2011) Strigolactones interact with ethylene and auxin in regulating root-hair elongation in *Arabidopsis*. J Exp Bot 62:2915–2924

Kariali E, Kuanar SR, Mohapatra PK (2008) Individual tiller dynamics of two wild *Oryza* species in contrasting habitats. Plant Prod Sci 11:355–360

Kebrom TH, Spielmeyer W, Finnegan EJ (2013) Grasses provide new insights into the regulation of shoot branching. Cell Press 18:1–8

Kende H, van der Knaap E, Cho H-T (1998) Deepwater rice: a model plant to study stem elongation. Plant Physiol 118:1105–1110

Li C, Bangerth F (1999) Auto-inhibition of indole acetic acid transport in the shoot of two-branched pea (*Pisum sativum*) plants and its relationship to correlative dominance. Physiol Plant 106:415–420

Li X, Qian Q, Fu Z, Wang Y, Xiong G, Zeng D, Wang X, Liu X, Teng S, Hiroshi F, Yuan M, Luo D, Han B, Li J (2003) Control of tillering in rice. Nature 422:618–621

Ling F, Su Q, Jiang H, Cui J, He X, Wu Z, Zhang Z, Liu J, Zhao Y (2020) Effects of strigolactone on photosynthetic and physiological characteristics in salt-stressed rice seedlings. Sci Rep 10:1–8. https://doi.org/10.1038/s41598-020-63352-6

Liu Y, Wang Q, Ding Y, Li G, Xu J, Wang S (2011a) Effects of external ABA, GA3, and NAA on the tiller bud outgrowth of rice is related to changes in endogenous hormones. Plant Growth Regul 65:247–254

Liu Y, Ding YF, Wang QS, Meng DX, Wang SH (2011b) Effects of nitrogen and 6-benzyl amino purine on rice tiller bud growth and changes in endogenous hormones and nitrogen. Crop Sci 51:786–792

Liu X, Hu Q, Yan J, Sun K, Liang Y, Jia M, Meng X, Fang S, Wang Y, Jing Y, Liu G, Wu D, Chu C, Smith SM, Chu J, Wang Y, Li J, Wang B (2020) z-Carotene isomerase suppresses tillering in rice through the coordinated biosynthesis of strigolactone and abscisic acid. Mol Plant 13:1784–1801

Luo L, Pan S, Liu X, Wang H, Xu G (2017) Nitrogen deficiency inhibits ell division-determined elongation, but not initiation, of rice tiller buds. Israel J Plant Sci 64:32–40

Luo L, Wang H, Liu X, Hu J, Zhu X, Pan S, Qin R, Wang Y, Zhao P, Fan X, Xu G (2018) Strigolactones affect the translocation of nitrogen in rice. Plant Sci 270:190–197

Matsushima S (1957) Analysis of developmental factors determining yield and yield prediction in lowland rice. Bull Natl Agric Sci A 5:1–271

Mohapatra PK, Panda BB, Kariali E (2011) Plasticity of tiller dynamics in wild rice *Oryza rufipogon* Griff. A strategy for ensuring resilience in suboptimal environments. Int J Agron 543237, 9 pages

Mubeen K, Sarwar N, Shehzad M, Ghaffar A, Aziz M (2022) Irrigation management in rice. In: Sarwar N, Atique-ur-Rehman AS, Hasanuzzaman M (eds) Modern techniques of rice crop production. Springer, Singapore. https://doi.org/10.1007/978-981-16-4955-4_8

Myung E-J, Known Y-W, Lee B-W (2004) Ethylene production and accumulation in the leaf sheath and its relation to tillering suppression of deep-irrigated rice plants. Korean J Crop Sci 49:5

Nawaz A, Farooq M (2017) Rice physiology, Chapter 17. In: Chauhan BS et al (eds) Rice production worldwide. Springer, Cham, pp 455–485. https://doi.org/10.1007/978-3-319-47516-5_17

Oikawa T, Kyozuka J (2009) Two-step regulation of LAX PANICLE1 protein accumulation in axillary meristem formation in rice. Plant Cell 21:1095–1108

Ongaro V, Leyser O (2008) Hormonal control of shoot branching. J Exp Bot 59:67–74
Qi W, Sun F, Wang Q, Chen M, Huang Y, Feng YQ, Luo X, Yang J (2011) Rice ethylene-response AP2/ERF factor *Os*EATB restricts internode elongation by down-regulating a gibberellin biosynthetic gene. J Plant Physiol 157:216–228
Raskin I, Kende H (1984) Regulation of growth in stem sections of deep-water rice. Planta 160:66–72
Reddy MM, Ulaganathan K (2015) Nitrogen nutrition, it's regulation and biotechnological approaches to improve crop productivity. Am J Plant Sci 6:2745–2798. https://doi.org/10.4236/ajps.2015.618275
Sachs T, Thimann K (1967) The role of auxins and cytokinins in the release of buds from dominance. Am J Bot 54:136–144
Saeed W, Naseem S, Ali Z (2017) Strigolactones biosynthesis and their role in abiotic stress resilience in plants: a critical review. Front Plant Sci 8:1487. https://doi.org/10.3389/fpls.2017.01487
Sakakibara H, Takei K, Hirose N (2006) Interactions between nitrogen and cytokinin in the regulation of metabolism and development. Trends Plant Sci 11:440–448
Sasaki T (2006) Rice in deep water. Nature 442:635–636. https://doi.org/10.1038/442635a
Shimizu-Sato S, Tanaka M, Mori H (2009) Auxin-cytokinin interactions in the control of shoot branching. Plant Mol Biol 69:429–435. https://doi.org/10.1007/s11103-008-9416-3
Shinohara N, Taylor C, Leyser O (2013) Strigolactone can promote or inhibit shoot branching by triggering rapid depletion of the auxin efflux protein PIN1 from the plasma membrane. PLoS Biol 11(1):e1001474. https://doi.org/10.1371/journal.pbio.1001474
Song X, Lu Z, Yu H, Shao G, Xiong J, Meng X, Jing Y, Liu G, Xiong G, Duan J, Yao X-F, Liu C-M, Li H, Wang Y, Li J (2017) IPA1 functions as a downstream transcription factor repressed by D53 in strigolactone signalling in rice. Cell Res 27:1128–1141. https://doi.org/10.1038/cr.2017.102
Sun H, Tao J, Liu S, Huang S, Chen S, Xie X, Yoneyama K, Zhang Y, Xu G (2014) Strigolactones are involved in phosphate- and nitrate-deficiency-induced root development and auxin transport in rice. J Exp Bot 65:6735–6746
Sun H, Li W, Burritt DJ, Tian H, Zhang H, Liang X, Miao Y, Golam Mostofa M, Phan Tran LS (2022) Strigolactones interact with other phytohormones to modulate plant root growth and development. Crop J 10(6):1517–1527. https://doi.org/10.1016/j.cj.2022.07.014
Taiz L, Zeiger E (2006) Plant Physiology, 4th edn. Sinauer Associates, Inc., Publishers, Sunderland, MA
Tanaka M, Takei K, Kojima M, Sakakibara H, Mori H (2006) Auxin controls local cytokinin biosynthesis in the nodal stem in apical dominance. Plant J 45:1028–1036
Umehara M, Hanada A, Magome H, Takeda-Kamiya N, Yamaguchi S (2010) Contribution of strigolactones to the inhibition of tiller bud outgrowth under phosphate deficiency in rice. Plant Cell Physiol 51:1118–1126
Wang B, Smith SM, Li J (2018) Genetic regulation of shoot architecture. Annu Rev Plant Biol 69:437–468
Wang Y, Shang L, Yu H, Zeng L, Hu J, Ni S, Rao Y, Li S, Chu J, Meng X, Wang L, Hu P, Yan J, Kang S, Qu M, Lin H, Wang T, Wang Q, Hu X, Chen H, Wang B, Gao Z, Guo L, Zeng D, Zhu X, Xiong G, Li J, Qian Q (2020) A strigolactone biosynthesis gene contributed to the green revolution in rice. Mol Plant 13:923–932
Waters MT, Brewer PB, Bussell JD, Smith SM, Beveridge CA (2012) The *Arabidopsis* orthologue of rice DWARF27 acts upstream of MAX1 in the control of plant development by strigolactones. Plant Physiol 159:1073–1085
Waters MT, Gutjahr C, Bennett T, Nelson DC (2017) Strigolactone signalling and evolution. Annu Rev Plant Biol 68:291–322
Wu F, Gao Y, Yang W, Sui N, Zhu J (2022) Biological functions of strigolactones and their crosstalk with other phytohormones. Front Plant Sci 13:821563. https://doi.org/10.3389/fpls.2022.821563
Xie X, Yoneyama K, Yoneyama K (2010) The strigolactone story, Annu. Rev. Phytopathol. 48, 93–117

Xu J, Li Y, Ding Y, Chen L, Chen L, Ding C, Wang S (2015) The interaction between nitrogen availability and auxin, cytokinin, and strigolactone in the control of shoot branching in rice (*Oryza sativa* L.). Plant Cell Rep 34:647–1662

Yang C, Lu X, Ma B, Chen S-Y, Zhang J-S (2015) Ethylene signaling in rice and Arabidopsis: conserved and diverged aspects. Mol Plant 8:495–505

Yin C-C, Zhao H, Ma B, Chen S-Y, Zhang J-S 2017. Diverse roles of ethylene in regulating agronomic traits in rice. Front. Plant Sci. 8:1676. https://doi.org/10.3389/fpls.2017.01676

Zha M, Imran M, Wang Y, Xu J, Ding Y, Wang S (2019) Transcriptome analysis revealed the interaction among strigolactones, auxin, and cytokinin in controlling the shoot branching of rice. Plant Cell Rep 38(3):279–293. https://doi.org/10.1007/s00299-018-2361-y

Zha M, Zhao Y, Wang Y, Chen B, Tan Z (2022) Strigolactones and cytokinin interaction in buds in the control of rice tillering. Front Plant Sci 13:837136. https://doi.org/10.3389/fpls.2022.837136

Zhang YX, van Dijk AD, Scaffidi A, Flematti GR, Hofmann M, Charnikhova T, Verstappen F, Hepworth J, van der Krol S, Leyser O, Smith SM, Zwanenburg B, Al-Babili S, Ruyter-Spira C, Bouwmeester HJ (2014) Rice cytochrome P450 MAX1 homologues catalyze distinct steps in strigolactone biosynthesis. Nat Chem Biol 10(12):1028–1033

Genetics of Tiller Development in Rice

9

Abstract

Rice has been accepted as a model system for genomic studies of cereal crops. The genetic basis of shoot branching has received increased attention because of the agronomic importance of tillers for grain yield. Genes affecting tiller development also influence panicle architecture and thereby grain number. Several genetic mutants for tillering have been identified. The genes responsible for tiller number are expressed during apical and axillary meristem initiation. Often, they work in combination with hormones, which control tissue differentiation and exaggeration of new buds. In the last decade, several genes controlling tiller number have been identified in rice, such as *MOC1, TAD1, MIP1, LAX1, LAX2, OsPIN1, TLD1, OsIPT1, OsGA2ox, OsTB1,* etc. Environmental factors overwhelm the genotypic permissibility of rice tillering.

Keywords

Axillary meristem · Brassinosteroid signalling · Effective tiller · Ideal plant architecture · LAX genes · Monoculm mutant · MOC genes · Shoot apical meristem · Strigolactone related genes · Tiller angle · Tiller dynamics

9.1 Introduction

The tillering pattern of rice has been one of the most important agronomic traits for grain production. It is also related to crop adaptation to target environments. Therefore, rice breeders have been interested in controlling tillering to identify genotypes better adapted to the environment in which they are grown. Among the monocots of economic importance, rice tillering has been preferred as a model system for studying branching. There are two types of branching in the above-ground plant parts. First, the pattern of shoot branching accounts for the tiller number;

© The Author(s), under exclusive license to Springer Nature Singapore Pte Ltd. 2025

P. K. Mohapatra et al., *Tillering Behavior of Rice Plant*,
https://doi.org/10.1007/978-981-97-5235-5_9

second, panicle branching is responsible for the fixation of panicle architecture, i.e. critical for the numbers of branches, spikelets, and grains. The kingdom of rice genotypes is vast, and the varietal diversity of tiller numbers is extremely high. Some varieties bear a numerous tiller number compared to uniculm varieties, bearing only the main shoot bereft of any subsidiary branches. Crop adaptation to target environments depend on the tillering type, and high-tillering cultivars are the preferred choice for farmers growing crops in the most favourable environments. In contrast, low-tillering rice cultivars are cultivated in stress-prone environments where crop growth is threatened by periodic exposure to suboptimal conditions. Subject to deficiency and deprivation of essential resources, plants tend to fold up vegetative growth early and switch to a reproductive mode for quick completion of their life cycle fore-closing tiller number increase.

9.1.1 Tillering: An Indicator of Phenotypic Adaptation

Changes in the phenotype of rice plants (phenotypic plasticity) play an important role in determining plant responses to resource availability, with the magnitude and direction of plasticity of individual traits being critical for grain yield. According to the analyses of magnitude and hierarchy in response to yield-related traits based on biomass partitioning and grain yield components, shoot and panicle branching, which determines the numbers of tillers and panicle branches, respectively, are the most plastic. In comparison, seed size is the least plastic attribute. Because panicle branching and spikelet development from tillers specify the grain number per panicle and determine grain yield, rice breeders have given foremost attention to studying the mechanisms involved in rice tillering and panicle branching (Liang et al. 2014).

In pursuance, rice breeders have chosen high-tillering rice genotypes for cultivation in lowland irrigated ecosystems and take advantage of the bounties of resources available. The profusely tillering high-yielding semi-dwarf rice plants are preferred in this environment, and they made an immense contribution to the success of the Green Revolution in the late 1960s. However, an excessive tiller number produces several sterile and unproductive tillers, a minimized leaf area, and a reduction in leaf photosynthesis via mutual shading. In contrast, under wet direct-seeded conditions, where the seed rate is high for successful competition against weeds, the most desirable rice genotypes are low tillering because of high allelopathic competition among plants. Similarly, in stress-prone environments inviting episodes of drought, high temperature, and saline water flooding during crop growth, low-tillering cultivars become the preferred choice for cultivation. A minimized branching pattern increases the robustness of a limited number of tillers produced and enhances resilience for survival and completion of the life cycle. In conditions where stressful events are expected to occur episodically during crop growth, farmers use low-tillering genotypes to reduce tiller production and conserve water and other essential resources for survival in exigency. However, too few tillers result in meagre biomass production and a consequential decline in carbohydrate production and grain filling. Thus, tillering becomes a key component of the phenotypic plasticity of rice plants in response to transient changes in growing environments (Mohapatra et al. 2011).

9.1.2 Tiller Development in Rice

In rice, tillers are specialized shoot-siding branches that grow from the nodes of unelongated internodes. The number of tillers produced determines the overall shoot architecture of the plant. Tillers, dependent on the main shoot for resources, may gain independence later by developing adventitious roots. Tillers develop on the main shoot through the continual organogenesis process during the rice's vegetative growth period. This process involves the activity of pluripotent stem cells located in meristems, such as shoot apical meristems and axillary meristems. The activity and determinacy of both of these meristems ultimately determine the shoot architecture of the rice plant, similar to the system proposed for eudicots by Wang and Li (2008). The development of a tiller in rice and other grasses involves three important stages (i) establishment of an axillary meristem, as indicated by the formation of a stem cell population in the leaf axil; (ii) production of leaf primordia from the axillary meristem to form an axillary bud; and (iii) the emergence of the axillary bud to form a tiller (Hussien et al. 2014). The development of the main shoot or tiller is calibrated in units of phytomers, with each phytomer comprising an internode (stem segment), a node, a leaf, and an axillary bud.

In rice, the first axillary meristem is formed during embryogenesis when a mature rice embryo possesses one axillary bud in the axil of the coleoptile leaf. Following germination, young rice plants continue to produce more leaves and axillary meristems in an ordered and coordinated progression terminating with the initiation of the reproductive stage (Itoh et al. 2005; Oikawa and Kyozuka 2009). Leaves originate from the tissue on the flanks of the shoot apical meristem, where the cells of the tissue transit from indeterminate to determinate cell fate as they are converted into nascent leaf primordia. According to Itoh et al. (2005), the time interval between the formation of two successive leaf primordia is called the plastochron, and leaf developmental stages are calibrated by their plastochron number (Pi). The youngest visible leaf primordium is identified as P1, the next youngest is P2, and so on. The axillary meristem develops in the leaf axil, and the Pi of the subtending leaves identifies the developmental stages of axillary meristem formation. The first visible symptom of axillary meristem development in rice is a protrusion on the stem surface by cellular proliferation directed towards the P2 leaf primordium. Cell proliferation continues unabated through the P3 and P4 stages when the cells gain a meristematic fate. The process ends at the P5 stage. At this stage, the new axillary meristem initiates its leaf primordia bearing an axillary bud primordium (Oikawa and Kyozuka 2009). Subject to environmental conditions and intrinsic factors, the axillary bud may remain dormant or grow into a new tiller. Each tiller retains a growth-oriented new axis, similar to the phytomer units of the main shoot.

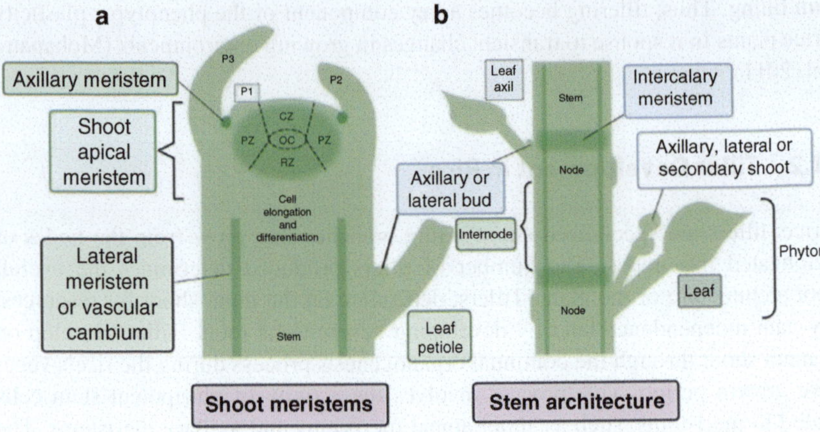

Fig. 9.1 Shoot meristems and architecture. (**a**) Meristems at the shoot apex, P1, P2, and P3 are leaf primordia. (**b**) Stem architecture comprising phytomers and lateral shoots. *CZ* central zone, *PZ* peripheral zone, *OC* organizing centre, *RZ* rib zone. (Adapted from Wang et al. 2018)

The body plan created by the shoot apical meristem that produces new cells and organs has been defined by Wang et al. (2018) depicted in a model (Fig. 9.1). Accordingly, shoot apical meristem (SAM) determines plant phyllotaxy and cues axillary meristem (AM) formation. There are three zones of tissues in the SAM. The central zone (CZ) comprises pluripotent stem cells that divide to form initials for the peripheral zone (PZ). PZ generates axillary or lateral buds at the flanks. The rib zone (RZ) is responsible for stem tissue formation. The organizing centre (OC) is located below the CZ. It regulates stem cell proliferation in the CZ and differentiation in the PZ. Thereby, SAM organization and functions are maintained. The PZ generates both the leaf primordia and AM. AM later gives rise to vegetative branches or inflorescence (Fig. 9.1).

9.1.3 Genes Controlling Axillary Meristem Development and Tiller Dynamics

Adaptation of rice to a wide variety of environmental conditions, such as normal, unfavourable, and over-favourable conditions worldwide, has led to relevant genetic changes and species diversity. In such adaptation, rice plants have acquiesced several environment-specific genetic factors as derivatives of spontaneous mutations associated with variations in the tiller number. In addition to natural mutations, mutations induced through radiation and ethyl sulphonate methods have also contributed to the identification of several genes involved in rice tillering. The elucidation and characterization of QTLs for tiller number has continued in the past few decades with endeavours to increase rice's genetic yield ceiling because each tiller can produce a seed-bearing inflorescence that contributes to increasing grain

Table 9.1 Gene action to determine tiller formation during mutation and over-expression in rice

Genes	Genes expression/mutant	Phenotype
Axillary meristem maintenance		
MONOCULM 1 (MOC1)	*MOC1* mutant	Predominantly main culm, tiller numbers reduce drastically
	Over-expression	Plant height decreases, tiller numbers increase
Tillering and dwarf 1 (*TAD1*)	*TAD1* mutant	Increased tillers and reduced plant height
	Over-expression	Plant height increases, tiller numbers reduce
MOC1 interacting proteins (MIP1)	Over-expression of MIP1	Plant height decreases, tiller numbers increase
Lax panicle1 (*LAX1*) encode basic helix-loop-helix (bHLH) transcription factors	*LAX1-1* single mutant	Formation of tiller does not affect
	LAX1-2 and LAX1-3 mutant	Reduction of tillers
Lax panicle2 (*LAX2*) encodes a novel nuclear protein	*LAX2-1* mutant	Reduction of tillers
LAX1 × *LAX2* interaction	*LAX1*, *LAX2* double mutant	The effect is synergistic, tiller numbers reduce drastically
LAX1 × *MOC1* interaction	Double mutation in both *LAX1* and *MOC1* point	Greater reduction of tiller
MORE PANICLES 3 (MP3)	Single gene	Promote tillering to increase panicle number moderately

number. Approximately, 100 rice tiller number genes have been identified thus far, but the genetic basis of tiller number has remained mostly elusive because of its complex traits and control by both major genes and QTLs (Jiang et al. 2019). In addition to the efficacy of grain yield, a balance between tiller number and vigour is essential, as late-formed tillers consume photosynthetically assimilated nutrients but do not produce grains. In pursuance, breeders have focused their prime attention on the initiation and establishment of axillary buds controlled by a suite of genetic factors; while a complex network of genetic, hormonal, and environmental factors regulates bud outgrowth. Several genes and mutants involved in tillering and their homologues, phenotypes, molecular functions, and pathways are described in Table 9.1 (Oikawa and Kyozuka 2009; Tabuchi et al. 2011; Xu et al. 2012; Hussien et al. 2014).

9.1.4 GWAS for Rice Tiller Number

In the exploration of genes controlling rice tiller number, a genome-wide association study (GWAS) based on high-density SNP maps and widely contrasting germplasm collections has been used as a new strategy for efficient and high-throughput gene identification. Jiang et al. (2019) investigated the genetic architecture underlying rice tiller number variation through application of this technique involving next-generation genome sequencing and phenotyping. They identified five new tiller number-associated genes at 15 loci that were significantly associated with tiller number variation. The results obtained could be useful for dissecting the genetic architecture of rice tiller numbers and molecular breeding of high-yielding rice plants. The five novel genes associated with tiller number were *Os01g28690*, *Os05g32120*, *Os07g28890*, *Os11g15130*, and *Os11g15210*. All these genes exhibited SNPs, which were associated with tiller number variation. In addition, the expression levels of two other genes, *Os01g28690* and *Os05g32120*, were closely correlated with the tiller number. Similarly, Zhao et al. (2020) sequenced 219 accessions of Korean rice to construct a high-quality SNP dataset for evaluating the genetic basis of the tiller number. By GWAS, 20 significant association signals for tiller-related traits were detected, 5 of which were identified close to genomic regions of known candidate genes involved in the phase transition of the shoot apical meristem from the vegetative to reproductive stages, such as *HD1* (*HEADING TIME1*). Multiple linear regression models for lead SNPs showed similar coefficients of determination for maximum and productive tiller numbers. Taken together, these findings imply that the genetic basis for tiller number in rice can be revealed via GWAS and that a formulated prediction model can be constructed via linear regression analysis. The expression and function of several of the important genes involved in axillary meristem formation are discussed as follows.

9.2 Genes Regulating Tiller Number and Tiller Dynamics

9.2.1 MOC Genes

In rice plants, three important monoculm genes have been identified, namely, *MOC1*, *MOC2*, and *MOC3*. *MOC1* (*Os06g40780*) was the first gene identified as the key regulator controlling rice tillering and branching (Li et al. 2003).

9.2.1.1 MOC1 Gene

MOC1 is located on chromosome 6 and encodes a putative GRAS family nuclear protein that is expressed in the axillary buds of rice. This protein functions to initiate axillary buds and promote their outgrowth. Loss-of-function *moc1*mutant plants display only a main shoot bereft of any tiller because of a defect in the formation of axillary tiller buds. The single main culm phenotype is displayed in the *moc1*mutant because of a 1.9 kb retrotransposon inserted in *MOC1* (Li et al. 2003). The authors convincingly proved that *MOC1* functions as a positive regulator of rice tillers and panicle branches by studying the expression of the *OSH1* and *OsTB1* genes involved in shoot and axillary meristem development, respectively. *OSH1* controls shoot meristem initiation and maintenance, while *OsTB1* is expressed in axillary buds and regulates its outgrowth. In the *moc1* mutants, the expression of both *OSH1* and *OsTB1 was* significantly downregulated. Li et al. (2003) showed that wild-type rice plants exhibit *OSH1* RNA in situ hybridization in the axillary meristem and shoot apical meristem of tiller buds, but there was no expression in the *moc1* leaf axil. Similarly, *OsTB1* RNA in situ hybridization revealed expression in the axillary meristem and the entire tiller bud in the wild type, but there was no expression in the *moc1* leaf axil. Thus, the rice mutant *moc1* completely loses its tillering ability and behaves as a monoculm.

The mechanism by which *MOC1* controls the branching pattern of rice was explored further. A novel gene, *Tillering and Dwarf 1* (*TAD1*) was shown to interact with *MOC1* to control the cell cycle machinery of meristems (Xu et al. 2012). The *TAD1* protein interaction leading to the degradation of *MOC1* was identified from analysis of the allelic mutant *TAD1* (Xu et al. 2012) and *tillering enhancer* (*te*) (Lin et al. 2012). Both of these mutants are characterized by reduced stem height and enhanced tillering. *TAD1* encodes a coactivator of the anaphase-promoting complex (APC/C), a multi-subunit E3 ligase. By interacting with *MOC1*, *TAD1* forms a complex with OsAPC10 and functions as a co-activator of APC/C to target *MOC1* for degradation in a cell cycle-dependent manner. TE is a substrate recognition and binding factor for APC/C. It is involved in the formation of the APC/C^{TE} complex, which interacts with *MOC1* and *OsCDC27* to induce the degradation of *MOC1* via the ubiquitin-26 proteasome pathway and reduces the expression of the meristem identity gene *OSH1*. Moreover, *MOC1* over-expresses in both wild-type plants and cultured *TAD1* mutant cells, but its dynamic degradation is comparatively low in the latter. The over-expression of *MOC1* leads to increased tillering of rice plants. Sun et al. (2010) characterized the *MOC1 Interacting Protein 1* (*MIP1*) gene involved in rice tillering; subject to over-expression, the *MIP1* gene also causes enhanced tillering. Liang et al. (2014) confirmed that *MOC1* is the key regulator in controlling rice tillering and branching in a review article. A working model for the degradation of *MOC1* has been proposed by its interaction with APC/C$^{TAD1/TE}$ in a cell-cycle-dependent manner, which is responsible for the tillering of rice (Fig. 9.2). This exploration elucidates the genetic mechanism controlling the determination of rice yield (Lin et al. 2012; Xu et al. 2012).

9.2.1.2 MOC1-Related Genes

As discussed above, *MOC1* encodes the GRAS family of transcription factors. However, its target has not been specifically investigated, although the MOC1 protein has been shown to interact significantly with other proteins, such as MIP1, TAD1, and SLR1, in post-translational regulation. Therefore, it is imperative to screen the natural alleles of *MOC1* and its cofactors for use in the molecular design and breeding of rice. Zhang et al. (2021b) reported the characterization of a novel natural *moc1* allele called *gnp6*, which comprises a thymine insertion in the coding sequence of the SAW motif in the GRAS domain. The SAW motif is needed for the nuclear localization of GNP6/MOC1, where it functions as a transcription factor or coregulator. Loss-of-function mutations in *gnp6* reduce axillary branch formation.

Fine mapping of the reduced tillering *MOC1*-related genes was explored further by Yu et al. (2017). In pursuance, these workers characterized the *low-tillering mutant 1* (*lt1*), which is a single recessive mutation on chromosome 6. The *lt1* mutant exhibited a phenotype of fewer tillers, thicker culms, fewer panicle branches,

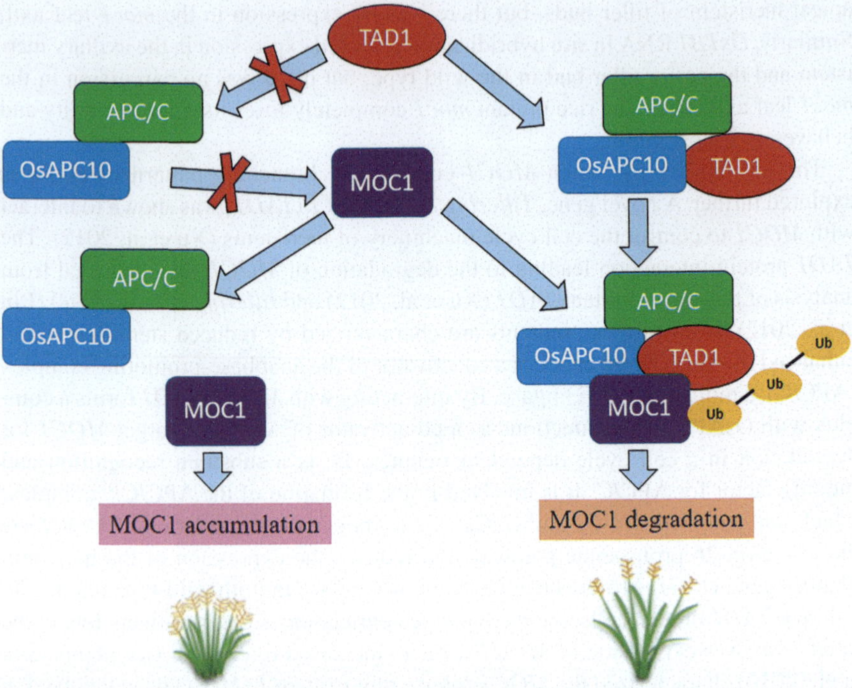

Fig. 9.2 The AP\C/C^{TAD1} complex-mediated degradation of MOC1 and MOC1 accumulation in absence of TAD1 leading to low- and high-tillering response in rice. (Adapted from Xu et al. 2012)

abnormal spikelets, and larger grains. It was reported that the transcript levels of cell proliferation- and expansion-associated genes are affected by the *LT1* gene in grains because of a size reduction. Two candidate genes, *LOC_Os0640780/MOC1* and *LOC_Os6g40480*, are involved in the molecular mechanism of tillering and panicle branching, and their decreased expression could be responsible for the *lt1* phenotype. In a similar investigation, Uddin et al. (2016) reported the identification of a single recessive low-tillering (*ltn2*) gene located on chromosome 7 from a new plant type cultivar of rice.

9.2.1.3 MOC2 Gene

Koumoto et al. (2013) characterized the rice monoculm mutant *moc2*, which exhibited a dwarf phenotype, possessing only a few tillers, a reduced growth rate, and pale green leaves. Although the mutant produced tiller buds, there was a significant deficiency in their outgrowth. The inconsistent change observed in the expression of genes involved in tiller bud outgrowth reveals the *moc2* mutant has the defective function needed for tiller bud outgrowth. *MOC2* (*Os01g64660*) is located on chromosome 1. This gene encodes fructose 1,6-biphosphatase 1 (FBP1), which is imperative for bud outgrowth. The insertion of the Tos 17 retrotransposon into exon 4 of *MOC2* leads to a significant decrease in tiller number compared to that in wild plants. In this *moc2* mutant, defective FBP activity resulted in a fall in the sucrose supply, causing the inhibition of tiller bud outgrowth. Hence, sucrose supply could be an important cue for tiller bud outgrowth in rice.

9.2.1.4 MOC3 Gene

A rice monoculm mutant, *monoculm 3* (*moc3*), was isolated and functionally characterized by Lu et al. (2015). *MOC3* (*Os04g56780*)/*TAB1*/*OsWUS* is located on chromosome 4, and this gene is needed for axillary bud formation on the shoot apical meristem. The normal specification of stem cell identity in the apical meristem is regulated by homeobox genes, *WUSCHEL* (*WUS*) and *WUSCHEL-RELATED HOMEOBOX* (*WOX*). Like its orthologue *WUS* in *Arabidopsis*, rice *MOC3* encodes a nucleus-localized transcription factor, and its loss-of-function mutation decreases tiller bud formation. The expression of *MOC3* is induced by cytokinin and the expression of two-component cytokinin response regulators, such as *OsRRs* and *ORRs*. Thus, *MOC3* is necessary for tiller bud initiation in rice through a complex relationship with cytokinin. In the rice monoculm mutant *moc3*, a point mutation causes premature termination of *Oryza sativa WUS* (*OsWUS*) expression, precluding axillary tiller formation. *WUS* expression is essential for tiller development in rice; it encourages tiller bud release by establishing weak apical dominance in the plant (Xia et al. 2020). Loss of function of *WUS* in the low-tillering mutant *decreased culm number 1* (*dc1*), which led to greater apical dominance and a greater reduction in tiller number than in the wild type. In the *dc1* mutant, the auxin response is enhanced, and knocking out the auxin-associated gene *ABERRANT SPIKELET AND PANICLE 1* (*ASP1*) releases the inhibition of tiller bud growth, suggesting that both *WUS* and *ASP1* are involved in this process.

Shao et al. (2019) showed that *MOC1* and *MOC3* (*MONOCULM 3/TILLERS ABSENT 1/STERILE AND REDUCED TILLERING 1* (*MOC3/TAB1/SRT1*) are two vital regulators for tiller formation in rice and they physically interact to regulate tiller bud outgrowth through over-expression of *FON1* (*FLORAL ORGAN NUMBER 1*) gene. In rice, *FON1* is homologous to *CLAVATA 1* and over-expression is activated in the axillary meristems when *MOC1* binds to its promoter region (Fig. 9.3). Loss of function mutation in *moc1* and *moc3* significantly reduces *FON1* function. Functionless *FON1* mutants exhibit defective bud growth and reduced tiller number.

9.2.2 LAX Genes

9.2.2.1 LAX1 Gene

The aerial architecture of rice plants is determined by shoot branching initiated by lateral bud growth in leaf axils. The *LAX1* gene is prominently responsible for the control of axillary meristem initiation and maintenance. By positional cloning, the gene was identified from the rice *lax1* (*lax panicle1*) mutant (Komatsu et al. 2003). The defects in the *lax1* mutant indicate that the *LAX1* gene is responsible for axillary bud initiation. The gene encodes a putative transcription factor that participates

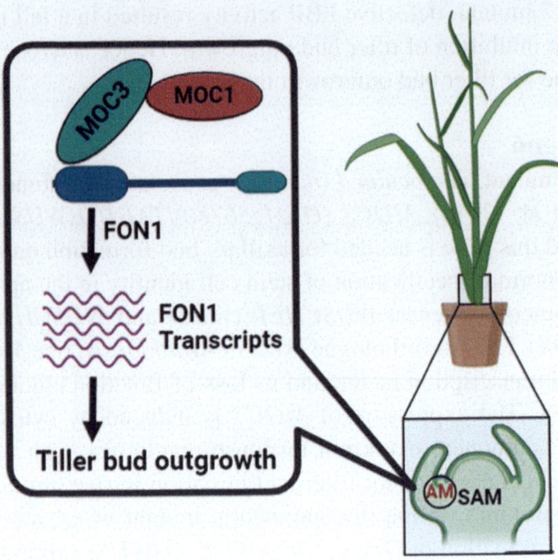

Fig. 9.3 A proposed model of rice tillering by *MOC3* and *MOC1* via activation of *FON1* in the axillary meristem. *SAM* shoot apical meristem, *AM* axillary meristem. (Adapted from Shao et al. 2019)

in the control mechanism. During the period of tiller initiation, *LAX1* expression is needed for the generation of the axillary meristem in rice. In addition, *LAX1* also acts as a floral meristem identity gene that determines the terminal spikelet of the rice panicle. *LAX1* expression is disrupted in plants harbouring a loss-of-function mutation in *LAX1*. Hence, *lax1* strongly reduces the number of rachis branches and spikelets on panicles and the number of tillers (Komatsu et al. 2001; Oikawa and Kyozuka 2009; Wang et al. 2020).

According to Oikawa and Kyozuka (2009), *LAX1* expression is initiated in the leaf axil of rice at the plastochron 4 (P4) stage, and two-step regulation of the spatial and temporal distribution of *LAX1* activity is needed for its complete function. *LAX1* mRNA accumulates in the boundary region between the axillary and the shoot apical meristems specifically. The LAX1 protein synthesized therein is subsequently trafficked towards the axillary meristem. Thus, the spatial distribution of the LAX1 protein is regulated by two consecutive steps, i.e. cell-specific mRNA expression and protein trafficking (Fig. 9.4). A similar two-step regulation also controls the temporal activity of LAX1. LAX1 mRNA expression is initiated at the P4 stage and continues thereafter, while LAX1 protein movement is confined to the P4 stage. This phenomenon reveals that stage- and cell-type-specific LAX1 mRNA expression is refined by stage- and direction-specific LAX1 protein movement. The Japanese authors concluded that during the axillary meristem initiation at the P4 stage, the LAX1 protein accumulates transiently through strict regulation of mRNA expression and subsequent protein trafficking, and this two-step regulatory process is critical for the establishment of new axillary meristems.

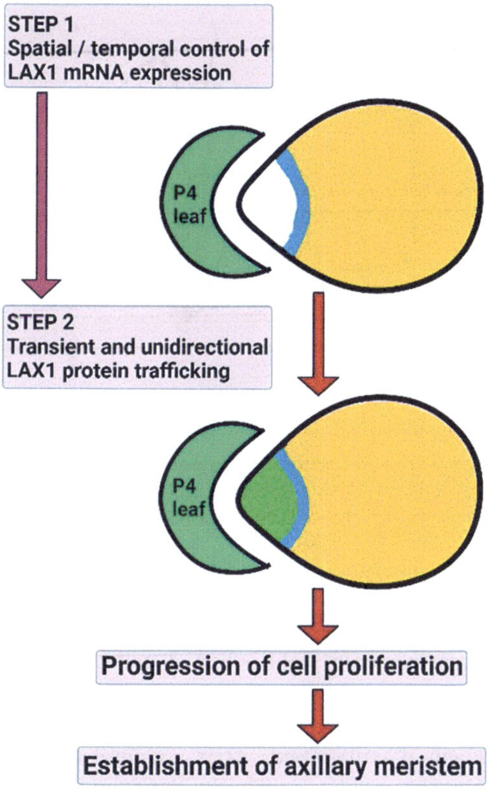

Fig. 9.4 A schematic model of the function of *LAX1* in the axillary meristem formation of rice. LAX1 mRNA is specifically expressed in the boundary region when the subtending leaf is at the P4 and P5 stages. At P4, the LAX1 protein synthesized in the boundary region moves towards the position of the yet-to-develop axillary meristem and promotes cell proliferation. In the absence of *LAX1* function, cell proliferation does not occur, and the axillary meristem is not established. LAX1 mRNA accumulation is shown in blue, and the LAX1 protein distribution is shown in green. (Adapted from Oikawa and Kyozuka 2009)

9.2.2.2 LAX2 Gene

Like in the case of *lax1* mutants, plants harbouring mutations in the *LAX2* gene exhibit reduced branching in the vegetative and reproductive phases. The *lax2* mutant lacks an axillary meristem in most of the lateral branches of the panicle and possesses a reduced number of axillary meristems in the vegetative stage. According to Tabuchi et al. (2011), *LAX2* encodes a nuclear protein with a conserved plant-specific domain that interacts with *LAX1*. These authors performed double mutant analyses to show that *MOC1*, *LAX1,* and *LAX2* work in a partially independent but overlapping pathway to control axillary meristem formation and development. A model summarizing the interaction between different genes in the regulation of tiller development in rice was proposed by Hussien et al. (2014) (Fig. 9.5).

9.2.3 MP3 Gene

Most recently a QTL called *MORE PANICLES 3* (*MP3*) has been cloned as a single gene that promotes tillering to increase panicle number moderately in rice (Takai 2024). This gene is a natural allele of *OsTB1/ FC1*, which is an ortholog of maize domestication gene *Teosinte Branched 1* (*TB1*) and holds immense value in

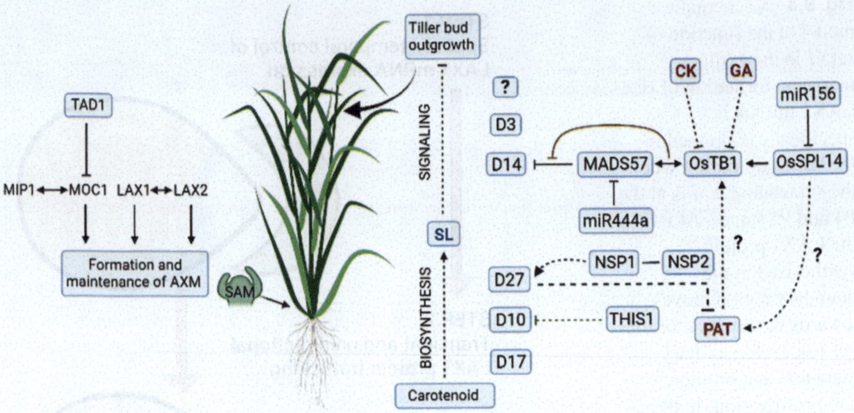

Fig. 9.5 Key interactions among genes controlling rice tiller number. Left: genes involved in axillary meristem (AXM) initiation and establishment. Right: genes involved in axillary bud outgrowth through the strigolactone (SL) pathway and its interactions with auxin (Aux), gibberellic acid (GA), and cytokinin (CK). Arrows indicate positive regulation, blunt-ended lines indicate negative regulation, dashed lines indicate multiple steps or indirect effects, and double-headed arrows represent protein–protein interactions. (Adapted from Hussien et al. 2014)

increasing grain yield potential under challenging situations encountered by the plants under climate change and nutrient deficiencies. In rice *OsTB1/ FC1* acts downstream of strigolactone biosynthesis and signalling pathways to inhibit outgrowth of axillary buds. *MP3*, a natural allele of *OsTB1/ FC1,* exhibits action similarly and regulates tiller development. Strigolactones are a group of terpenoid hormones synthesized in roots and shoots and transported acropetally in the xylem; they reach axillary buds and inhibit their growth. Interactions between genetic factors and plant hormones in the formation and growth of axillary tillers in rice have been revised in a new model (Takai 2024) (Fig. 9.6).

9.2.4 Amino Acid Transporter Genes

Inorganic nitrogen is absorbed by plants in the form of nitrate and ammonium ion in the roots. In the assimilation process, the absorbed nitrogen is converted into amino acids before they enter proteins. The amino acids synthesized in the roots or leaves require transporter proteins to traffic them from the source to the sink. These amino acid transporter proteins are membrane-bound, and they transport specific amino acids across the membrane barrier; their role is critical for various developmental processes of plants, including tiller production. Whole-genome sequence analysis revealed the presence of 79–85 amino acid homologue genes in rice (Lu et al. 2012; Zhao et al. 2012), and the biomass and yield of the plant changed when the *OsAAT* (rice amino acid transporter) genes were knocked out. In transport of

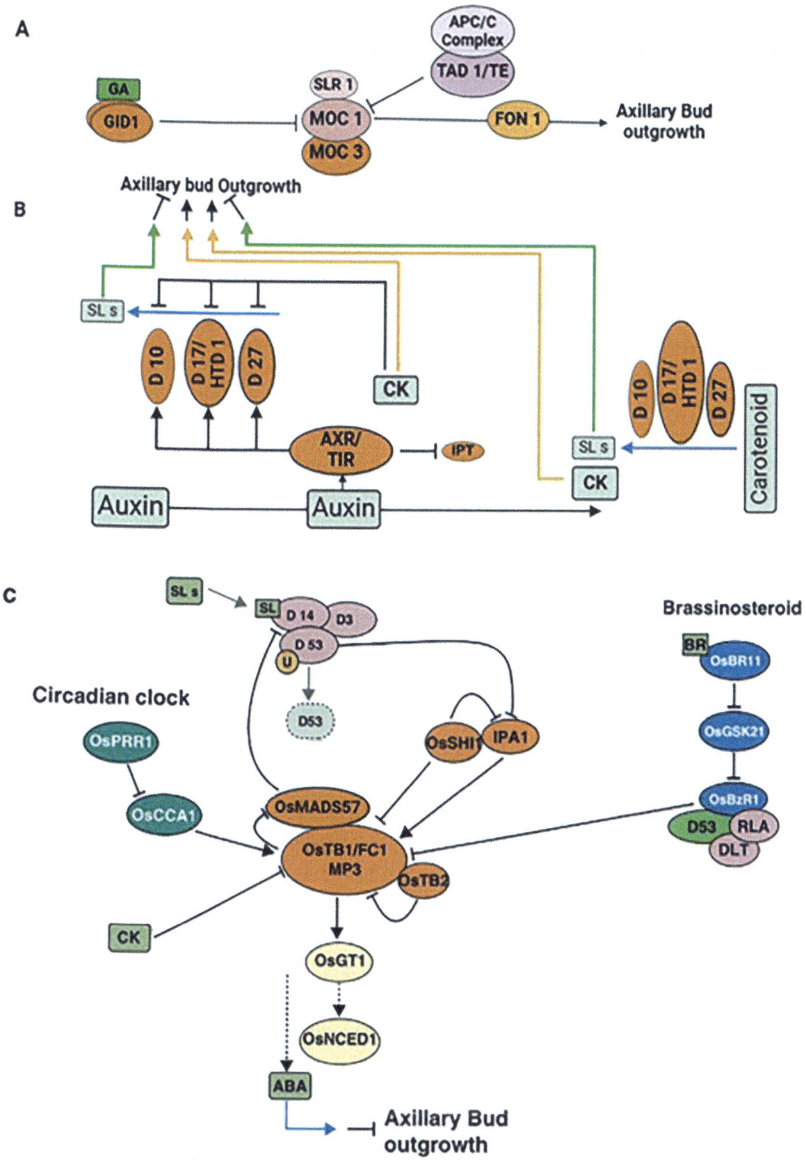

Fig. 9.6 The molecular mechanism for formation of a tiller bud outgrowth in rice. (**A**) Axillary bud outgrowth controlled by gibberellic acid (GA) and MONOCULM1 (MOC1); (**B**) Cross-talk among auxin, cytokinin (CK), and strigolactones (SLs) in shoots and roots; (**C**) Strigolactone-signalling pathway involving OsTB1/FC1 (MP3). Pointed arrows indicate promotion and flat-ended arrows indicate inhibition. (Adapted from Takai 2024)

rice-AATs, OsAAT3 ferries amino acids, namely, Ser, Met, Lys, Leu, His, Gln, Arg, Ala, and Gly (Taylor et al. 2015). Lu et al. (2018) showed that the expression of *OsAAP3 is* negatively correlated with the tiller number of rice and that blocking *OsAAP3* has the potential to improve grain yield by promoting tiller bud outgrowth and tiller number, especially in *japonica* rice, by regulating the accumulation of Lys, Arg, His, Asp, Ala, Gln, Tyr, and Thr amino acids. Accordingly, these authors proposed a model in which *OsAAP3* regulates rice tillering and grain yield (Fig. 9.7).

9.2.5 Genes Controlling Effective Tiller Number

In rice, all tillers produced in the vegetative stage do not survive to maturity or make yield contributions. Thus, instead of the whole tiller number, the effective tiller number is designed to determine the final grain yield. The effective tiller number per plant is an essential yield component of rice, and this trait is determined by both genetic and environmental factors (Xing and Zhang 2010). By using mutants of high- and low-tillering rice, many genes controlling rice tillers have been cloned, but seldom have they been utilized as breeding materials for the improvement of the rice phenotype. Although effective tiller number per plant is more pertinent to final yield, this parameter has received less attention in molecular breeding because it is a complex trait with relatively low heritability and is substantiated by multiple QTLs (Liu et al. 2010). Very recently, Ren et al. (2021) conducted a genome-wide association study for effective tiller number (ETN) per plant utilizing an exploratory panel of 490 rice accessions. In brief, the authors identified 38 ETN-associated QTLs in the GWAS, four of which co-existed with the *OsAAP1, DWL2, NAL1,* and *OsWRKY74* genes involved in rice tillering regulation. This exploration included the mining of genes homologous to known tillering-related genes and genes involved in hormone synthesis and signal transduction. Overall, 25 novel candidate genes related to ETN were discovered, including ten genes involved in auxin synthesis or signal transduction, three genes involved in cytokinin synthesis, and two genes involved in carotenoid synthesis (a precursor for strigolactones?), one gene involved in the regulation of GA signal transduction, two genes encoding ammonium transporters, two genes encoding peptide/amino acid transporters, and five genes encoding sugar/glucose transporters.

9.3 Integration of Hormonal Signals Controlling Rice Tillering and Tiller Dynamics

Some plant hormones, such as the classical hormones auxin and cytokinin and the new hormone strigolactone, are involved mainly in rice tillering and branching. Typically, hormones influence various aspects of plant development, including tillering, through the influence of various genetic factors. The unifying principles of hormonal and genetic control of shoot branching in deciding the number of tillers produced have been the subject of great discussion in laying a foundation for

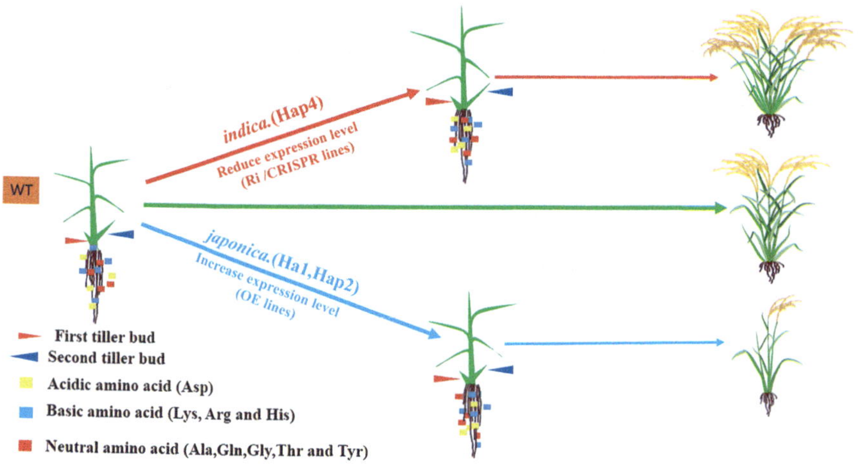

Fig. 9.7 The *OsAAP3* gene modulates tiller initiation and grain yield in rice. In the *OsAAP3-over-expressing* line (*OE*) and *japonica* line, elevated expression of *OsAAP3* results in the accumulation of basic amino acids (Lys, Arg, and His), acidic amino acid (Asp), and neutral amino acids (Ala, Gln, Gly, Thr, and Tyr). Amino acid accumulation is unfavourable for rice tillering. However, the RNAi/CRISPR line or the *indica* line of *OsAAP3* had the opposite effects. The small red or blue arrows located at the base of the plants indicate the first or second tiller buds, the long blue arrows indicate the development of rice *OE* lines or *japonica* plants (such as KY131) from seedlings to mature plants, the longer green arrow indicates the development of rice wild-type plants (such as *japonica* ZH11) from seedlings to mature plants, and the longer red arrows indicate the development of the rice RNAi/CRISPR line or *indica* plants from seedlings to mature plants. The blue squares indicate basic amino acids, the yellow squares indicate acidic amino acids (Asp), and the red squares indicate neutral amino acids. (Adapted from Lu et al. 2018)

rational approaches to breeding and the genesis of novel designer crops for improved performance and productivity (Wang et al. 2018). In addition to the effects of these hormones, several downstream genes were found to be activated, and some of these genes have been isolated. In rice plants, initiated axillary meristems contributing to tiller number far outnumber fully developed tillers because tiller bud growth is under the control of various phytohormones. The roles of auxin and cytokinin, in particular, are most critical for axillary bud outgrowth. Auxin synthesized at the shoot apical meristem is transported basally, and this hormone inhibits the growth of axillary meristems. In contrast, cytokinin promotes bud outgrowth and reverses this process. The hormone is synthesized in roots and exported to the axillary meristem through the xylem, which induces bud release (Muller and Leyser 2011). It is proposed that the concentration of these two hormones, as well as the ratio between them in the tillering node, are critical factors for the release of tiller buds (Cai et al. 2018). In addition, the third class of hormones that affects tillering in rice is strigolactones, which are negatively correlated with tiller number. This hormone is an inhibitor of axillary meristem outgrowth; it acts downstream of auxin or independently of polar auxin transport (Gomez-Roldan et al. 2008). It is also possible that microRNAs and their targeted transcription factors regulate tillering and branching patterns in rice plants (Wang et al. 2015). Concerning the mechanism underlying the hormonal regulation of tiller bud release, Liu et al. (2011) assessed the endogenous contents of auxin, cytokinin, and ABA in the tiller buds and nodes of rice plants treated with exogenous growth regulators such as GA, NAA, and ABA. The results suggested that auxin, cytokinin, and ABA are closely associated with tiller bud outgrowth; of these hormones, the first two play key roles, while the third affects the growth velocity of tiller buds.

9.3.1 Auxin-Associated Genes

The auxin distribution pattern inside plants is a major factor recognized to alter plant phenotypes. In rice, auxin is synthesized in the shoot primordia and young leaves around the shoot apical meristem and directionally transported to subtending organs extending up to the roots. The basipetal transport of auxin at the cellular level results from the asymmetric distribution of various membrane auxin carriers, such as the auxin influx carrier AUX1 family protein, the efflux-facilitating PIN FORMED (PIN) proteins, and several p-glycoprotein ABC transporters. Among these transporters, PIN proteins perform a rate-limiting role in stimulating the efflux of auxins from one cell to the cell positioned immediately below. The asymmetric cellular localization of PIN proteins on the basal side of the cell determines the direction of cell-to-cell flow, which is essential for auxin movement and distribution among plant organs and for their development. Hence, PIN protein-encoding gene expression has received increased amounts of attention about auxin-regulated growth processes.

A unique feature of auxin in rice is its ability to control tillering, tiller angle, and plant height by modifying the expression of *PIN* genes (Chen et al. 2012). In the rice genome, Wang et al. (2009) reported the presence of 12 putative *PIN* genes, which included four *PIN1* (*OsPIN1a-d*), three *PIN5* (*OsPIN5a-c*), three monocot-specific *PINs* (*OsPIN9, OsPIN10a,* and *b*), and one each of the *OsPIN2* and *OsPIN8* genes. The expression of these genes was upregulated by exogenous auxin, while their response to other hormones was uneven. An investigation of tissue-specific expression using RT–PCR analysis showed that the *OsPIN* genes exhibit constitutive expression in the stem, leaf, young panicle, and roots, but there is no specific target for axillary meristem-producing tiller buds in rice. Using a transgenic approach, Chen et al. (2012) showed that the over-expression of *OsPIN2* in *japonica* cv. *Nipponbare* rice resulted in a shorter plant height, wider tiller angle, and more tillers than those of the wild type. The over-expression of *OsPIN2* suppressed the expression of the gravitropism-related gene *OsLazy1* in the shoots while maintaining the expression of the tiller angle-controlling genes *OsPIN1b* and *OsTAC1*. Thus, *OsPIN2* has been shown to participate in a distinct auxin-dependent regulatory pathway together with the *OsPIN1b* and *OsTAC1* genes, which are responsible for designing the shoot architecture of rice. Compared with the wild-type control plants, the *OsPIN2*-over-expressing transgenic plants had shorter panicles, fewer grains per panicle, and a lower grain weight per panicle at the grain ripening stage. The over-expression of the *OsPIN2* significantly improved the total and effective tiller number per plant, but the plant height decreased (Chen et al. 2012, Table 9.2).

The pathway for auxin movement from one cell to the other is termed polar auxin transport (PAT), and fluctuations in nitrogen supply strongly impact auxin distribution in tissues. An imbalance in IAA distribution influences the nitrogen-regulatory mechanism involved in tiller development in rice. The mode of action of ammonium treatment on auxin-induced tillering in rice plants is greater than that of nitrate application (Hou et al. 2020). Based on GeneChip data, these authors revealed that *OsPIN9* is the candidate gene responsible for ammonium-regulated tillering of rice plants and that the OsPIN2 transporter co-ordinately controls tiller bud elongation in the presence of ammonium. *OsPIN2* is abundantly expressed in the roots and junctions from 2 weeks of age to the maturity stage in response to overt symptoms.

9.3.2 Auxin Genes Controlling Tiller Angle

The shoot architecture of rice plants is determined not only by tiller number but also by tiller angle. The tiller angle is the angle between the main shoot and its subordinate tillers. In recent years, the tiller angle has been considered one of the main target traits in rice breeding for achieving ideal plant phenotype and optimized yields. The tiller angle substantially influences grain yield because of its strong effect on plant density, leaf photosynthetic efficiency, and disease resistance. Rice

Table 9.2 Agronomic traits and yields of wild type (WT) and OsPIN2 transgenic seedlings (O1 and O2) in the pot experiment, Numbers are presented as mean ± SE

Genotype	WT	O1	O2
PH (cm)	71.60 ± 1.10^{a}	57.00 ± 2.80^{b}	52.80 ± 1.90^{b}
TTN/plant	37.33 ± 2.81^{b}	53.80 ± 2.48^{a}	49.33 ± 2.50^{a}
ETN/plant	31.84 ± 3.21^{b}	43.04 ± 3.53^{a}	39.47 ± 2.30^{a}
PL (cm)	21.24 ± 0.54^{a}	17.73 ± 0.64^{b}	17.84 ± 0.43^{b}
GW (g/panicle)	01.10 ± 0.06^{a}	00.74 ± 0.04^{b}	00.67 ± 0.06^{b}
SSR (%)	73.21 ± 2.31^{a}	75.84 ± 1.19^{a}	74.21 ± 1.40^{a}
GN/ panicle	59.47 ± 4.01^{a}	44.93 ± 3.49^{b}	36.38 ± 1.57^{b}
GL (mm)	07.25 ± 0.09^{a}	06.91 ± 0.09^{b}	06.82 ± 0.07^{b}
GB (mm)	03.19 ± 0.03^{a}	02.80 ± 0.06^{b}	02.73 ± 0.04^{b}
L/B	02.28 ± 0.02^{b}	02.47 ± 0.05^{a}	02.45 ± 0.03^{a}
1000-GW (g)	23.20 ± 0.41^{a}	20.81 ± 0.27^{b}	20.30 ± 0.09^{b}
Y (g/plant)	15.76 ± 0.54^{b}	18.33 ± 0.63^{a}	17.30 ± 1.02^{ab}

Means in the same column followed by the same letter are not significantly different ($P < 0.05$, LSD test). *PH* plant height (cm), *TTN* total tiller number per plant, *ETN* effective tiller number per plant, *PL* panicle length (cm), *GW* grain weight (g/panicle), *SSR* seed setting rate (%), *GN* grain number per panicle, *GL* grain length (mm), *GB* grain breadth (mm), *L/B* length/breadth, *1000-GW* 1000-grain weight (g), *Y* yield (g/plant). (Adapted from Chen et al. 2012)

plants with large tiller angles obtain easier access to resources, and they experience fewer pathogen attacks. However, a prostrate orientation occupies more space, decreasing yield per unit area. Several genes related to rice tiller angle have been characterized in the rice functional genomics project. These genes are *LAZY1* (*LA1*), *DWARF2*, *TILLER ANGLE CONTROL1* (*TAC1*), *TAC3*, *TAC4*, *PROSTRATE GROWTH 1* (*PROG1*), *PROG7*, *LOOSE PANICLE ARCHITECTURE1* (*LPA1*), *ONAC106* of the NAC protein family, *PLANT ARCHITECTURE AND YIELD1* (*PAY1*), and *TILLER INCLINED GROWTH1* (*TIG1*) (Chen et al. 2022).

The developing organs of plants sense gravity to control their orientation, including tiller angle (Wang et al. 2022). Among the genes described above, *LAZY1* (*LA1*) is the first gene identified to play a prominent role in the gravitropic response during tiller growth and to control tiller angle (Wang and Li 2008). Gravistimulation by dynamic reallocation of the LAZY1 (LA1) protein between the nucleus and plasma membrane controls the tiller angle. A mutation in *LA1* changes the IAA distribution in aerial plant parts, resulting in a tiller-spreading phenotype in rice plants (Li et al. 2007; Yoshihara and Iino 2007). According to Chen et al. (2022), a non-synchronous mutation in the transmembrane domain-encoding region of the allele *la1* is responsible for the spreading phenotype because the mutation causes a loss of shoot gravitropism, where prostrate growth of the shoot overtakes vertical alignment. Here, *LAZY1* acts as a negative regulator of polar auxin transport and redistributes auxin after it perceives a gravitropic stimulus. The stimulus reallocates the LA1 protein between the plasma membrane and nucleus for impact. Loss of function mutation in *LA1* expedites polar auxin transport and alters the asymmetric distribution of auxin in shoots, resulting in a spreading phenotype in rice plants. Li et al. (2019) reported that the interaction between Brevis Like 4 (OsBRXL4) and LAZY1 at the plasma membrane determines the relative proportion of the LA1 protein in the plasma membrane and nucleus, leading to an asymmetric redistribution of auxin and thereby a reorientation of the tiller angle. The most recent investigation on this subject marginalizes the role of LA1 localization to the plasma membrane in shoot gravitropism and instead suggests that LA1 could localize to the endoplasmic reticulum, which controls the gravitropic response of rice plants through triggering Ca^{2+} release.

Zhang et al. (2018) described the existence of a core regulatory pathway underlying the rice tiller angle mediated by the *LAZY1*-dependent asymmetric distribution of auxin. These findings demonstrated that HEAT STRESS TRANSCRIPTION FACTOR2D (HSFA2D) is an upstream positive regulator of the *LAZY1*-induced asymmetric auxin distribution pathway. *HSFA2D* induces the transcription of *LAZY1*, resulting in asymmetric auxin distribution. Auxin, in turn, induces asymmetric expression of two functionally redundant transcription factor genes, namely, *WUSCHEL-RELATED HOMEOBOX6* (*WOX6*) and *WOX11*, to determine the rice tiller angle. Compared to the compactly packed tillers in the wild type, the *hsfa2d* mutant exhibited widespread tillers occupying a larger space (Fig. 9.8, Table 9.3).

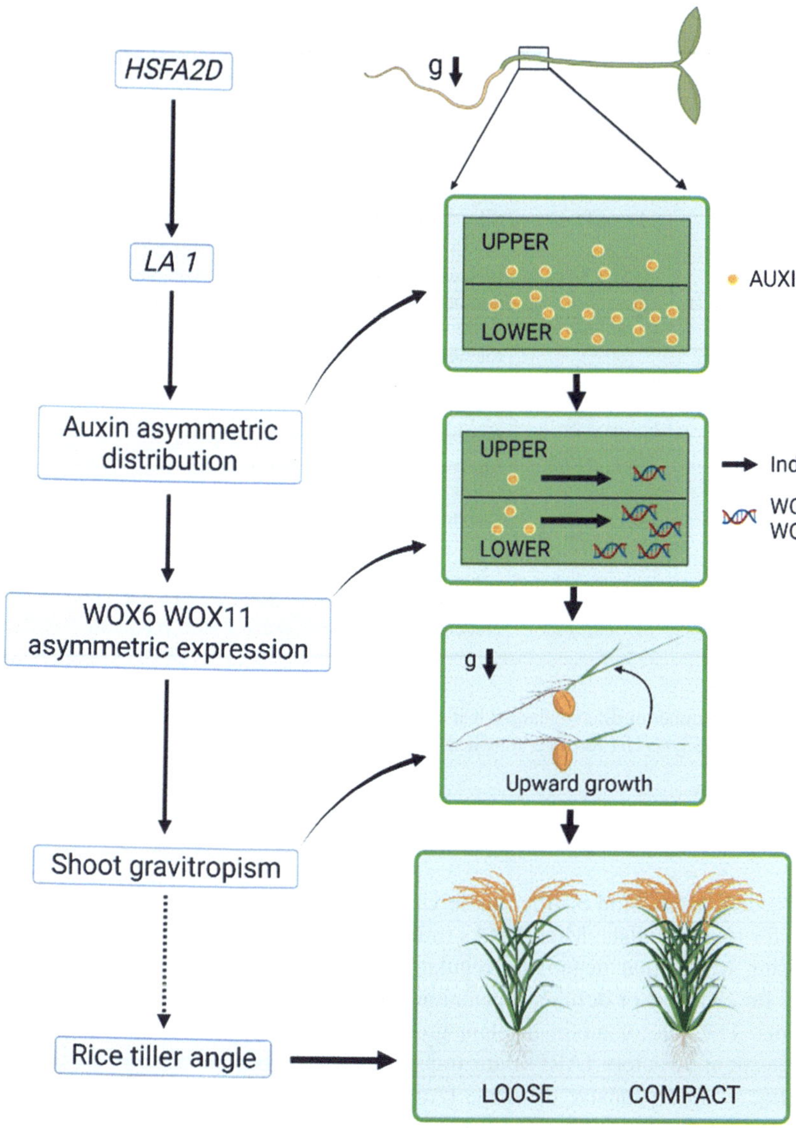

Fig. 9.8 A regulatory pathway of rice tiller angle via *LA1*-mediated asymmetric auxin distribution. *HSFA2D* regulates asymmetric auxin distribution by regulating the expression of *LA1*, which in turn modulates the asymmetric expression of the auxin downstream regulators *WOX6* and *WOX11* to determine rice shoot gravitropism and tiller angle. The arrows marked with 'g' indicate the direction of gravity. (Adapted from Zhang et al. 2018)

Table 9.3 Auxin-related genes *OsPIN1,OsPIN2, PINFORMED* and encoded proteins (Adapted from Hussien et al. (2014)

Name of the gene, protein or gene family	Homologue genes	Plant phenotype	Functions or pathway
OsPIN1-PIN1 family	*AtPIN1* in *Arabidopsis*	High tiller number	Auxin efflux carrier/ transports auxin
OsPIN2-PIN2 family	*AtPIN2* in *Arabidopsis*	High tiller number and tiller angle and reduced plant height	Auxin efflux carrier/ transports auxin and downregulation of *Oslazy1*
OsGH3.13-GH3 (*Gretchen Hagen3*)	*WSE1/GH3.5* in *Arabidopsis*	Increased tiller number, more leaf angle, higher drought tolerance, reduced panicle length, plant height and spikelet numbers	Multifunctional acetyl-amino synthetase conjugates IAA to amino acids, *Arabidopsis* homologues respond to IBA and salicylic acid pathways
ASP1/Ramosa1 Enhanced locus2-transcriptional corepressor	*TPL, REL2*	Increased primary tillers and decreased secondary tillers	It is involved in auxin signalling, depression of axillary bud growth and acts as a transcriptional co-repressor
OsmiR393, MIR393 Non-coding *miRNA*	*miR393* in *Arabidopsis*	Increased tiller number and early flowering	Represses *OsTIR1, OsAFB2* that further represses *OsAUX1* and auxin transport to axillary bud

TLD1 increased number of tillers/enlarged leaf angles/dwarfism, *ASP1* aberrant spikelet and panicle1, *TPL* TOPLESS, *OsTB1* teosinte branched1

9.3.3 Epigenetic Regulation of Tiller Angle

The epigenetic regulation of tiller angle via microRNAs has been a new subject of interest for rice scientists. MicroRNAs (miRNAs) are approximately 21 nucleotide non-coding RNAs that negatively regulate gene expression by binding to target mRNAs for cleavage or defunct translation at the post-translational level in various organisms. The role of auxin in plant growth and development is indispensable. Auxin Response Factors (ARFs) are transcription factors that bind to TGTCTC-containing auxin-responsive elements (AuxREs) in the promoter regions of early auxin response genes and mediate the signal transduction of auxin to the nucleus. Hence, infringement of *ARF* gene expression becomes critical for growth and development. In plant systems, several ARF genes have been identified as cleavage targets for miR160, miR390, and miR167. In rice, there are 25 *OsARFs,* of which *OsARF6, OsARF12, OsARF17,* and *OsARF25* are promising target genes of OsmiR167a. Li et al. (2020) reported that OsmiR167a represses the expression patterns of *OsARF12, OsARF17,* and *OsARF25* to modulate the tiller angle of rice plants by fine-tuning the asymmetric distribution of auxin in shoots. Over-expressed OsmiR167a in plants strongly represses the expression of *OsARF12, OsARF17,* and *OsARF25,* increasing the tiller angle, similar to the pattern observed in *osarf12/*

osarf17 and *osarf12/osarf25* plants. These plants exhibited abnormal auxin distributions and were less sensitive to gravity. The authors concluded that the *OsARFs* function redundantly and become involved in the *HSFA2D*- and *LAZY1*-directed asymmetric auxin distribution pathways to regulate the rice tiller angle (Fig. 9.9). In rice, both the *OsARF12/17/25* and *HSFA2D-LAZY1* systems specify tiller angles by regulating the asymmetric distribution of auxin and subsequently inducing asymmetric expression of *WOX6/11*.

9.3.4 Cytokinin-Associated Genes

In rice plants, initiated axillary meristems far outnumber the fully developed tillers because many of the axillary meristems do not grow after initiation. Axillary bud growth is normally under hormonal regulation. Auxin synthesized at the shoot tip is transported basally to repress tiller bud growth, whereas cytokinin synthesized in roots and stem nodes acts antagonistically to reverse this process and promote bud release. In plants, cytokinins regulate growth and development by promoting cell division and tissue differentiation. Therefore, genes controlling cytokinin homeostasis are critical for organ growth and development. The intrinsic concentration of the hormone in the tissue is determined by the balance between the amount synthesized and the amount inactivated. Cytokinin synthesis begins from tRNA, ATP, or ADP catalyzed by different adenosine phosphate isopentenyl transferases; the LONELY GUY enzyme activates the enzymes. The biodegradation of the hormone occurs by specific enzymes, such as cytokinin oxidase/dehydrogenases (CKXs), which lead to irreversible oxidative cleavage. Compared to the genes encoding adenosine phosphate isopentenyl transferase, more emphasis has been given to the expression of *CKX* genes in deciding on cytokinin homeostasis for tissue-specific growth regulation.

Fig. 9.9 A schematic pathway regulating the rice tiller angle via *OsmiR167a-OsARFs* and *HSFA2D-LAZY1*. (Adapted from Li et al. 2020)

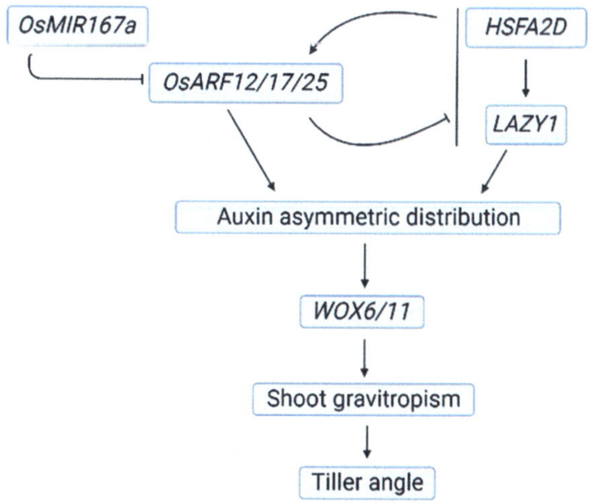

The number of *CKX* genes varies among species. In rice, 11 *CKX* family genes have been identified, although their specific functions have not been elucidated (Rong et al. 2022). In the Rice Annotation Project database (https://rapdb.dna.affrc.go.jp), these genes were assigned accession numbers as follows. *OsCKX1* (Os01g0187600), *OsCKX2* (Os01g0197700), *OsCKX3* (Os10g0483500), *OsCKX4* (Os01g0940000), *OsCKX5* (Os01g0775400), *OsCKX6* (Os02g0220000), *OsCKX7* (Os02g0220100), *OsCKX8* (Os04g0523500), *OsCKX9* (Os05g0374200), *OsCKX10* (Os06g0572300), and *OsCKX11* (Os08g0460600). These genes exhibit functional divergence and overlap in growth and development, particularly in the control of tillering. Using CRISPR/Cas9 technology, Rong et al. (2022) constructed mutants of these 11 genes and found that the *ckx* single- or double-mutant lines exhibited functional overlap and different organ-specific expression patterns. The *OsCKX* phylogenetic tree consisted of five clades (Rong et al. 2022) as follows (Table 9.4):

Among the *CKX* genes, eight were expressed at the shoot base and in the inflorescence meristem at different stages and positions. *CKX2*, *CKX3*, and *CKX11* were not expressed in flag leaf primordia. In axillary buds, *CKX1* is expressed at the top, *CKX2*, *CKX3*, *CKX4*, and *CKX5* are expressed at the base, and *OsCKX9* is expressed throughout the entire bud. *CKX* expression in rice plants is influenced by different plant hormones, affecting the homeostasis of cytokinin in tissues. According to Duan et al. (2019), the expression of most rice *CKXs* is upregulated by exogenous treatment with trans-zeatin (tZ), *N*6-(Δ2-isopentenyl) adenine (iP), or 6-benzylaminopurine (6-BA). Individually, *OsCKX4* is induced by auxin; this gene integrates auxin and cytokinin signalling to control crown root initiation (Gao et al. 2014). The over-expression of *OsCKX4* reduces the tiller number, panicle grain number, and plant height but increases the root number. *OsCKX9* interacts with cytokinin to determine tiller development in rice. The gene is activated by strigolactone. Compared to those of the WT plants, the *osckx9* mutants had more tillers but fewer panicles. Moreover, plants over-expressing *OsCKX9* also exhibit more tillers, smaller panicles, shorter plant heights, and lower grain-setting rates (Duan et al. 2019). Conversely, *OsCKX11* elicits an antagonistic relationship between cytokinin and ABA (Zhang et al. 2021a). Cytokinin levels in the flag leaves of rice *osckx11* mutants significantly increase, while the expression of ABA-biosynthesizing and ABA-degrading genes decreases and increases, respectively. Ultimately, compared

Table 9.4 Tissue-specific expression of *OsCKX* genes

Clade	Gene	Tissue-specific expression pattern
First	*OsCKX1*	Shoot base and at the top axillary buds
	OsCKX2	Leaf sheath and inflorescence
Second	*OsCKX6*	Very low in all tissues
	OsCKX7	Very low in all tissues
	OsCKX10	Very low in all tissues
Third	OsCKX4	Strongly affects vegetative organs, particularly roots
	OsCKX5	Strong in all tissues, particularly roots and leaves
	OsCKX9	Low in all tissues, high in the leaf blade and axillary bud
Fourth	*OsCKX3*	Shoot base and young panicle
	OsCKX8	Shoot base, flag leaf primordia, and panicle
Fifth	*OsCKX11*	Shoot base, roots, and young panicle

with those in the WT plants, the ABA level in the mutant plants was lower, and leaf photosynthesis and grain and tiller numbers were greater. *CKX2/Gn1a* is the first cytokinin gene identified in rice. Reduced expression of *CKX2* or *ckx2* in rice plants leads to an increase in panicle grain number, tiller number, and grain weight (Ashikari et al. 2005; Yeh et al. 2015).

To elucidate the expression patterns and functions of these genes comprehensively, Rong et al. (2022) used CRISPR/Cas9 technology to construct mutants of all 11 *OsCKXs*. The *ckx* single mutants and higher-order *ckx4 ckx9* mutant lines exhibited functional overlap. There was considerable phenotypic difference between the *osckx1 osckx2* and *osckx4 osckx9* double mutants and the WT. The *osckx1 osckx2* double mutant had significantly fewer tillers, while the double knockouts of *OsCKX4* and *OsCKX9* increased tillering. Thus, *OsCKX4* and *OsCKX9* inhibited tillering, and *OsCKX1* and *OsCKX2* promoted tillering in rice. The *osckx4 osckx9* double mutant also presented shorter and fewer roots. These findings imply that *OsCKX4* and *OsCKX9* are involved prominently in the biodegradation of cytokinin and they regulate cytokinin levels to exhibit an impact on plant phenotype. In the study by Rong et al. (2022), the expression of the *OsRRs* (response regulator genes induced by cytokinin), namely, *OsRR4*, *OsRR6*, *OsRR7*, *OsRR9*, and *OsRR10*, was significantly depressed in the roots, but except *OsRR2*, the expression of all detectable *RR* genes was enhanced in the axillary buds of *ckx4 ckx9* double mutants. The functions of various *OsCKX* genes in rice plants are illustrated in Fig. 9.10.

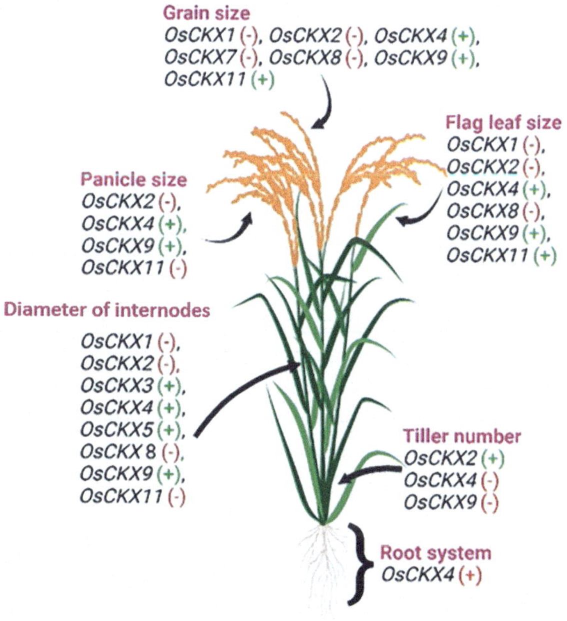

Fig. 9.10 Functions of CKX genes in rice. The function of each gene in regulating the growth and development of different organs is summarized based on the phenotypes of all the mutants observed. '+' indicates positive regulation, and '–' indicates negative regulation. (Adapted from Rong et al. 2022)

9.3.5 Adenosine Phosphate Isopentenyl Transferase (IPT) Genes

Another candidate gene family in the rice genome responsible for the production of bioactive cytokinin and the subsequent control of tillering is the adenosine phosphate isopentenyl transferase (IPT) gene. IPT catalyzes the rate-limiting step of cytokinin biosynthesis. According to Sakamoto et al. (2006), there are eight *IPT* genes, *OsIPT1*-*OsIPT8*, in rice, including the pseudogene *OsIPT6*. Over-expression of the *IPT* gene in transgenic rice induces axillary bud growth but inhibits root growth, and the phenotype of these plants resembles that of KNOX (KNOTTED-like homeobox)-overproducing rice plants. KNOX proteins encoded by *Knox* genes accumulate in indeterminate cells of the shoot apical meristem. It promotes the synthesis of bioactive cytokinins, such as trans-zeatin (tZ) and isopentenyl adenine (iP), but suppresses the expression of *GA20 oxidase* genes to the detriment of GA biosynthesis. Thus, the KNOX protein plays a crucial role in promoting meristem activity at the shoot apex through the simultaneous activation of cytokinin and suppression of gibberellin biosynthesis. The expression of a weak allele of the *KNOX* gene, *SHOOTMERISTEMLESS* (*STM*), provided evidence that a combination of constitutive GA signalling and low cytokinin levels is unfavourable for shoot apical meristem function (Jasinski et al. 2005). A BLAST search performed by Sakamoto et al. (2006) using predicted amino acid sequences encoded by *Arabidopsis IPT* genes as probes against the rice DNA database revealed previously reported higher expression of the *OsIPT2* and *OsIPT3* genes subject to induction of KNOX protein function. The results confirmed that the KNOX protein decreases GA biosynthesis and promotes cytokinin biosynthesis via induction of *OsIPT2* and *OsIPT3* expression, leading to the high-cytokinin and low-GA conditions necessary for the initiation and maintenance of meristem activities. Multiple clumps of shoots were found in the *OsIPT*-transformants of rice, although they did not form tillers.

9.3.6 Strigolactone-Related Genes

As discussed before, shoot branching in rice is one of the most crucial morphological traits and is physiologically modulated by auxins, cytokinins, and strigolactones cumulatively. Therefore, the interplay of these three hormones is highly important for understanding the complex phenomenon of axillary branching initiation and bud release. In this complex interplay, auxin moving basipetal from shoot apices acts indirectly to inhibit axillary bud release by regulating the synthesis of strigolactones and cytokinins in the stem nodes. In addition, root-synthesized strigolactones and cytokinins are directly transported to buds and affect axillary bud outgrowth, where the hormones interact antagonistically because of their opposite effects on a common target, i.e. *BRC1* expression. The mechanism of the antagonistic effects of cytokinins and strigolactones on axillary bud release has been recently elucidated by studying the transcription levels of *OsCKXs* (Zha et al. 2022). The authors reported that strigolactones function antagonistically with cytokinins in controlling axillary bud release at node 2 of rice culms by enhancing *OsCKX* gene expression

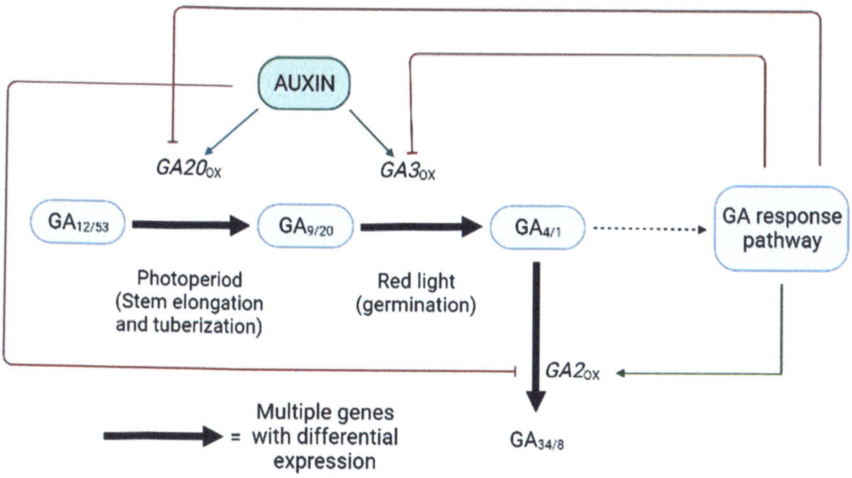

Fig. 9.11 The pathway of gibberellin biosynthesis in plants showing the identities of the metabolic enzyme-encoding genes and the way that their transcription is regulated by feedback, the environment, and other endogenous hormones. (Adapted from Taiz and Zeiger 2002)

in the bud and eventual degradation of the cytokinin pool. Thus, strigolactones inhibit bud outgrowth by reducing local cytokinin levels. The cytokinin contents exhibit a sub-optimal response to both endogenous and exogenous strigolactones in the axillary buds.

Like that of strigolactones, bud outgrowth is also suppressed by nitrogen and phosphorus deficiencies at the tillering stage. In cytokinin homeostasis, type-A response regulator (*RR*) genes are considered the targets of cytokinin signalling for primary transcription and rapid response. In Zha et al. (2022) study, a direct supply of the synthetic strigolactone rac-GR24 reduced benzyl adenine (BA)-stimulated axillary bud outgrowth on node 2 of the culm. The transcript levels of most of the ten *OsRR* genes examined increased upon treatment with benzyl adenine (BA) but significantly decreased following the application of synthetic strigolactone. Thus, specific suppression of *OsCKX* by strigolactones could increase the growth and yield of rice plants by increasing the tiller number. It is possible that the induction of *OsCKX* expression occurred via a cytokinin-independent pathway. The molecular interaction between cytokinin and strigolactone-related genes that modulate the tillering and tiller dynamics of rice has been enlisted and described by Hussien et al. (2014).

The most recent investigation on how SL signalling regulates tiller bud outgrowth in rice via *D14* has been elucidated by Liu et al. (2022). The authors described a rice mutant, *dwarf and high-tillering1* (*dht1*), which exhibits pleiotropic phenotypes of dwarfism along with high tiller numbers similar to those of their counterparts defective in SL-signalling. *DHT1* encodes a monocotyledon-specific hnRNP-like protein that works as an intron-splicing factor for many precursor mRNAs (pre-mRNAs), inclusive of the SL-receptor gene *D14*. The *dht1* mutant protein loses stability and RNA-binding activity, resulting in defective splicing of

D14 pre-mRNA and reduced *D14* expression. This leads to an SL-defective phenotype.

9.3.7 Gibberellin-Related Genes

Tillering in rice is inhibited by gibberellin. This concept has been utilized in the Green Revolution to decrease plant height and increase the number of tillers and achieve higher grain yield. The inverse relationship between plant height and tillering was elucidated by the exploration targeting GA biosynthesis gene *sd1*, which resulted in the generation of semi-dwarf stature of the plant. The *sd1* of rice is a recessive allele that resides on chromosome 1. It is a dwarf gene of the Taiwanese cultivar Dee-geo-woo-gen. The dwarf gene bestows lodging resistance through the shortened stem, and the plant becomes profuse tillering and high-yielding (Gaur et al. 2020). The dwarfs occur due to a low amount of GA caused by mutation of the *GA20ox-2* gene. The expression of the gene infringes upon GA biosynthesis, which limits the availability of the chemical for culm elongation. The mechanism by which gibberellin overwhelms tiller production for stem elongation has been elucidated through its control on *MOC1*. *MOC1* is the tiller regulator gene responsible for the axillary meristem formation. It is expressed in the axillary buds. It was previously discussed that *TAD1* interacts with MOC1 and anaphase-promoting complex 10 (APC 10) targeting *MOC1* for destruction in a cell cycle-dependent manner. MOC1 binds to rice DELLA protein SLENDER RICE 1 (SLR1) and protects the degradation of MOC1 with a reduced tiller number. According to (Liao et al. 2019, GA promotes the degradation of SLR1 and MOC1 for increased stem height and reduction of tiller number.

In land plants, IAA promotes the biosynthesis of active gibberellin by increasing the expression of *GA20ox* and *GA3ox* and blocking the expression of *GA2ox*. (O'Neill and Ross 2003; Taiz and Zeiger 2002). C19- and C20-GA2 oxidases catalyze the 2-β-hydroxylation reaction to inactivate the intrinsic level of bioactive GA, and thereby affect GA-regulated development. In rice, there are two classes of *GA2oxs* gene numbering ten that inactivate bioactive GA through 2b-hydroxylation: *C19 GA2oxs* are the larger class, whereas *C20 GA2oxs* are the smaller class (Lo et al. 2008). Rice lines over-expressing *GA2oxs* exhibit early and increased tillering and adventitious root growth. Lo et al. (2008) reported that, in rice, GA negatively regulates the expression of two transcription factors, *O. sativa* homeobox 1 and TEOSINTE BRANCHED1, which in turn control meristem initiation and axillary bud outgrowth, respectively, leading to the inhibition of tillering. Compared to the *C19 GA2oxs* class, the *C20 GA2oxs* class causes a less severe GA-defective phenotype. The improvements in rice phenotypes, such as semi-dwarfism, greater root growth, and greater tillering, could be induced by the over-expression of modified *C20 GA2oxs*. The pathway of gibberellin biosynthesis in plants has been described in Fig. 9.11.

Approximately, 60 transcription factors have been identified in plants so long, and they include factors such as AP2/ERF, bZIP, C2H2, MYB, MADS, NAC, and

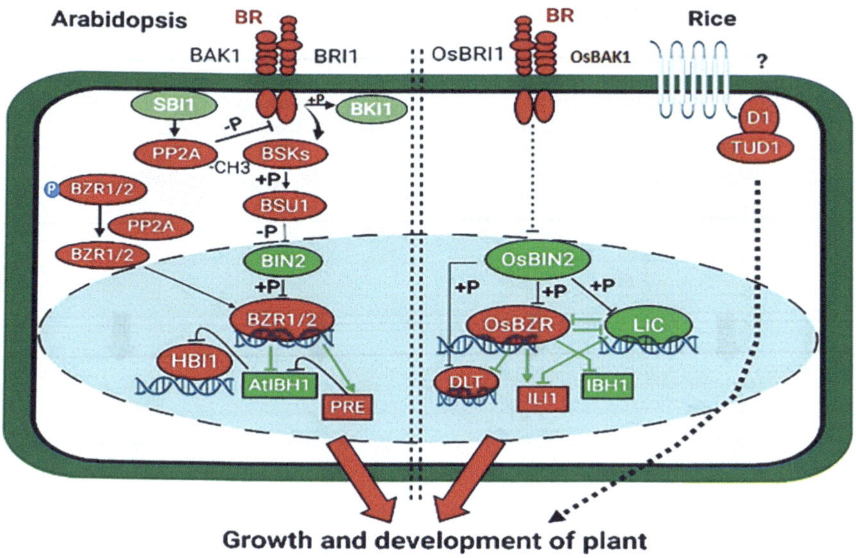

Fig. 9.12 A schematic model for BR-signalling pathway in rice and *Arabidopsis*. The left panel (*Arabidopsis*) shows the direct interaction of BR with BRI1 (BRASSINOSTEROID-INSENSITIVE1) and BAK1 (BRI1-ASSOCIATED RECEPTOR KINASE 1) to form the BRI1-BR-BAK1 complex. The sequential transphosphorylation between BRI1 and BAK1 activates BRI1, which phosphorylates BSKs (BR-SIGNALLING KINASEs)/CDG1 (CONSTITUTIVE DIFFERENTIAL GROWTH1). The active BSKs/CDG1 phosphorylates BSU1 (BRI1-SUPPRESSOR1), which dephosphorylates and inactivates BIN2 (BRASSINOSTEROID-INSENSITIVE2. BZR1(BRASSINAZOLE-RESISTANT1) and BES1(BRI1-EMS-SUPPRESSOR1)/BZR2 are dephosphorylated by PP2A (PROTEIN PHOSPHATASE2A) and move into the nucleus to induce expression of PRE1 (PACLOBUTRAZOL RESISTANT1), but repress expression of IBH1 (INCREASE LEAF INCLINATION1 BINDING HLH1). PRE1, IBH1, and HBI1 form an antagonistic cascade to regulate plant growth. SBI1 (SUPPRESSOR OF BRI1) is involved in the deactivation of BRI1 through methylation of PP2A. The right panel shows the BR-signalling pathway in rice. BR binding to OsBRI1 promotes association with OsBAK1 and inactivates OsBIN2 via an unknown mechanism. OsBIN2 phosphorylates OsBZR1, LIC (TILLER ANGLE INCREASED CONTROLLER), and DLT (DWARF and LOW-TILLERING), and inhibits their activity. OsBZR1 induces the expression of ILI1 (Increase Leaf Inclination 1), but represses the expression of IBH1, LIC, and DLT; whereas LIC represses the expression of OsBZR1. The heterotrimeric G-protein α (alfa) subunit known as D1(DWARF1)/RGA1 in rice is involved in another BR-signalling pathway in which D1/RGA1 and TUD1 (TAIHU DWARF1) act together to mediate BR signalling. The red colour indicates components responsible for positive roles in BR signalling, whereas green colour denotes the components that play negative roles. (Adapted from Zhang et al. 2014)

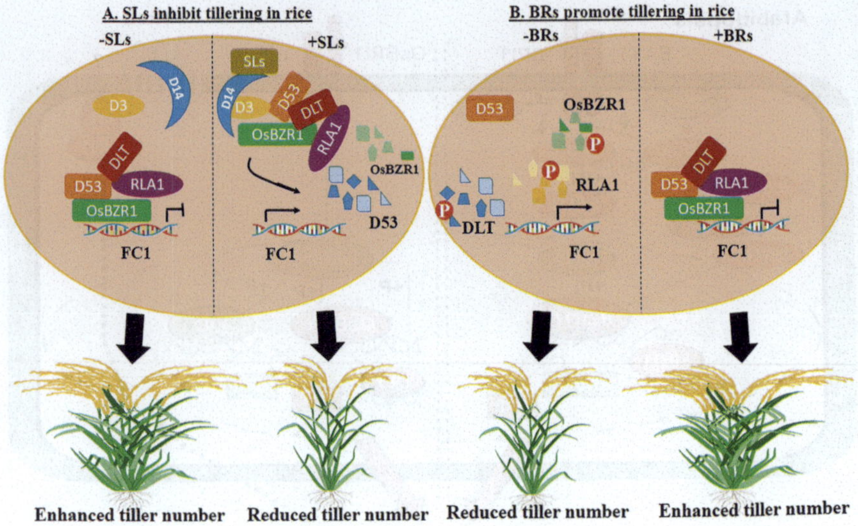

Fig. 9.13 SL and BR signalling in rice for regulation of tiller number. (**A**) SL-mediated tillering is dependent on the protein level of the D53–OsBZR1 complex in rice. The degradation of D53 and OsBZR1 induced by SL signalling promotes the expression of FC1 to inhibit rice tillering. (**B**) BR-mediated tillering is dependent on the OsBZR1–RLA1–DLT complex in rice. The induced accumulation of OsBZR1 facilitates the formation of the D53–OsBZR1 complex, which in turn inhibits the expression of FC1 to promote rice tillering. (Adapted from Fang et al. 2020)

WRKY. Transcription factors associated with DNA-binding domains control various developmental processes. APETALA2/ethylene response factor (AP2/ERF) is widely distributed among plants, including rice; the AP2/ERF family of transcription factors has various physiological and developmental functions in rice and holds immense prospects for breeding excellent rice varieties with good grain yield (Xie et al. 2022). The *AP2/ERF* gene *OsEATB* (rice *ERF protein associated with tillering and panicle branching*) was cloned from an *indica* rice variety by Qi et al. (2011), who showed that the putative interaction between ethylene and gibberellin is mediated by this gene, leading to a varietal difference in stem elongation. *OsEATB* gene expression underrates ethylene-induced enhancement of stem elongation by down-regulating the gibberellin biosynthesis-related gene ent-kaurene synthase A. Thus, *OsEATB* decreases plant height and panicle length and enhances the branching potential of tillers and panicles of rice. This evidence confirms the negative correlation between gibberellins and tillering in rice.

9.3.8 Brassinosteroid-Related Genes

In plants, gibberellins and brassinosteroids are major hormones that exhibit similar patterns of growth and development. Like gibberellins, a deficiency of brassinosteroids reduces plant height, which could be beneficial for tillering in rice. In general, tiller number and height are negatively correlated in rice, and dwarf plants usually have more tillers than control plants. In contrast to the negative relationship between plant height and tiller number, dwarf rice plants defective in brassinosteroid biosynthesis pathways have more tillers than wild-type plants, except *dwarf and low-tillering (dlt)* plants (Tong et al. 2009). Furthermore, *dlt* has reduced sensitivity to brassinosteroids. OsDLT is a GRAS family protein. It downregulates the expression of several brassinosteroid synthesis genes, resulting in a reduction in tillering and height. Exogenous treatments of brassinosteroids repress the transcription of *OsDLT*. Tong et al. (2009) found that OsBZR1, a key transcription factor in the brassinosteroid signalling pathway, binds to the promoter region of *OsDLT* to inhibit its expression. Similarly, OsBZR1 functions as a positive regulator of brassinosteroid signalling, and its suppression leads to erect leaves, semi-dwarfism, and brassinosteroid-insensitive phenotypes in rice (Zhang et al. 2014). It is surmised that brassinosteroids acting via OsBZR1 (rice BRASSINAZOLE-RESISTANT1) determine rice tillering by regulating the expression of *OsDLT*. Furthermore, in brassinosteroid-deficient or brassinosteroid-insensitive mutants, the activity of OsBZR1 is repressed by OsBIN2 phosphorylation, resulting in high expression of OsDLT and a consequential decrease in tillering. The expression of *OsDLT* increases in the *osd2* (*osdwarf2*) and *osd11* mutants, which supports this finding. A model illustrating BR signalling in rice and *Arabidopsis* has been proposed by Zhang et al. (2014) (Fig. 9.12).

9.4 BR and SL Signalling Co-ordinately Regulate Rice Tillering

As discussed above, BR-signalling strongly accelerates rice tillering by promoting axillary bud release. However, genetic and biochemical analyses conducted by Fang et al. (2020) revealed that both BR and SL signalling co-ordinately determine rice tillering by regulating the stability of D53 and the OsBZR1-RLA-DLT module. In rice, *FC1* is a local inhibitor of tillering that acts downstream of SL and BR signalling. D53 interacts with OsBZR1 to repress *FC1* expression. The level of inactivation depends on direct DNA binding by OsBZR1, which involves D53 binding to the *FC1* promoter in axillary buds. The results revealed that the SL and BR-signalling pathways interact antagonistically to control rice tillering through the OsBZR1-D53 complex via the early responsive gene *FC1*. Furthermore, these authors observed that OsBZR1 functions as a substrate of D3 in the rice SL-signalling pathway. A schematic model showing the coordinated interaction of the SL- and BR-signalling pathways for the regulation of rice tillering has been proposed in Fig. 9.13 (Fang et al. 2020).

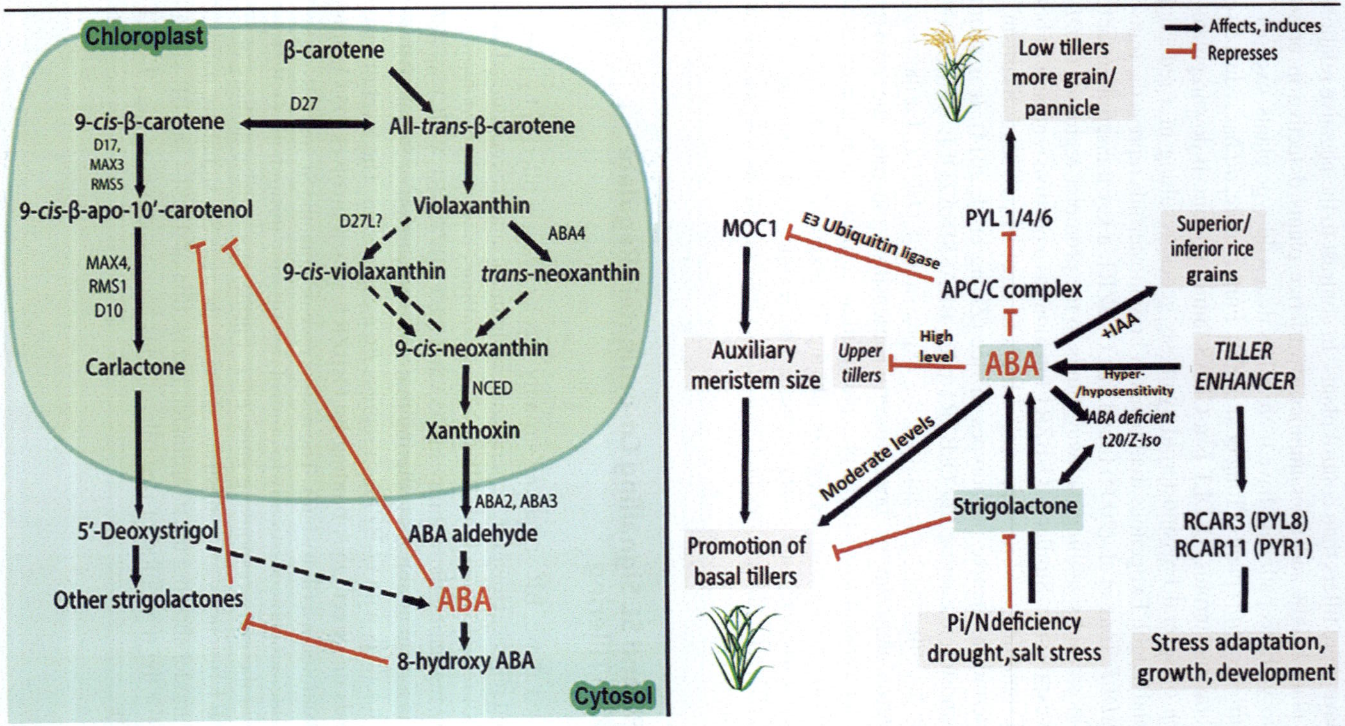

Fig. 9.14 The role of ABA signalling in plant development and yield formation. *ABA* abscisic acid, *APC* anaphase-promoting complex, *NCED* 9-cis-epoxy carotenoid dioxygenase, *PYL* pyrabactin resistance-like, *PYR* pyrabactin resistance, *RCAR* Regulatory component of ABA receptor, *MOC1* monoculm 1. (Adapted from Kavi Kishor et al. 2022 with permission)

9.5 ABA Co-ordinates with SL to Modulate Tillering

In rice plants, ABA mediates important physiological events ultimately targeting grain-yielding attributes. The concentration of the hormone fluctuates with the variation of environmental conditions; it remains low (basal) level in a normal ambient situation but rises briskly under stress-prone conditions. The functions of the hormone-modulating plant development differ significantly between basal versus elevated concentrations. The dynamic function of ABA in regulating the tiller formation of rice plants has been elucidated in a recent review (Kavi Kishor et al. 2022). The impact of the hormone on tillering is not independent and coordinated with strigolactones (SL). Both ABA and SL share a common metabolite source from the carotenoid biosynthesis pathway. The enzyme ζ-carotene isomerase, which is involved in the carotenoid biosynthesis pathway, also accounts for the biosynthesis of both ABA and SL (Liu et al. 2020). Thus, ABA also exhibits a negative impact on the tiller formation of rice similar to SL. The authors found that treatment of an SL analogue stimulated expression of *Oryza sativa 9-CIS-EPOXYCAROTENOID DIOXYGENASE 1 (OsNCED1)*, which generates a precursor for ABA in the shoot base, but the application of ABA was not productive for SL biosynthesis. It was concluded that high tiller formation occurs in rice mutant *tillering20* because of defects in *ζ-carotene isomerase* gene expression responsible for limitations in ABA and SL biosynthesis. Kavi Kishor et al. (2022) opined that the effect of endogenous ABA on rice tillering is site-specific; a high concentration negatively affects the formation of upper tillers, while moderate levels of ABA promote basal tillers. Conversely, SL inhibits the promotion of basal tillers. The molecular biology showing ABA and SL biosynthesis and their interaction for regulation of various aspects of development in rice plants is given in Fig. 9.14.

9.6 Other Factors Controlling Tillering

9.6.1 Ideal Plant Architecture 1 (IPA1)

In addition to the genes discussed above, there are several major and minor genes involved in the determination of the number of tillers, spikelets, and panicle branches in rice. In rice plants, tiller growth and their number depend on expression profile and loss of function of some genetic factors. Since the discovery of the *semi-dwarf* gene (*sd1*), several attempts have been made to break the yield barrier in rice. The ideal plant architecture (IPA) concept has been proposed to overcome the ceiling of high-yielding rice cultivars created based on the *sd1* background. Jiao et al. (2010) reported that a semi-dominant quantitative trait locus, *IPA1* (*Ideal Plant Architecture 1*), strongly influenced the rice plant architecture and greatly enhanced rice grain yield. *IPA1 QTL* encodes OsSPL14 (SQUAMOSA PROMOTER BINDING PROTEIN-LIKE14), which is regulated by OsmiR156, a non-coding miRNA. Jiao et al. (2010) found that a point mutation in *OsSPL14* disturbed the OsmiRNA156 regulation of *OsSPL14*. Perturbation of the function of *OsSPL14* was found to be associated with IPA, i.e. reduced tiller, greater resistance to lodging, and improved yield.

9.6.2 FZP Gene

APO2/RFL and *FZP* play crucial roles in regulating the floral organ identity genes (Bai et al. 2016). Bai et al. (2016) did a unique experiment where they used a novel mutant *fzp-11*, *FZP*-over-expressing plants, and wild type. *FZP* through the regulation of OsMADS-box controls the floral organ identity. FZP is the most important negative regulator of APO2/RFL and also determines the transition from panicle branching to spikelet differentiation. Over-expression of *FZP* severely restricts the AXM formation in both the vegetative and reproductive stages and thus impairs tiller as well as spikelet formation. Both losses of function and over-expression of *FZP* perturb normal growth leading to unwanted AXM differentiation, which impacts greatly tiller growth as well as spikelet formation. In contrast to *FZP*, rice *aberrant panicle organization 1* (*apo1*) positively controls spikelet number. The *apo1* mutants have a smaller inflorescence meristem in contrast to *APO1* over-expression plants (Ikeda-Kawakatsu et al. 2009). In *APO1*-over-expression plants, a higher rate of cell proliferation occurs in inflorescence meristem even as compared to wild types. *APO1* and *APO2/RFL* both work jointly on inflorescence development. So, the transition from vegetative to the reproductive stage, panicle differentiation, and spikelet formation is a coordinated action of several genes.

Table 9.5 Other genes/factors influencing tillering, spikelet number, and branching (Adapted from Ikeda-Kawakatsu et al. 2009; Jiao et al. 2010; Qi et al. 2011; Schertl et al. 2012; Hussien et al. 2014; Bai et al. 2016; Zhao et al. 2019)

Genes	Function	Phenotype
OsmiR156	Perturbation of function	Reduce effective tiller, increase spikelet and grain number
	Over-expression	Tiller numbers increase, and spikelet and grain numbers decrease
FRIZZY PANICLE (*FZP*)	fzp mutant, loss of function of *FZP*	AXM formation, increased secondary branches, panicle length normal however, spikelets formation ceased to '0'
	Over-expression	Reduced tiller number, panicle length and spikelet number
Aberrant panicle organization (*APO1*)	Mutant, loss of function	Increased tillers and leaf number, decreased panicle branches and spikelets
Aberrant panicle organization 2 (*APO2*)	Mutant, loss of function	Reduced panicle branching, increased tillers and leaf number
ABCG (*OsABCG5*/*rcn1*-)	abcg mutant, loss of function	Reduced tillers, reduced panicle branches
APETALA2 (AP2)	Over-expression	Reduced plant height and panicle length, increased tiller and panicle branches
l-Galactono-1,4-lactone dehydrogenase (GLDH)	Downregulation	Premature senescence, reduced plant height and tillers

9.6.3 ERF Gene

Tiller formation in rice is also affected by ethylene-gibberellin signal interaction. The rice *APETALA2 (AP2)/Ethylene-Responsive Element-Binding Factor* (*ERF*) gene *OsEATB* influences greatly in determining the plant height, tiller number, and grain yield. Its over-expression restricts ethylene-induced augmentation of GA responsiveness through downregulation of the GA biosynthetic gene, *ent-kaurene synthase*. Its action thus determines plant height, tiller number, and grain yield (Qi et al. 2011). In addition to plant hormones, antioxidant like ascorbic acid (Asc) plays an important role in plant growth and determines the length of the plant and the tiller number (Schertl et al. 2012). GLDH (l-Galactono-1,4-lactone dehydrogenase) enzyme catalyzes the last step of ascorbic acid biosynthesis in plants. The transgenic plant with partial loss of GLDH activity greatly reduces the ascorbic acid formation compared to wild types. The number of tillers is significantly lower in Asc-deficient plants. Lack of sufficient quantities of Asc induces oxidative stress with hormonal disbalances resulting in premature senescence and reduction of tiller number (Schertl et al. 2012). Thus, tiller development is a complex event where several genes act in a coordinated fashion involving hormonal regulation to antioxidant involvement suppressing oxidative damage.

9.7 Other Genes/Factors Influence Tillering, Spikelet Number, and Branches

Several genes are involved in rice architecture. The action of many minor and major genes determines tillering, spikelets number, and panicle branching. As shown in Table 9.5. tiller growth and number depend on the expression profile and loss of function of genes. Since the discovery of the *semi-dwarf* gene (*sd1*), several attempts have been made to break the yield barrier in rice. The ideal plant architecture (*IPA*) concept has been proposed to overcome the genetic ceiling of high-yielding rice cultivars based on the *sd1* background. Jiao et al. (2010) reported that a semi-dominant quantitative trait locus, *IPA1* (*Ideal Plant Architecture 1*), strongly influenced rice plant architecture and greatly enhanced rice grain yield. The *IPA1 QTL* encodes OsSPL14 (SQUAMOSA PROMOTER-BINDING PROTEIN-LIKE 14), which is regulated by a non-coding miRNA, OsmiR156. Jiao et al. (2010) reported that a point mutation in *OsSPL14* disrupted the OsmiRNA156-mediated regulation of *OsSPL14*. Perturbation of the function of *OsSPL14* was shown to be associated with IPA, i.e. reduced tillers, greater resistance to lodging, and improved yield.

APO2/RFL and *FZP* play critical roles in regulating floral organ identity genes (Bai et al. 2016). Bai et al. (2016) performed a unique experiment using a novel mutant *fzp-11*; *FZP*-over-expressing plants; and the wild type. FZP, through the regulation of OsMADS-box, control the floral organ identity. FZP is the most important negative regulator of APO2/RFL and determines the transition from panicle branching to spikelet differentiation. Over-expression of FZP strongly restricts axillary meristem (AXM) formation in both the vegetative and reproductive stages and thus impairs tiller and spikelet formation. Both loss of function and over-expression of FZP perturb normal growth, leading to unwanted AXM differentiation, which greatly impacts tiller growth and spikelet formation. In contrast to the *FZP* gene, rice *aberrant panicle organization 1* (*apo1*) positively controls spikelet number. Compared to *APO1*-over-expressing plants, *apo1* mutants have a smaller inflorescence meristem (Ikeda-Kawakatsu et al. 2009). In *APO1*-over-expressing plants, a greater rate of cell proliferation occurs in the inflorescence meristem than in the wild type. *APO1* and *APO2/RFL* both work jointly in inflorescence development. Therefore, the transition from the vegetative to the reproductive stage, panicle differentiation and spikelet formation are coordinated actions of several genes.

Antioxidant levels and variations in the quantities of plant hormones have profound effects on general growth as well as on determining tiller numbers and spikelet formation, leading to grain yield. Hormones such as cytokinin, ethylene, gibberellin, and antioxidant ascorbic acid are highly important for determining plant height and tiller number. It was observed that very frequently severe short-stature plants produce tillers profusely; however, these types of plants are often not highly efficient at providing good yields. A balance of plant height and tiller number is needed to obtain maximum yield. A plant that is too tall is also not optimal for obtaining yield due to its weakness in lodging. In most cases, the site of formation and the site of action of a plant hormone are quite different. Cytokinin produced in roots must be transferred to shoots for tiller development. The ATP-binding cassette transporter OsABCG18 transports root-derived cytokinin to the shoot. In *osabcg18* mutants, the transport of cytokinin is severely impaired, leading to a drastic reduction in grain yield. However, its over-expression improved cytokinin transport and grain yield (Zhao et al. 2019). Ethylene-gibberellin signalling plays a vital role in plant growth. The AP2/ERF transcription factor is found only in planta. The rice APETALA2 (AP2)/Ethylene-Responsive Element-Binding Factor (ERF) gene *OsEATB* strongly influences plant height, tiller number, and grain yield. *OsEATB* over-expression restricts ethylene-induced augmentation of GA responsiveness through downregulation of the GA biosynthesis-related gene ent-kaurene synthase. Its activity thus determines plant height, tiller number, and grain yield (Qi et al. 2011).

In addition to plant hormones, antioxidants such as Asc play important roles in plant growth and determine the height of plants as well as tiller number (Schertl et al. 2012). The enzyme GLDH (l-Galactono-1,4-lactone dehydrogenase) catalyzes the terminal step of ascorbic acid biosynthesis in plants. Compared with wild-type plants, transgenic plants with partial loss of GLDH activity greatly reduce Asc formation. The number of tillers is significantly lower in Asc-deficient plants. A lack of sufficient quantities of Asc induces oxidative stress, and hormonal imbalances result in premature senescence and a reduction in tiller number (Schertl et al. 2012). In rice, tiller development is a complex process in which several genes act in a coordinated fashion involving hormonal regulation and antioxidant activity to suppress oxidative damage.

References

Ashikari M, Sakakibara H, Lin S, Yamamoto T, Takashi T, Nishimura A, Angeles ER, Qian Q, Kitano H, Matsuoka M (2005) Cytokinin oxidase regulates rice grain production. Science 309(5735):741–745

Bai X, Huang Y, Mao D, Wen M, Zhang L, Xing Y (2016) Regulatory role of FZP in the determination of panicle branching and spikelet formation in rice. Sci Rep 6:19022. https://doi.org/10.1038/srep19022

Cai T, Meng X, Liu X, Wang H, Jia Z, Yang D, Ren X (2018) Exogenous hormonal application regulates the occurrence of wheat tillers by changing endogenous hormones. Front Plant Sci 9:1886. https://doi.org/10.3389/fpls.2018.01886

Chen Y, Fan X, Song W, Zhang Y, Xu G (2012) Over-expression of OsPIN2 leads to increased tiller numbers, angle, and shorter plant height through suppression of *OsLAZY1*. Plant Biotechnol J 10(2):139–149

Chen S, Huang Y, Han J, Zhang S, Yang Q, Li Z, Zhang Y, Mao R, Fan L, Liu Y, Chen Y, Xie X (2022) Blocking rice shoot gravitropism by altering one amino acid in LAZY1. Int J Mol Sci 23(16):9452

Duan J, Yu H, Yuan K, Liao Z, Meng X, Jing Y, Liu G, Chu J, Li J (2019) Strigolactone promotes cytokinin degradation through transcriptional activation of CYTOKININ OXIDASE/DEHYDROGENASE 9 in rice. Proc Natl Acad Sci USA 116(28):14319–14324

Fang Z, Ji Y, Hu J, Guo R, Sun S, Wang X (2020) Strigolactones and Brassinosteroids antagonistically regulate the stability of the D53–OsBZR1 complex to determine FC1 expression in rice tillering. Mol Plant 13(4):586–597

Gao S, Fang J, Xu F, Wang W, Sun X, Chu J, Cai B, Feng Y, Chu C (2014) CYTOKININ OXIDASE/DEHYDROGENASE4 integrates cytokinin and auxin signalling to control rice crown root formation. Plant Physiol 165(3):1035–1046

Gaur VS, Channappa G, Chakraborti M, Sharma TR, Mondal TK (2020) 'Green revolution' dwarf gene sd1 of rice has gigantic impact. Brief Funct Genom 19:390–409

Gomez-Roldan V, Fermas S, Brewer PB, Puech-Pages V, Dun EA, Pillot JP, Letisse F, Matusova R, Danoun S, Portais JC, Bouwmeester HJ, Becard G, Beveridge CA, Rameau C, Rochange SF (2008) Strigolactone inhibition of shoot branching. Nature 455(7210):189–195. https://doi.org/10.1038/nature07271

Hou M, Luo F, Wu D, Zhang X, Lou M, Shen D, Yan M, Mao C, Fan X, Xu G, Zhang Y (2020) OsPIN9, an auxin efflux carrier, is needed for the regulation of rice tiller bud outgrowth by ammonium. New Phytol 229(2):935–949

Hussien A, Tavakol E, Horner DS, Munoz-Amatriain M, Muehlbauer GJ, Rossini L (2014) Genetics of tillering in rice and barley. Plant Genome 7(1):1–20. https://doi.org/10.3835/plantgenome2013.10.0032

Ikeda-Kawakatsu K, Yasuno N, Oikawa T, Iida S, Nagato Y, Maekawa M, Kyozuka J (2009) Expression level of ABERRANT PANICLE ORGANIZATION1 determines rice inflorescence form through control of cell proliferation in the meristem. Plant Physiol 150(2):736–747

Itoh J-I, Nonomura K-I, Ikeda K, Yamaki S, Inukai Y, Yamagishi H, Kitano H, Nagato Y (2005) Rice plant development: from zygote to spikelet. Plant Cell Physiol 46(1):23–47

Jasinski S, Piazza P, Craft J, Hay A, Woolley L, Rieu I, Phillips A, Hedden P, Tsiantis M (2005) KNOX action in *Arabidopsis* is mediated by coordinate regulation of cytokinin and gibberellin activities. Curr Biol 15:1560–1565

Jiang S, Wang D, Yan S, Liu S, Liu B, Kang H, Wang G-L (2019) Dissection of the genetic architecture of rice tillering using a genome-wide association study. Rice 12:43. https://doi.org/10.1186/s12284-019-0302-1

Jiao Y, Wang Y, Dawei Xue D, Wang J, Yan M, Liu G, Guojun Dong G, Dali Zeng D, Lu Z, Zhu X, Qian Q, Li J (2010) Regulation of *OsSPL14* by OsmiR156 defines ideal plant architecture in rice. Nat Genet 42(6):541–544. https://doi.org/10.1038/ng.591

Kavi Kishor PB, Tiozon RN Jr, Fernie AR, Sreenivasulu N (2022) Abscisic acid and its role in the modulation of plant growth, development, and yield stability. Trends Plant Sci 27(12):1283–1295

Komatsu M, Maekawa M, Shimamoto K, Kyozuka J (2001) The LAX1 and FRIZZY PANICLE 2 genes determine the inflorescence architecture of rice by controlling the rachis branch and spikelet development. Dev Biol 231(2):364–373

Komatsu K, Maekawa M, Ujiie S, Satake Y, Furutani I, Okamoto H, Shimamoto K, Kyozuka J (2003) LAX and SPA: major regulators of shoot branching in rice. Proc Natl Acad Sci U S A 100(20):11765–11770. https://doi.org/10.1073/pnas.1932414100

Koumoto T, Shimada H, Kusano H, She K-C, Iwamoto M, Takano M (2013) Rice monoculm mutation *moc2*, which inhibits outgrowth of the second tillers, is ascribed to lack of a fructose-1,6-bisphosphatase. Plant Biotechnol 30(1):47–56

Li X, Qian Q, Fu Z, Wang Y, Xiong G, Zeng D, Wang X, Liu X, Teng S, Hiroshi F, Yuan M, Luok D, Han B, Li J (2003) Control of tillering in rice. Nature 422(6932):618–621

Li P, Wang Y, Qian Q, Fu Z, Wang M, Zeng D, Li B, Wang X, Li J (2007) LAZY1 controls rice shoot gravitropism by regulating polar auxin transport. Cell Res 17(5):402–410

Li Z, Liang Y, Yuan Y, Wang L, Meng X, Xiong G, Zhou J, Cai Y, Han N, Hua L, Liu G, Li J, Wang (2019) OsBRXL4 regulates shoot gravitropism and rice tiller angle by affecting LAZY1 nuclear localization. Mol Plant 12(8):1143–1156

Li Y, Li J, Chen Z, Wei Y, Qi Y, Wu C (2020) OsmiR167a-targeted auxin response factors modulate tiller angle via fine-tuning auxin distribution in rice. Plant Biotechnol J 18(10):2015–2026

Liang WH, Shang F, Lin QT, Lou C, Zhang J (2014) Tillering and panicle branching genes in rice. Gene 537(1):1–5

Liao Z, Yu H, Duan J, Yuan K, Yu C, Meng X, Kou L, Chen M, Jing Y, Liu G, Smith SM, Li J (2019) SLR1 inhibits MOC1 degradation to coordinate tiller number and plant height in rice. Nat Commun 10:2738. https://doi.org/10.1038/s41467-019-10667-2

Lin Q, Wang D, Dong H, Gu S, Cheng Z, Gong J, Qin R, Jiang L, Li G, Wang JL, Wu F, Guo X, Zhang X, Lei C, Wang H, Wan J (2012) Rice APC/C(TE) controls tillering by mediating the degradation of MONOCULM1. Nat Commun 3:752

Liu GF, Zhu HT, Liu SW, Zeng RZ, Zhang ZM, Li WT, Ding XH, Zhao FM, Zhang GQ (2010) Unconditional and conditional QTL mapping for the developmental behaviour of tiller number in rice (*Oryza sativa* L.). Genetica 138(8):885–893

Liu Y, Ding YF, Wang QS, Li GH, Xu JX, Liu ZH, Wang SH (2011) Effect of plant growth regulators on the growth of rice tiller bud and changes of endogenous hormones. Acta Agron Sin 37(4):670–676

Liu X, Hu Q, Yan J, Sun K, Liang Y, Jia M, Meng X, Fang S, Wang Y, Jing Y, Liu G, Wu D, Chu C, Smith SM, Chu J, Wang Y, Li J, Wang B (2020) z-Carotene isomerase suppresses tillering in rice through the coordinated biosynthesis of strigolactone and abscisic acid. Mol Plant 13:1784–1801

Liu T, Zhang X, Zhang H, Cheng Z, Liu J, Zhou C, Luo S, Luo W, Li S, Xing X, Chang Y, Shi C, Ren Y, Zhu S, Lei C, Guo X, Wang J, Zhao Z, Wang H, Zhai H, Lin Q, Wan J (2022) *Dwarf and High Tillering1* represses rice tillering through mediating the splicing of *D14* pre-mRNA. Plant Cell 34(9):3301–3318

Lo S-F, Yang S-Y, Chen K-T, Hsing Y-I, Zeevaart JAD, Chen LJ, Yu S-M (2008) A novel class of Gibberellin 2-Oxidases control semidwarfism, tillering and root development in rice. Plant Cell 20(10):2603–2618

Lu Y, Song Z, Lu K, Lian X, Cai H (2012) Molecular characterization, expression and functional analysis of the amino acid transporter gene family(*OsAATs*) in rice. Acta Physiol Plant 34:1943–1962

Lu ZF, Shao GN, Xiong JS, Jiao YQ, Wang J, Liu GF, Meng XB, Liang Y, Xiong GS, Wang YH (2015) MONOCULM 3, an orthologue of WUSCHEL in rice, is needed for tiller bud formation. J Genet Genomics 42(2):71–78

Lu K, Wu B, Wang J, Zhu W, Nie H, Qian J, Huang W, Fang Z (2018) Blocking amino acid transporter *OsAAP3* improves grain yield by promoting outgrowth buds and increasing the tiller number in rice. Plant Biotechnol J 16(10):1710–1722

Mohapatra PK, Panda BB, Kariali E (2011) Plasticity of tiller dynamics in wild rice *Oryza rufipogon* Griff. A strategy for ensuring resilience in suboptimal environments. Int J Agron 543237:9

Muller D, Leyser O (2011) Auxin, cytokinin and the control of shoot branching. Ann Bot 107(7):1203–1212. https://doi.org/10.1093/aob/mcr069

O'Neill DP, Ross JJ (2003) Auxin regulation of the gibberellin pathway in pea. Plant Physiol 130(4):1974–1982. https://doi.org/10.1104/pp.010587

Oikawa T, Kyozuka J (2009) Two-step regulation of LAX PANICLE1 protein accumulation in axillary meristem formation in rice. Plant Cell 21(4):1095–1108

Qi W, Sun F, Wang Q, Chen M, Huang Y, Feng Y-Q, Luo X, Yang J (2011) Rice ethylene-response AP2/ERF factor OsEATB restricts internode elongation by downregulating a gibberellin biosynthesis-related gene. Plant Physiol 157(1):216–228

Ren M, Huang M, Qiu H, Chun Y, Li L, Kumar A, Fang J, Zhao J, He H, Li X (2021) Genome-Wide Association Study of the genetic basis of effective tiller number in rice. Rice 14(1):56

Rong C, Liu Y, Chang Z, Liu Z, Ding Y, Ding C (2022) Cytokinin oxidase/dehydrogenase family genes exhibit functional divergence and overlap in rice growth and development, especially in control of tillering. J Exp Bot 73(11):3552–3568

Sakamoto T, Sakakibara H, Kojima M, Yamamoto Y, Nagasaki H, Inukai Y, Sato Y, Matsuoka M (2006) Ectopic expression of KNOTTED1-like homeobox protein induces expression of cytokinin biosynthesis genes in rice. Plant Physiol 142(1):54–62

Schertl P, Sunderhaus S, Klodmann J, Grozeff GEG, Bartoli CG, Braun H-P (2012) L-Galactono-1,4-lactone dehydrogenase (GLDH) forms part of three subcomplexes of mitochondrial complex I in *Arabidopsis thaliana*. J Biol Chem 287(18):14412–14419

Shao G, Lu Z, Xiong J, Wang B, Jing Y, Meng X, Liu G, Ma H, Liang Y, Chen F, Wang Y, Li J, Yu H (2019) Tiller bud formation regulators MOC1 and MOC3 cooperatively promote tiller bud outgrowth by activating *FON1* expression in rice. Mol Plant 12:1090–1102

Sun F, Zhang W, Xiong G, Yan M, Qian Q, Li J, Wang Y (2010) Identification and functional analysis of the MOC1 interacting protein1. J Genet Genomics 37(1):69–77. https://doi.org/10.1016/S1673-8527(09)60026-6

Tabuchi H, Zhang Y, Hattori S, Omae M, Shimizu-Sato S, Oikawa T, Qian Q, Nishimura M, Kitano H, Xie H, Fang X (2011) *LAX PANICLE2* of rice encodes a novel nuclear protein and regulates the formation of axillary meristems. Plant Cell 23(9):3276–3287

Taiz L, Zeiger E (2002) Plant physiology, 3rd edn. Sinauer Associates, Sunderland, MA

Takai T (2024) Potential of rice tillering for sustainable food production. J Exp Bot 75:708–720

Taylor MR, Reinders A, Ward JM (2015) Transport function of rice amino acid permeases (AAPs). Plant Cell Physiol 56(7):1355–1363. https://doi.org/10.1093/pcp/pcv053

Tong H, Jin Y, Liu W, Li F, Fang J, Yin Y, Qian Q, Zhu L, Chu C (2009) DWARF AND LOW-TILLERING, a new member of the GRAS family, plays positive roles in brassinosteroid signaling in rice. Plant J 58(5):803–816

Uddin MN, Tomita A, Obara M, Yanagihara S, Fukuta Y (2016) Identification of a low tiller gene from a new plant type cultivar in rice (*Oryza sativa* L.). Breed Sci 66(5):790–796

Wang Y, Li J (2008) Molecular basis of plant architecture. Annu Rev Plant Biol 59:253–279. https://doi.org/10.1146/annurev.arplant.59.032607.092902

Wang JR, Hu H, Wang GH, Li J, Chen JY, Wu P (2009) Expression of PIN genes in rice (*Oryza sativa* L.): tissue specificity and regulation by hormones. Mol Plant 2(4):823–831

Wang L, Sun S, Jin J, Fu D, Yang X, Weng X, Xu C, Li X, Xiao J, Zhang Q (2015) Coordinated regulation of vegetative and reproductive branching in rice. Proc Natl Acad Sci U S A 112(50):15504–15509

Wang B, Smith SM, Li J (2018) Genetic regulation of shoot architecture. Annu Rev Plant Biol 69:437–468

Wang Y, Wei S, He Y, Yan L, Wang R, Zhao Y (2020) Synergistic roles of *LAX1* and *FZP* in the development of rice sterile lemma. Crop J 8(1):16–25

Wang W, Gao H, Liang Y, Li J, Wang Y (2022) Molecular basis underlying rice tiller angle: current progress and future perspectives. Mol Plant 15(1):125–137

Xia T, Chen H, Dong S, Ma Z, Ren H, Zhu X, Fang X, Chen F (2020) *OsWUS* promotes tiller bud growth by establishing weak apical dominance in rice. Plant J 104(6):1635–1647

Xie W, Ding C, Hu H, Dong G, Zhang G, Qian Q, Ren D (2022) Molecular events of rice AP2/ERF transcription factors. Int J Mol Sci 23(19):12013. https://doi.org/10.3390/ijms231912013

Xing YZ, Zhang QF (2010) Genetic and molecular bases of rice yield. Annu Rev Plant Biol 61(1):421–442. https://doi.org/10.1146/annurev-arplant-042809-112209

Xu C, Wang Y, Yu Y, Duan J, Liao Z, Xiong G, Meng X, Liu G, Qian Q, Li J (2012) Degradation of MONOCULM 1 by APC/C(TAD1) regulates rice tillering. Nat Commun 3:750

Yeh SY, Chen HW, Ng CY, Lin CY, Tseng TH, Li WH, Ku MS (2015) Downregulation of cytokinin oxidase 2 expression increases tiller number and improves rice yield. Rice 8(1):36

Yoshihara T, Iino M (2007) Identification of the gravitropism-related rice gene LAZY1 and elucidation of LAZY1-dependent and -independent gravity signalling pathways. Plant Cell Physiol 48(5):678–688

Yu H, Qiu Z, Xu Q, Wang Z, Zeng D, Hu J, Zhang G, Zhu L, Gao Z, Chen G, Guo L, Qian Q, Ren D (2017) Fine mapping of *LOW TILLER 1*, a gene controlling tillering and panicle branching in rice. Plant Growth Regul 83:93–104

Zha M, Zhao Y, Wang Y, Chen B, Tan Z (2022) Strigolactones and cytokinin interaction in buds in the control of rice tillering. Front Plant Sci 13:837136. https://doi.org/10.3389/fpls.2022.837136

Zhang C, Bai M, Chong K (2014) Brassinosteroid-mediated regulation of agronomic traits in rice. Plant Cell Rep 33(5):683–696

Zhang N, Yu H, Yu H, Cai Y, Huang L, Xu C, Xiong G, Meng X, Wang J, Chen H, Liu G, Jing Y, Yuan Y, Liang Y, Li S, Smith SM, Li J, Wang Y (2018) A core regulatory pathway controlling rice tiller angle mediated by the LAZY1-dependent asymmetric distribution of auxin. Plant Cell 30:1461–1475

Zhang W, Peng KX, Cui FB, Wang D, Zhao J, Zhang Y, Yu N, Wang Y, Zeng D, Wang Y, Cheng Z, Zhang K (2021a) Cytokinin oxidase/dehydrogenase OsCKX11 coordinates source and sink relationship in rice by simultaneous regulation of leaf senescence and grain number. Plant Biotechnol J 19(2):335–350

Zhang Z, Sun X, Ma X, Xu B, Zhao Y, Ma Z, Li G, Khan NU, Pan Y, Liang Y, Zhang H, Li J, Li Z (2021b) GNP6, a novel allele of MOC1, regulates panicle and tiller development in rice. Crop J 9:57–67

Zhao H, Ma H, Yu L, Wang X, Zhao J (2012) Genome-wide survey and expression analysis of amino acid transporter gene family in rice (Oryza sativa L.). PLoS ONE 7(11):e49210. https://doi.org/10.1371/journal.pone.0049210

Zhao J, Yu N, Ju M, Fan B, Zhang Y, Zhu E, Zhang M, Zhang K (2019) ABC transporter OsABCG18 controls the shoot ward transport of cytokinins and grain yield in rice. J Exp Bot 70(21):6277–6291

Zhao S, Jang S, Lee YK, Kim D-G, Jin Z, Koh H-L (2020) Genetic basis of tiller dynamics of rice revealed by Genome-Wide Association Studies. Plants 9(12):1695. https://doi.org/10.3390/plants9121695

Manipulation of Tiller Number in Designer Rice

10

Abstract

Among monocotyledonous plants, especially cereals, rice is a model system for the study of branching patterns. The plasticity of tiller dynamics is associated with agronomic management in complex ways and is essential for not only survival but also the determination of the grain yield of rice plants. However, how plant architecture changes with tiller dynamic plasticity and how plant architecture interacts with agroecological adaptation are unclear. Controlling tiller dynamics to increase rice yield by manipulating agronomic inputs in various environments has been tested very often. Still, the objective of obtaining designer rice plants with tiller numbers appropriate for the environment has rarely been achieved. The grain yield margin can improve by as much as 40% via the regulation of tiller dynamics. In rice agroecology, dry and wet conditions of the paddy fields alternately increase tiller number and yield significantly. Also, the application of organic and inorganic fertilizers improves tiller number and yield. However, tiller production above resource capacity is corrected by the senescence of late tillers. Identification of genetic markers of tiller production and survival is essential for designer rice.

Keywords

Designer rice · Phenotypic plasticity · Phytogeography · *Oryza rufipogon* · Tiller classes · Hormone cross-talk

10.1 Phenotypic Plasticity and Dynamics of Rice Tillers

Tillering is a critical agronomic trait in rice crops for grain production. Additionally, among monocotyledonous plants, especially cereals, rice is a model system for the study of branching. Each tiller produced is genetically competent to bear a panicle

© The Author(s), under exclusive license to Springer Nature Singapore Pte Ltd. 2025

P. K. Mohapatra et al., *Tillering Behavior of Rice Plant*,
https://doi.org/10.1007/978-981-97-5235-5_10

at maturity. A grain-bearing panicle contributes to overall plant yield; hence, the greater the number of grains is, the higher the opportunity for grain yield. Rice plants subjected to various field conditions exhibit an amazing capacity to adapt and adjust tiller dynamics appropriately. The plants produce more tillers in thin stands and fewer in dense stands. In addition to density, which determines tiller number, other factors, such as temperature, water depth, nutrients, allelopathic competition, light intensity, and soil physical conditions, such as pH, salinity, and moisture content, have distinct impacts on tillering and tiller development. Under suboptimal conditions, plants become resource-deficient and tend to restrict tiller production and vice versa. Thus, tillering is one of the most important stages of rice plant development that can be influenced by management practices for harnessing optimum yields from crops.

10.2 *Oryza rufipogon* Griff. A Model System for Tiller Dynamics

Among monocot plants, rice is construed as a model system for the study of branching patterns. However, genetic changes accumulated through thousands of years of cultivation in their respective habitats have made plant phenotypes recalcitrant or obligatory to the ecosystem concerned, and they have lost the capacity for growth in any environment other than their own. Thus, cultivated rice does not abide by environmental changes to support new adaptations. The survival of these plants in stress-prone environments is often critically threatened much to the misery of the farming community. Almost every year, rice cultivation in the tropical and subtropical areas of Asia is prone to drought and flooding because of uncertain weather conditions, and the need for the exploration of stress-resistant genes is a high priority. Under these circumstances, rice breeders do not have any option to explore genetic factors to evaluate stress resistance within cultivated species of rice, which otherwise display no resilience to stress-prone environments. In this perilous situation, with null prospects for designer rice, the wild rice *Oryza rufipogon* Griff. holds achievable opportunity as a genetic repository of traits for survival under challenging environments (Mohapatra et al. 2011). Further, it is recognized as a prominent component of the rice agroecosystem and an integral contributor to crop evolutionary dynamics in Southeast Asia (Vigueira et al. 2019). As a conspecific wild relative of a model crop like *O. sativa*, *O. rufipogon* provides a productive system for exploring the dynamics of tillering patterns under varying environments. The species is native to Asia, but during diffusion, it has migrated to rice-growing agroecosystems in different parts of the world. This species grows in close sympatric association with cultivated rice in various widely contrasting habitats across the globe, and because of its ability to complete its life cycle without human force in the natural ecosystem, it has achieved the ability to adapt to climate-resilient genetic characteristics. The species adapts according to the level of soil moisture, high or low, and manifests changes in phenology, biomass production, and yield; these features positively correlate with tillering.

10.3 Beneficial Traits of *Oryza rufipogon* Griff.

In the *Poaceae* family of monocot plants, species belonging to the genus *Oryza* are of great economic importance. This genus comprises two cultivated species, *O. sativa* L. and *O. glaberrima* and 20 other non-domesticated species. Genetically, *Oryza* species are more closely related to bamboo than to other cereal plants, such as wheat, barley, or maize (Tzvelev 1989; Watson et al. 1985). Hence, for cultivated rice, the wild species of *Oryza* are the only source of genetic recombination. Among the wild species, *O. rufipogon* possesses an immense advantage in gene exchange with cultivated rice because of its similarity in terms of its genome complex (AA genome complex), chromosome number (n = 12), and existence in compatible growth habitats. Hence, the species has been assessed for the enrichment of the gene pool in cultivated rice for resilience to biotic and abiotic stresses and the derivation of appropriate yields.

The other wild species of the genus are less attractive for scientific manipulation of gene flow from wild to domesticated rice because of their incompatibility with genome complexes such as BB, CC, DD, and EE, which makes them unsuitable for gene pyramiding. In addition, the incidence of natural intercrossing between *O. rufipogon* and cultivated rice is high leading to the establishment of numerous morphological intergrades in the rice agroecosystem (Chang et al. 1982). The existence of spontaneous interspecific hybrids in *O. sativa* confirms this proposition (Rao et al. 1997). The hybrids diverge across a variety of agricultural niches and establish themselves into new ecotypes. Therefore, several authors have considered *O. rufipogon* to be the progenitor of *O. sativa* (Khush 1997; Takeoka 1962; Oka 1991). During evolution, the perennial *O. rufipogon was* transformed into the annual ecotype *O. nivara* while living in summer-parched dry land conditions; it subsequently gave rise to *O. sativa* (Sharma and Shastry 1965). In addition, the distribution of the species is extraordinarily large. It is native to ecosystems ranging from summer-parched dry upland areas to permanent water holes in pools or ditches. The ditches hold stagnant or mild flow of water. The altitude of the habitats ranges from 0 m at sea-plains to 1000 m on mountain tops.

10.4 Superiority in Tiller Production

The variation in tiller dynamics of *O. rufipogon* between contrasting environments provides an estimate of the ability of the plant to adapt to a wide range of environments. This capacity for adaptation is rarely seen in cultivated rice. Hence in the rice agro-ecosystem, *O. rufipogon* is a superior competitor compared to cultivated rice (Kim et al. 2016). Besides genetic limitation, a low tiller number per plant in a genotype indicates the adverse influence of a sub-optimal environment, while profuse tillering accounts for optimum growth under favourable environments. In rice agroecosystems, *O. rufipogon* is a superior competitor compared to cultivated rice. The species possesses early vigour, establishes more tillers, and achieves greater plant height to access efficient irradiance perception. These traits are essential for

Fig. 10.1 Photograph of *O. rufipogon* in a deep-water ecosystem (**A**) and *O. nivara* (often described as annual ecotype of *O. rufipogon*) in a dryland habitat (**B**) (Source: Photo gallery of E. Kariali)

adaptation in natural ecosystems (Fig. 10.1). The superior vegetative growth helps the production of more tillers in *O. rufipogon* and outpaces the growth of cultivated rice. Given the advantage of the abundance of resources in irrigated field conditions, a profuse tillering feature of the species helps to better utilize space for carbon assimilation and out-manoeuvre *O. sativa* in tiller and biomass production. Thus, the tiller dynamics of rice plants in a stable environment are species-specific and an important agronomic factor not only for the nature of adaptation to the environment and for biomass production but also for reproductive growth and grain yield.

10.5 Phytogeography of *Oryza rufipogon*

In the undulating flood plains of Asia, the nativity of *O. rufipogon* is confined to permanent water holes. It grows unabated through the years to become perennial. However, its ecotype counterpart, *O. nivara*, lives in the upland area, where it survives only with monsoon seasonal runoff water for 2–3 months. Compared to *O. nivara* the perennial form of *O. rufipogon* lives in a stable and more favourable environment with minimal risk of environmental perturbation (Fig. 10.1). Distinctive features of *O. rufipogon* and *O. nivara* are given below.

Oryza rufipogon

Perennial, photosensitive (short photoperiod), cross-pollinated, found in permanent water ditches, aquatic, elongated stem, prostrate, long spreading panicles.

Oryza nivara

Annual, short height, upright, photoperiod insensitive, self-pollinated, native of seasonally wet habitats with periodic flooding, panicles relatively compact and short.

In swampy marshlands, *O. rufipogon* plants elongate quickly in response to monsoonal flood waters. The elongation continues until the water level stabilizes at the end of the monsoon rainfall period. A rhizomatous stem submerged under water expresses a strong ratooning ability by producing new tillers from various nodes. Moreover, fibrous roots emerge from the submerged stem nodes, which support the ability of putative tillers to gain independence. Reproduction by seed production being rare, the plant multiplies in number by the production of tillers. The culm sleeps over the ground when the field water depth reaches zero during the post-monsoon period. The plant is strongly photoperiodic and flowers at the advent of the winter season. Only a few seeds develop on the inflorescence. The period of seed development is very short. Endosperm starch filling is completed in only approximately 10 days (Chen 2001). The premature seeds are shed to moist ground, where the sharp awn provides anchorage, and remain viable and dormant for up to 3 years. Physiological adaptation helps the seeds to remain embedded for a year, not showing germination. The change in phenotypic expression leading to possible ecotypic divergence of *O. rufipogon* in response to different water regimes across a graded-topo sequence has been illustrated by Vaughan (1994).

10.6 Tiller Production as an Indicator of Adaptation

In cultivated rice, tillering is an agronomic trait that provides a measure of grain production. In addition, tillering can allow plants to adjust their growth relative to the plant density and quality of the growing conditions. The tillering pattern determines the overall plant architecture and canopy development for the capture of incident light for primary production. As discussed before, tillers are branches that develop from leaf axils on nodes of the main culm or other tillers during vegetative growth. The initiation of reproductive growth coincides with the increase in distance between nodes. The elongated internodes do not support tiller generation. Thus, tiller formation is restricted to unelongated basal internodes. The formation of new tillers depends on the main shoot for photo assimilatory root-derived nutrients until the plants expand their leaves and initiate roots. These plants grow independently of the main culm using their adventitious roots.

The cultivated rice *O. sativa* originated from landraces that are intermediate between the annual and perennial forms of the wild rice *O. rufipogon* Griff. (Vaughan 1994). It diffused into a wide spectrum of habitats across the globe varying in soil moisture regimes, such as aerobic drylands, rainfed lowlands, rainfed deep water, irrigated systems, and very deep-water agro-ecosystems (Huke and Huke 1990), where the plant phenotype remained specific and unique to the culture conditions. Subject to environmental fluctuations, rice plants strongly regulate their sink–source relationships (Yoshida 1981), which dictates the nature of assimilate partitioning to different organs. The tillering and tiller dynamics of plants are morphological features that sustain assimilate provision from source organs, and they are strongly influenced by changes in habitat-induced alterations in assimilate partitioning. Thus, the variation in assimilate partitioning between tillers of rice plants across a range of habitats could be a reliable physiological index reflecting the permissible limits of variation according to the situation.

Because cultivated rice has been recalcitrant to adaptation to a new environment, assimilates partitioned into different classes of tillers of *Oryza rufipogon* Griff. was used as an index of adaptation to sub-optimal environments (Kariali et al. 2010). The species is more amenable to exhibiting tiller dynamics appropriate for the environment, but the lack of such studies precludes knowledge on the genotypic permissibility of cultural landraces for phenotypic plasticity in response to challenging environments. The authors examined the margin of plasticity in assimilate partitioning between individual tillers of three ecotypes of *O. rufipogon* living in different habitats, such as dryland, lowland, and deep-water habitats, both in natural and simulated conditions resembling the respective natural habitats (Table 10.1). The ecotypes of the species survived in all three habitats by changing the distribution pattern of soluble carbohydrates and free amino acids that are unparalleled in cultivated rice. The growth duration, biomass, tiller number, grain yield, and assimilate concentration of reproductive organs were low in the dryland ecotype but high in the other two ecotypes. Switching of dryland ecotypes to either low- or deep-water conditions improved biomass and tiller production, but the effects of a change in habitat were inconsistent for the other two ecotypes (Table 10.1). Thus, the difference in the morphological attributes of the ecotypes was mostly determined by the variation in environmental conditions, irrespective of the variation accrued in the genotype during its adaptation to the habitat.

10.7 Tiller Dynamics: Low Versus High Tillering Rice

Rice plants are rhizomatous monocots that have immense potential for producing tillers subject to the availability of resources in the growth media. While the abundance of resources leads to a profuse tillering phenotype, deficiencies in essential nutrients and moisture in the habitat support the growth of one- or no-tiller. A rice plant exhibits a high-tillering rate in the early stage of growth due to an abundance of resources (Dingkuhn and Kropff 1996). Thus, it is important to study the physiological processes and functionality of morphological changes in response to environmental factors to explore the adaptive plasticity of plants and their evolution (Blom 1999; Sultan 2000). In pursuance, Sultan (2004) suggested four promising directions: examining dynamic, anatomical/architectural, and cross-generational plasticity along with simpler growth traits and carefully assessing the adaptive significance of those plasticity patterns. The authors investigated the intricate transduction pathways that lead from environmental signals to phenotypic responses while considering the rich environmental context of natural systems. Furthermore, the tillering capacity of rice can be altered by changes in the environment and agronomic practices (Yoshida 1973). Accordingly, Kariali et al. (2008) and Kariali (2014) used two differentially expressed tillering wild-rice species, *O. rufipogon* and *O. nivara* (considered the annual ecotype of *O. rufipogon*), which are native to two contrasting habitats, deep-water and dryland, respectively. The water depth in the natural habitats and meteorological observations during the experiment are presented in Fig. 10.2 to determine the stability of the tiller hierarchy among the

Table 10.1 Morphological and phenological features of three ecotypes of *Oryza rufipogon* grown in natural and three different simulated (*) habitats, namely, dryland, lowland, and deep-water conditions

Wild rice eco-types	Growing habitats	No. of tillers	Dry wt. of main shoot (g)	Dry wt. of panicle on the main shoot (g)	Grain number of panicles on main shoot			100 grain Weight (g)	Panicle wt/total shoot wt	Growth duration (days)	
					Fertile	Sterile	Total			Vegetative	Reproductive
Dry	Natural	1.66 ± 0.47	0.85 ± 0.11	0.433 ± 0.03	21.33 ± 2.05	09.33 ± 0.471	30.670 ± 1.672	1.220 ± 0.011	0.337	44	27
Dry	Dry*	4.33 ± 0.42	1.16 ± 0.40	0.493 ± 0.012	22.33 ± 2.054	08.33 ± 1.632	30.667 ± 0.816	1.616 ± 0.069	0.298	51	27
Dry	Lowland*	9.67 ± 1.24	1.78 ± 0.23	0.519 ± 0.021	23.67 ± 0.942	07.67 ± 0.471	31.333 ± 1.247	1.680 ± 0.057	0.226	59	32
Dry	Deep*	6.67 ± 0.47	2.47 ± 0.37	0.580 ± 0.032	28.00 ± 2.449	09.00 ± 0.816	37.000 ± 2.160	1.760 ± 0.131	0.190	67	39
Lowland	Natural	4.67 ± 0.47	1.78 ± 0.21	0.610 ± 0.04	28.67 ± 4.49	10.33 ± 2.052	38.990 ± 2.451	1.513 ± 0.082	0.255	55	30
Lowland	Dry*	4.67 ± 0.64	1.45 ± 0.50	0.503 ± 0.012	18.00 ± 1.632	07.67 ± 0.471	25.667 ± 1.247	1.523 ± 0.061	0.258	60	32
Lowland	Lowland*	6.33 ± 0.71	3.97 ± 0.72	0.580 ± 0.032	25.00 ± 1.414	08.67 ± 0.471	33.667 ± 1.632	1.740 ± 0.048	0.128	63	34
Lowland	Deep*	4.67 ± 0.82	5.27 ± 0.98	0.603 ± 0.021	30.00 ± 1.632	09.67 ± 1.247	39.667 ± 2.247	1.843 ± 0.069	0.103	68	39
Deep	Natural	4.33 ± 0.47	8.03 ± 0.11	1.041 ± 0.11	61.67 ± 6.18	14.00 ± 1.411	75.670 ± 4.780	1.401 ± 0.011	0.115	85	44
Deep	Dry*	7.67 ± 1.02	1.51 ± 0.40	0.469 ± 0.092	19.67 ± 1.128	07.67 ± 1.247	27.333 ± 3.599	1.540 ± 0.032	0.237	65	34
Deep	Lowland*	6.33 ± 0.91	4.75 ± 0.87	0.610 ± 0.040	34.00 ± 4.546	13.00 ± 0.816	47.000 ± 4.080	1.640 ± 0.089	0.114	70	39
Deep	Deep*	4.67 ± 0.47	5.0.36 ± 0.76	0.636 ± 0.065	38.00 ± 2.816	16.33 ± 1.247	54.333 ± 4.942	1.810 ± 0.073	0.106	79	42

Observations were recorded at maturity. ± Values are the means of three replicates (n = 3) (Kariali et al. 2010)

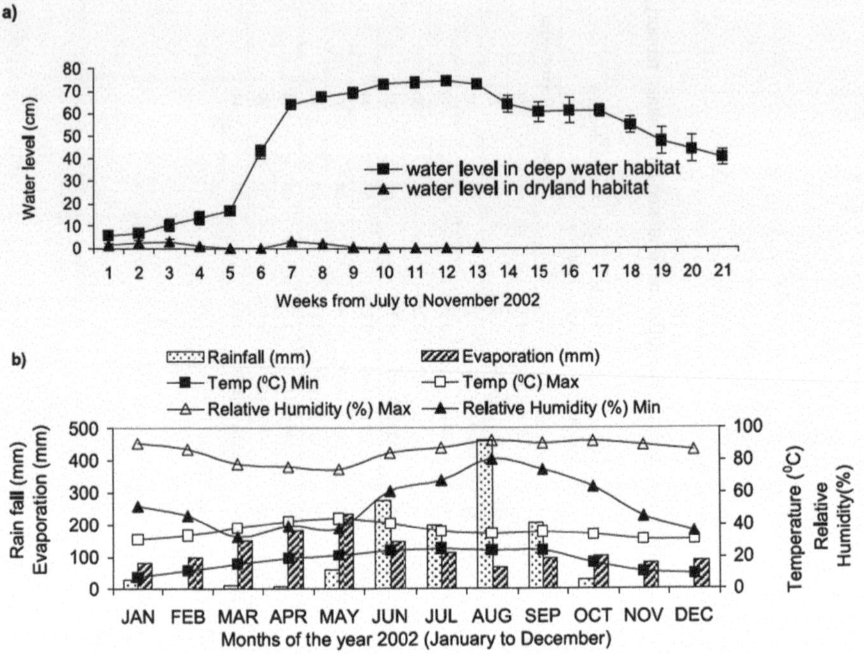

Fig. 10.2 Water depth in the deep-water and dry land habitats from July to November 2002 (**a**) and other meteorological observations (**b**) during the experiment (Kariali et al. 2008)

species and the relationship between tiller dynamics and hierarchy with the carbon assimilate concentration of panicles under natural and agronomically managed pot conditions.

In natural habitats, *O. nivara* produces one tiller and completes its life cycle much earlier than *O. rufipogon*, which produces four tillers and delays the duration of maturity. Under potted conditions, *O. nivara* and *O. rufipogon* exhibited a profuse tillering phenotype and produced 27 and 38 tillers, respectively. The maturity duration of the tillers and biomass accumulation decreased in an acropetal fashion, and the late-stage tillers were mostly sterile. Habitat-induced changes did not influence this hierarchy. Among the categories of tillers, the soluble sugar concentration in the panicle was high in the early-stage tillers at anthesis, but the trend reversed at maturity. The starch concentration in the panicles was always high in panicles of early-stage tillers, implying a preference for assimilate partitioning (Tables 10.2, 10.3, and 10.4). These observations indicate that the potential for tiller production remains un-influenced irrespective of genetic differences and that phenotypic expression is limited by the capacity for assimilate production due to a deficiency of essential resources in the environment. Poorer vegetative growth in a succeeding tiller compared to a preceding tiller subjugates vegetative growth in the partitioning of assimilates in favour of panicle growth. The late tiller becomes sterile or decimates, being targeted by discrimination in assimilate supply from the main culm.

Table 10.2 Phenological features of *Oryza nivara* under cultivated and natural conditions

Tillers	Days after tiller emergence				Dry weight at MA (g)		Panicle length at MA (cm)	Growth duration (days)		
	TE	PI	AN	MA	Shoot	Panicle		Veg.	Rep.	Total
Cultivated habitat										
M	–	61	74	90	5.50 ± 0.21	1.53 ± 0.02	27.22 ± 0.60	70	29	99
P1	0	62	75	90	4.39 ± 0.19	1.41 ± 0.03	27.01 ± 0.39	62	28	90
P3	5	62	75	90	3.80 ± 0.02	1.29 ± 0.09	26.24 ± 1.40	57	28	85
P6	10	63	76	90	2.46 ± 0.01	1.10 ± 0.01	25.53 ± 1.10	53	27	80
P1S1	8	62	75	90	3.60 ± 0.14	1.35 ± 0.01	26.91 ± 0.59	54	28	82
P1S3	14	65	78	92	3.09 ± 0.09	1.00 ± 0.01	23.11 ± 0.87	51	27	78
P2S1	10	62	75	90	3.40 ± 0.01	1.31 ± 0.02	26.05 ± 1.90	42	28	70
P2S3	16	66	78	92	2.90 ± 0.03	0.99 ± 0.01	23.01 ± 0.48	50	26	76
P3S1	14	65	77	92	3.06 ± 0.02	1.14 ± 0.02	25.92 ± 1.20	51	27	78
P6S1	17	66	78	92	2.40 ± 0.01	0.70 ± 0.01	23.32 ± 0.70	49	26	75
P1S1T1	23	68	79	93	1.30 ± 0.12	0.41 ± 0.02	22.11 ± 0.40	45	25	70
Natural habitat										
M	–	49	62	75	0.54 ± 0.11	0.54 ± 0.01	26.62 ± 1.57	55	26	81
P1	0	52	65	77	0.39 ± 0.07	0.51 ± 0.02	20.11 ± 0.92	52	25	77

TE tiller emergence, *PI* panicle initiation, *AN* 50% anthesis, *MA* maturity, *Veg* vegetative, *Rep* reproductive, *M* main shoot, *P1* first primary tiller, *P6* sixth primary tiller, *P1S1* first secondary tiller emerged from P1, *P2S1* first secondary tiller emerged from P2, *P6S1* first secondary tiller emerged from P6, *P1S1T1* first tertiary tiller emerged from P1S1, and so on. ± Values are the means of three replicates ($n = 3$) (Kariali et al. 2008)

Table 10.3 Phenological features of *Oryza rufipogon* under cultivated and natural conditions

Tillers	Days after tiller emergence				Dry weight at MA (g)		Panicle length at MA (cm)	Growth duration (days)		
	TE	PI	AN	MA	Shoot	Panicle		Veg.	Rep.	Total
Cultivated habitat										
M	–	84	97	132	13.23 ± 0.32	1.62 ± 0.01	27.31 ± 0.89	101	48	149
P1	0	85	97	132	11.12 ± 0.45	1.62 ± 0.02	27.13 ± 0.47	85	47	132
P3	7	86	98	132	09.12 ± 0.35	1.59 ± 0.02	26.41 ± 0.09	79	46	125
P6	14	89	100	133	06.85 ± 0.44	1.43 ± 0.01	25.84 ± 1.30	75	44	119
P1S1	11	86	98	132	09.15 ± 0.42	1.59 ± 0.01	26.52 ± 0.43	75	46	121
P1S3	20	90	101	133	05.12 ± 0.11	1.10 ± 0.01	23.12 ± 1.09	70	43	113
P2S1	12	87	99	132	08.15 ± 0.14	1.55 ± 0.02	26.15 ± 0.78	75	45	120
P2S3	22	91	102	134	04.89 ± 0.12	1.05 ± 0.01	23.91 ± 0.29	69	43	112
P3S1	15	90	100	133	07.88 ± 0.24	1.53 ± 0.01	26.02 ± 0.89	75	43	118
P6S1	19	91	102	134	03.18 ± 0.05	1.43 ± 0.02	25.13 ± 0.83	72	43	115
P1S1T1	21	93	104	134	02.45 ± 0.11	0.39 ± 0.01	24.91 ± 0.30	72	41	113
Natural habitat										
M	–	85	105	119	08.03 ± 0.12	1.37 ± 0.06	26.90 ± 1.20	98	44	142
P1	0	87	106	119	06.84 ± 0.10	1.36 ± 0.01	26.71 ± 1.40	87	42	129
P2	2	88	107	119	05.50 ± 0.06	1.24 ± 0.15	26.22 ± 1.05	86	41	127
P3	5	90	107	119	05.13 ± 0.07	1.22 ± 0.02	26.01 ± 1.50	85	39	124
P1S1	8	91	111	120	04.82 ± 0.10	1.20 ± 0.01	25.82 ± 1.10	83	39	122

The abbreviations used in the table are the same as those in Table 10.2. ± Values are the means of three replicates ($n = 3$) (Kariali et al. 2008)

10.8 Excessive Tillering Is Undesirable

For semi-dwarf rice, high and early tillering capacity is an essential trait for optimum grain yield (Chandler 1969). This trait is highly important because of its direct correlation with panicle number per unit land area, primary production, and grain yield (Moldenhauer and Gibbons 2003). Subject to the availability of resources, tiller initiation, defined as one tiller per culm node of the main stem or tillers, continues unchecked during the vegetative period of growth. The group of tillers produced by an individual plant constitutes a hill. The yield potential of rice can be characterized by the tillering capacity of hills to some extent but not entirely. In a hill, the percentage of panicle-bearing tillers influences yield (Wang et al. 2007). In semi-dwarf rice, the tiller number is high because of the initiation of secondary and tertiary tillers (Mohanan and Mini 2008), but most of these tillers are eliminated between the panicle initiation and anthesis stages (Badshah et al. 2014) or become sterile post-anthesis. Excessive tillering has been shown to lead to high tiller abortion, poor grain setting, small panicle size, and sub-optimal grain yield (Peng et al. 1994; Ahmad et al. 2005). Hence, excessive branching of rice culms is an investment loss in photo-assimilation (Dun et al. 2006) because the formation of un-/low-productive tillers in rice is a waste of photosynthetic energy (Mohapatra et al. 2011).

Table 10.4 Soluble sugar and starch concentrations in the panicles of different tiller categories of two wild rice species, *O. nivara* and *O. rufipogon*, at anthesis and maturity in cultivated and natural habitats

	O. nivara				*O. rufipogon*			
	Sugars (μg mg^{-1} dwt)		Starch (μg mg^{-1} dwt)		Sugars (μg mg^{-1} dwt)		Starch (μg mg^{-1} dwt)	
Tillers	Anthesis	Maturity	Anthesis	Maturity	Anthesis	Maturity	Anthesis	Maturity
Cultivated habitat								
M	229.6 ± 11.5	04.5 ± 0.3	360.9 ± 19.6	591.0 ± 11.9	166.9 ± 13.6	02.1 ± 0.1	334.1 ± 13.3	542.6 ± 13.0
P	166.4 ± 13.9	06.7 ± 1.6	269.4 ± 14.9	535.1 ± 27.8	108.3 ± 11.8	02.2 ± 0.8	276.8 ± 12.4	486.5 ± 20.6
S	104.6 ± 15.6	14.7 ± 4.4	207.0 ± 24.9	466.8 ± 16.4	086.5 ± 10.8	05.9 ± 1.8	246.8 ± 24.3	453.2 ± 36.4
T	037.5 ± 01.4	38.0 ± 2.8	083.5 ± 2.40	137.0 ± 11.5	040.0 ± 02.9	48.8 ± 4.2	073.6 ± 09.4	109.2 ± 11.4
Natural habitat								
M	134.3 ± 8.3	08.9 ± 1.1	109.7 ± 8.9	391.7 ± 13.9	138.1 ± 12.4	08.2 ± 0.8	186.3 ± 6.8	472.6 ± 11.2
P	111.5 ± 9.0	10.2 ± 0.6	089.3 ± 4.8	348.1 ± 08.0	093.1 ± 10.5	11.5 ± 1.7	161.6 ± 7.2	442.8 ± 09.9
S	–	–	–	–	083.9 ± 05.9	14.8 ± 0.6	154.5 ± 7.8	431.5 ± 10.1

M main shoot, *P* primary tiller, *S* secondary tiller, *T* tertiary tiller. ± Values are the means of three replicates ($n = 3$) (Kariali et al. 2008)

10.9 Senescence Patterns of Different Classes of Tillers

In a multi-tillering rice cultivar, the initiation of tillers is usually staggered and temporally spaced, but the event of maturity is mostly synchronous. Therefore, the duration of growth and development in a tiller that initiates later is shorter than that in a tiller formed earlier, and its contribution to biomass and grain yield is relatively smaller. Matching the ordered pattern of tiller dominance in a basipetal direction, the senescence of the leaves of individual tiller classes proceeds in a similar succession. Kariali et al. (2012) ascertained how tiller production capacity influences dry matter partitioning and tiller dynamics in two multi-tillering rice cultivars, Lalat (profuse tillering) and MGD-106 (medium tillering), by studying various physiological processes regulating leaf senescence. Senescence progressed among the different classes of tillers, as indicated by the gradual decrease in photosynthetic pigments (Fig. 6.2) and total nitrogen and protein concentrations (Fig. 6.3) and the increase in lipid peroxidation and peroxidase activity (Fig. 6.4) in the flag leaf; the effect was more pernicious in the newer tillers compared to the older ones. Biomass growth, grain yield, height, and leaf area were significantly low in the new tillers and they could not maintain longevity for long. The older tillers gained metabolic dominance due to greater photosynthetic activity, which decreased their vulnerability to senescence-induced oxidative stress. Both sink and source activities were maintained in these tillers for biomass production and grain filling. In late-formed tillers, the concentration of ethylene in the boot of the flag leaf sheath was greater than that in early-formed tillers just before anthesis, and a high ethylene concentration decreased grain filling, resulting in a significant loss of grain yield (Kariali and Mohapatra 2007, Table 10.5).

Exogenous application of kinetin often restored the fertility of spikelets. The ethylene concentration in the boot was positively correlated with spikelet sterility and lipid peroxidation in the flag leaf and negatively correlated with the and contents of chlorophylls and protein in the flag leaf and with the panicle grain yield, leaf area, and growth duration of the tillers (Kariali and Mohapatra 2007, Fig. 10.3).

Hence, the early induction of senescence and increased production of ethylene in the boot of late-developed tillers could be due to a deficiency of cytokinin. Furthermore, it was revealed that the diversity spectrum between the development of different classes of tillers increases with increasing tiller number; diversity is wider in the high-tillering cultivar Lalat than in the low-tillering cultivar IRON-104 (Kariali and Mohapatra 2007).

10.10 GWAS of Effective Tillers

As discussed forthwith, tiller formation is an important factor in determining grain yield in rice agronomy. However, late-formed tillers do not survive to maturity and do not contribute to the final grain yield. The number of surviving tillers, otherwise called effective tillers, directly controls grain yield. Thus, instead of the total tiller number, the effective tiller number per plant should be considered an essential yield

Table 10.5 The effect of Kinetin, IAA and GA_3 on morphological and phenological features of different type of tillers of rice cultivar Lalat

Tillers	Treatments	Panicle length (cm)	No. of grains	Grain yield (g)	Growth duration (day)	Maximum leaf area (cm^2)	% Sterility	Ethylene (nmol)
M	Control	28.00 ± 0.72	200.00 ± 7.00	3.89 ± 0.18	108.0 ± 1.0	220.00 ± 4.58	18.50 ± 3.16	6.709 ± 0.56
	Kinetin	28.47 ± 0.15	224.67 ± 4.93	4.59 ± 0.02	112.0 ± 1.0	234.08 ± 4.71	12.91 ± 1.95	5.421 ± 1.22
	IAA	28.90 ± 0.46	237.00 ± 3.61	4.82 ± 0.03	111.0 ± 2.0	252.00 ± 4.12	11.11 ± 1.23	4.701 ± 1.23
	GA_3	29.07 ± 0.06	225.00 ± 6.08	4.72 ± 0.05	110.0 ± 2.0	274.80 ± 1.85	15.26 ± 2.28	5.341 ± 0.30
	F (treatments)	3.623 NS	23.503***	57.211***	3.5 NS	105.924***	5.978*	2.499 NS
P1	Control	27.40 ± 0.35	179.00 ± 8.54	3.53 ± 0.20	93.0 ± 1.0	184.00 ± 3.61	17.89 ± 1.16	7.165 ± 0.37
	Kinetin	27.57 ± 0.32	205.67 ± 7.23	4.45 ± 0.05	97.0 ± 2.0	212.40 ± 2.69	13.61 ± 2.24	6.591 ± 0.60
	IAA	28.17 ± 0.15	223.67 ± 2.08	4.69 ± 0.12	98.0 ± 2.0	207.00 ± 2.65	12.07 ± 1.61	5.242 ± 0.30
	GA_3	28.13 ± 0.15	231.00 ± 4.58	4.63 ± 0.09	96.0 ± 1.0	221.67 ± 3.76	21.19 ± 0.83	6.479 ± 0.51
	F (treatments)	6.798*	42.676***	57.849***	5.6*	74.551***	21.285***	9.351**
P1S1	Control	26.00 ± 0.20	167.00 ± 1.73	2.94 ± 0.11	86.0 ± 1.0	154.67 ± 3.76	25.75 ± 1.20	7.771 ± 0.37
	Kinetin	26.40 ± 0.30	192.00 ± 1.73	3.94 ± 0.03	91.0 ± 1.0	157.00 ± 1.83	15.97 ± 1.94	7.069 ± 0.06
	IAA	24.83 ± 0.31	165.67 ± 7.23	2.89 ± 0.09	77.0 ± 2.0	131.00 ± 4.60	25.55 ± 1.98	9.250 ± 0.26
	GA_3	26.30 ± 0.27	196.67 ± 3.07	3.46 ± 0.07	82.0 ± 0.0	137.33 ± 1.62	17.93 ± 0.36	8.107 ± 0.86
	F (treatments)	21.227***	47.041***	110.538***	70.667***	47.649***	33.496***	10.527***
P1S1T1	Control	16.60 ± 0.80	62.00 ± 1.00	0.49 ± 0.09	67.0 ± 0.0	75.00 ± 2.77	68.04 ± 7.98	8.194 ± 0.60
	Kinetin	17.13 ± 0.31	77.67 ± 0.58	1.05 ± 0.07	79.0 ± 2.0	96.00 ± 3.23	32.62 ± 4.42	8.201 ± 0.32
	IAA	10.57 ± 1.10	69.33 ± 2.08	0.00 ± 0.00	55.0 ± 2.0	60.30 ± 3.90	100.0 ± 0.00	13.96 ± 0.74
	GA_3	09.80 ± 1.05	66.67 ± 0.58	0.00 ± 0.00	57.0 ± 1.0	77.54 ± 9.00	100.0 ± 0.00	13.09 ± 0.69
	F (treatments)	59.216***	86.426***	218.748***	161.333***	22.537***	148.803***	69.547***
	F (tillers)	46.219***	123.747***	82.106***	49.367***	76.834***	13.358**	9.801**

M mother shoot, *P1* first primary tiller emerged from M, *P1S1* first secondary tiller emerged from P1, *P1S1T1* first tertiary tiller emerged from P1S1 ± Values represent standard deviations ($n = 3$)

Data analysis was done by ANOVA (analysis of variance) in Microsoft excel computer program [Single factor ANOVA ($n1 = 3$, $n2 = 8$) to test the significance between treatments and double factor ANOVA without replication ($n1 = 3$, $n2 = 9$) between tillers] (Kariali and Mohapatra 2007)

$^{*}P \geq 0.05$, $^{**}P \geq 0.01$, $^{***}P \geq 0.001$

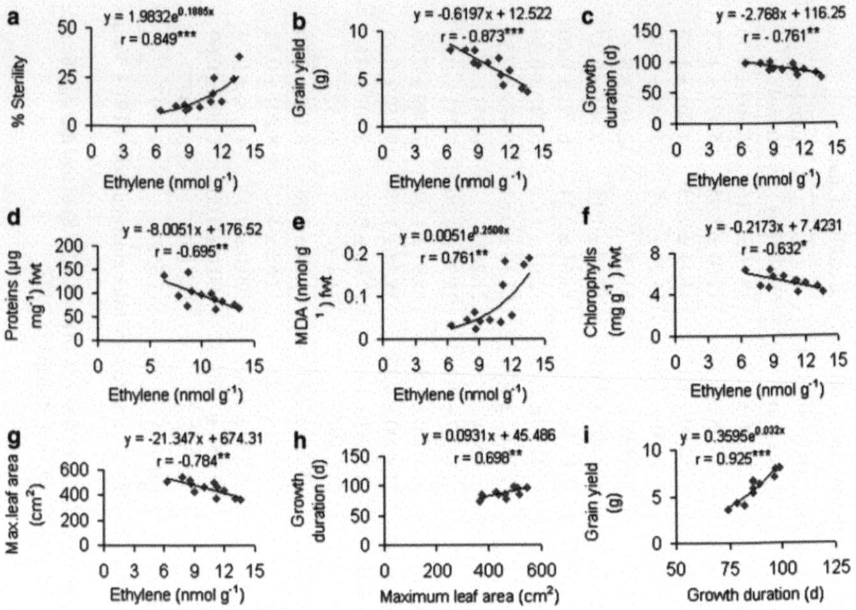

Fig. 10.3 Relationships between ethylene production inside the boot (nmol g^{-1} dwt of panicle) and (**a**) % sterility, (**b**) grain yield, (**c**) growth duration, (**d**) protein content, (**e**) lipid peroxidation, (**f**) chlorophyll content, and (**g**) maximum leaf area and between (**h**) maximum leaf area and growth duration and (**i**) growth duration and grain yield of different categories of tillers in the rice cultivar IRON-104 (Kariali and Mohapatra 2007)

component of rice. A genome-wide association study showed that effective tiller number (ETN) is a complex qualitative trait regulated by both genetic and environmental factors (Ren et al. 2021). These authors conducted a genome-wide association study (GWAS) for ETN using a panel of 490 rice accessions grown in two locations with the objective of (a) feasibility analysis of mining ETN genes by GWAS, (b) identifying novel genes affecting ETN, and (c) probing favourable alleles of ETN-associated genes. This exploration identified 38 ETN-associated QTLs, four of which co-localize with genes involved in the regulation of rice tillering, such as *OsAAP1*, *DWL2*, *NAL1*, and *OsWRKY74. Haplotype 5* (*Hap5* of *OsAAP1*, *Hap3* and 6 of *DWL2*, *Hap2* of *NAL1*, and *Hap3* and *Hap4* of *OsWRKY74*) are the favourable alleles for ETN, and they in combination enhance ETN. Additionally, Ren et al. (2021) identified 25 candidate genes involved in rice ETN regulation, and they were found to be located within QTL regions. These genes regulate ETN through plant hormone dynamics. Overall, ten genes are involved in auxin biosynthesis or signal transduction, three genes are associated with cytokinin biosynthesis, two genes are related to strigolactone homeostasis through the control of carotenoid biosynthesis, one gene participates in gibberellin signal transduction, two genes encode ammonium transporters, two genes encode peptide/amino acid transporters, and five genes encode sugar/glucose transporters. One of the candidate genes, *OsPILS6b* (*LOC_05g40330*), encodes an auxin transporter protein, and its expression levels correlate positively with ETN. The unique expression pattern of the ETN-associated genes contributes to the transition from vegetative to reproductive growth in tillers. According to Ren et al. (2021) experiment, most of the genes were expressed prominently between 21 and 49 days after transplantation, after which their expression decreased. This shows that the plant achieves competence in ETN production during the crucial period of vegetative growth.

10.11 Profuse Tillering Underrates Yield Potential

Unlike other important cereal crops of the family *Poaceae*, rice is semi-aquatic bearing a rhizomatous culm. It has immense potential to produce tillers. Although rice is an annual plant, many cultivars do not die at harvest. During the post-harvest period, the plant exhibits strong ratooning ability. Fibrous roots emerge from basal culm nodes and provide support for the initiation of new tillers. Regeneration of tiller production from the culm base perpetuates longevity and, thereby, the plant perennates through years as a ratoon crop, but grain yield becomes poorer in them. In rice, tillers, such as primary, secondary, and tertiary tillers, appear in a specific chronological sequence on the culm nodes. Subject to resource limitations, primary and secondary tillers bear grain-bearing panicles (ETNs) and become productive. The tillering potential of rice is a genetic trait. Wang et al. (2007) reported a significant difference in grain yield and yield between low-tillering *japonica* rice and its iso-allelic mutant, which has a high-tillering ability; the former has more grains per panicle and a greater filled grain percentage and grain weight than the latter for each type of tiller. Among all the yield components, the grain number per panicle exhibited the greatest variation among the tiller types. Contrary to expectations, the genetic potential of a cultivar is often overwhelmed by environmental factors. Under irrigated conditions, most rice cultivars produce primary and secondary tillers. However, tertiary tillers are produced only by profuse tillering high-yielding semi-dwarf rice towards the end of vegetative growth duration, i.e. just before the beginning of the reproductive growth period. The initiation of tertiary tiller development begins at the end of secondary tiller production and continues until the maximum tillering stage (De Datta 1981). Modern semi-dwarf rice cultivars produce 20–25 tillers under irrigated conditions, 14–15 of which bear panicles and the rest of them, in particular, tertiary tillers, become unproductive.

10.12 Heterogeneity of Tiller Development

Tiller development in rice is spatially and temporally spaced. This amounts to the heterogeneous distribution of tillers in plant architecture, where biomass partitioning among tillers is uneven and unsynchronized. Late-developed tertiary tillers being unsuccessful competitors in biomass partitioning, are victimized by growth discrimination, and do not gain competence in contributing to grain yield. The heterogeneity of tiller development, growth, and yield contributions among tillers increases with increasing tiller number; the higher the tiller number is, the greater the difference in growth dimensions (Fig. 10.4).

The bias against the development of less productive tillers was assessed by measuring morphological and phenological development, grain yield, and the assimilate content of panicles of different categories of tillers in the profuse tillering rice cultivar Lalat by Mohapatra and Kariali (2008). The duration of the vegetative and reproductive periods of the later-initiated tillers was short, as events such as tiller emergence, booting, and anthesis were delayed in them, but maturity was synchronized. Among the tillers, the panicle grain yield and assimilate concentration decreased with the decrease in the duration of vegetative and reproductive growth (Fig. 10.5).

The time taken for tiller emergence has a critical influence on the concentrations of soluble carbohydrates and starch in the panicle at the booting, anthesis, and maturity stages; a decrease in duration infringes assimilate partitioning to the detriment of biomass gain and yield formation (Fig. 10.6).

It was estimated that in the multi-tiller rice cultivar Lalat, a threshold biomass of 2.5 g/tiller is necessary for achieving at least 50% grain filling. A tiller enjoys photosynthetic independence after its release from the leaf sheath of the preceding leaf for primary production, but the time gap in the achievement of independence makes it subservient in growth and development concerning the preceding/early tillers. Hence, early tillering is highly essential for obtaining biomass and grain filling. Furthermore, the dominance of early tillers increases through the conservation of

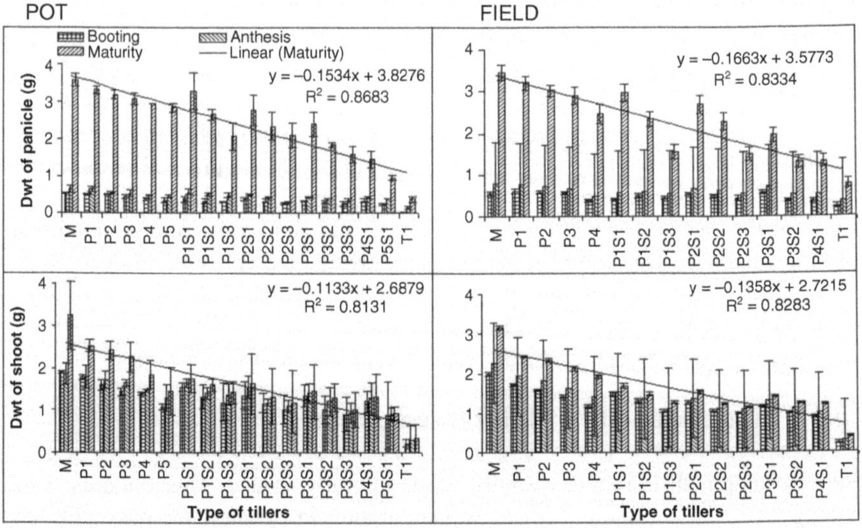

Fig. 10.4 The dry mass of panicles (top) and shoots (bottom) of different types of tillers at booting, anthesis, and maturity in the rice cultivar Lalat grown in pots (left) and in the field (right). The vertical bars represent the ± standard deviation ($n = 3$). The regression line was drawn at the maximum dry weight values (maturity) (Mohapatra and Kariali 2008)

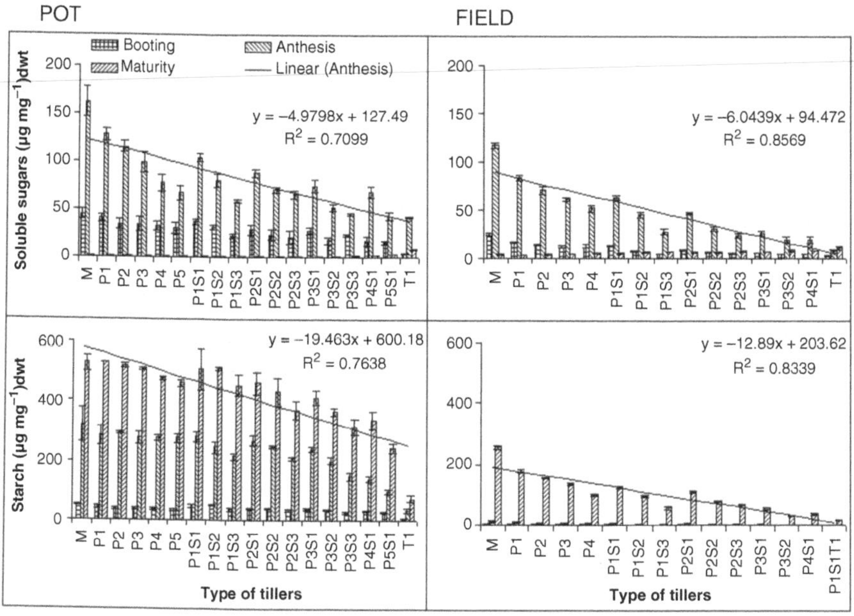

Fig. 10.5 The concentrations of soluble carbohydrates (top) and starch (bottom) in the panicles of different types of tillers at booting, anthesis, and maturity in the rice cultivar Lalat grown in pots (left) and in the field (right). The vertical bars represent ± standard deviation ($n = 3$). The regression line was drawn at the maximum values (carbohydrates at anthesis and starch at maturity) (Mohapatra and Kariali 2008)

resources when the development of late tillers is curtailed or restrained by agronomic management (Fig. 10.7).

Rice is cultivated in different ecosystems, yet ~70% of rice is produced in irrigated lowland ecosystems. The irrigated rice production system is highly water-dependent (Norton et al. 2017). To yield 1 kg of rice grain, 2500–3000 L of water is needed. In Asia, approximately 50% of the total freshwater area is used for different purposes, and rice consumes a major part of this area. In addition, water resources for the agricultural sector are declining rapidly due to demand for non-agricultural use and climate change. Because Asian people are primarily rice consumers, simply stopping rice cultivation for water conservation is not feasible. Therefore, rice agronomists have been trying to reduce water use in irrigated rice by improving water management techniques without curtailing production. Alternate wetting and drying (AWD) are a water management technique. It reduces water use by up to 30%. In addition to reducing water use, AWD is more environmentally friendly than continuous flooding (CF) for rice cultivation. Additionally, rice cultivated through CF is assumed to be a great source of methane, a highly potent greenhouse gas (Cheng et al. 2022). Methane has greater warming potential in the biosphere than carbon dioxide; on 20- and 100-year timescales, the warming potential is approximately 84 and 20 times greater, respectively. CF continuously maintains an anaerobic zone near the root hemisphere, whereas under AWD anaerobic-aerobic-anaerobic

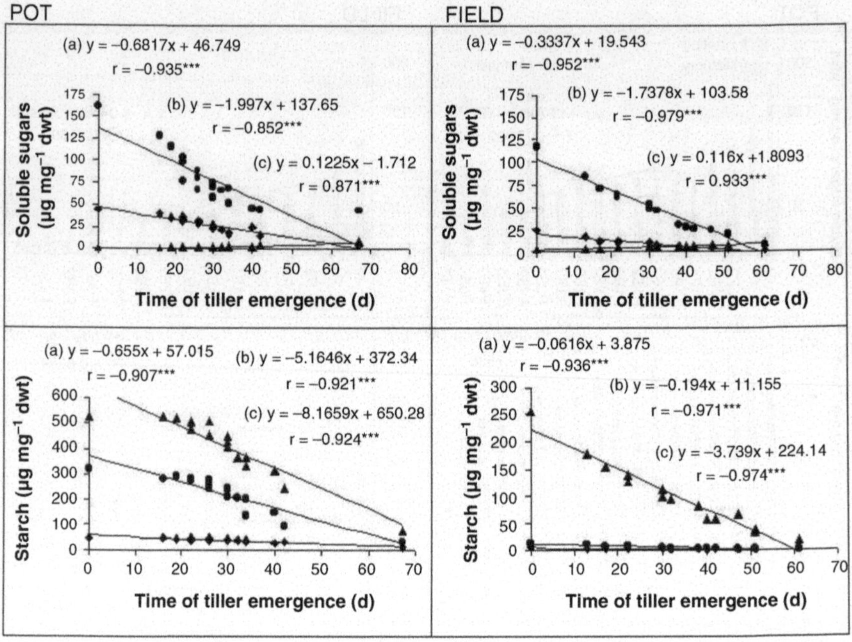

Fig. 10.6 Relationships between the time of tiller emergence and soluble carbohydrate concentration (top) and time of tiller emergence and starch concentration (bottom) for different types of tillers of the high-yielding rice cultivar Lalat cultivated in pots (left) and in the field conditions (right). The equations in parentheses (a), (b), and (c) represent the data for booting (rectangle), anthesis, (circle), and maturity (triangle), respectively (Mohapatra and Kariali 2008)

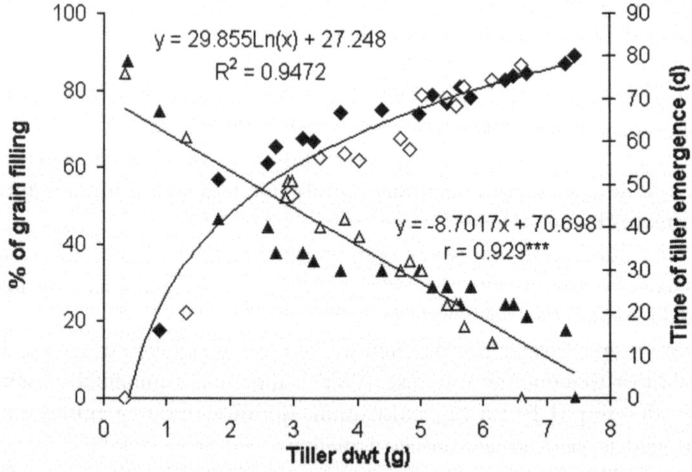

Fig. 10.7 Relationships between tiller dry weight and percentage of grain filling (rectangles) and time of tiller emergence and tiller dry weight (triangles) for different types of tillers of the rice cultivar Lalat under pot (closed symbols) and in the field (open symbols) conditions (Mohapatra and Kariali 2008)

conversion occurs periodically. This practice strongly impacts rice physiology. Obviously, the process also impacts tiller formation, survival, and productivity. The yield under AWD shows a mixed picture of promotion and demotion depending on the imposition of the severity of soil dryness. Flooding to a depth of 5 cm after 15 days of transplanting to maturity when water drops to 5 cm below the ground level was found to be appropriate (Mote et al. 2020). AWD yield is either greater or at par with CF due to an increase in tiller number or production of more ear-bearing tillers. AWD modifies shoot and root activity and alters the signalling of phytohormones such as ABA and CK (Yang and Zhang 2010). These interactions are, however, varietal or season-specific (Sriphirom et al. 2018). Improvement of effective and robust tillers caused by AWD has been reported by several investigators whereas both improvement and no impact on tiller number have also been reported by other investigators (Price et al. 2013). Kakehashi et al. (2021) reported that the dry matter production of basmati rice plants increases under CF, but this increase did not contribute to grain yield.

10.13 Cross-Talk Between Hormones During Tillering

In rice plants, hormonal regulation of tiller bud elongation or release has been undermined compared to that of reproductive structures, where plenty of evidence is available for hormones controlling grain number and grain filling. According to Dun et al. (2006), tiller buds at different locations in the main shoot exhibit different responses to auxin, cytokinin, and an unknown graft transmissible signal involved in branching inhibition. However, the homeostatic control that functions as a checkpoint for the feedback regulation of tiller number has not been studied. The constraint in exploration occurs because rice tillering is a complex multi-genic trait that influences grain yield; the regulatory effect of this molecular module is poorly understood (Guo et al. 2013). Liu et al. (2011) showed that the exogenous application of ABA, GA, and NAA influenced tiller bud outgrowth in rice, where GA and NAA completely inhibited outgrowth but ABA application did not. GA and NAA treatments increased the ABA content and decreased the zeatin + zeatin riboside content. Before the increase in ABA concentration, the IAA content increased in the GA3-treated plants, revealing the close association between these three hormones in tiller bud outgrowth. Thus, exogenous hormones regulate tiller bud growth by altering the endogenous levels of ABA, IAA, and Z + ZR in rice. The interplay of hormones also includes brassinosteroids (Bai et al. 2018). In a more recent study, Zha et al. (2022) showed that strigolactones and cytokinin play antagonistic roles in the control of shoot branching in rice. Strigolactones induce the transcription of *OsCKXs* in buds to the detriment of cytokinin content, thereby inhibiting the development of axillary buds in rice. The coordinated network of ABA in combination with other hormones for the determination of rice tillering and other yield attributes has been extensively described by Kavi Kishor et al. (2022) (Fig. 10.8). ABA fluxes maintained at the basal level regulate tillering, promote flowering, optimize cellularization in the endosperm, enhance sink strength capacity, and extend the duration of seed filling, thereby increasing grain yield.

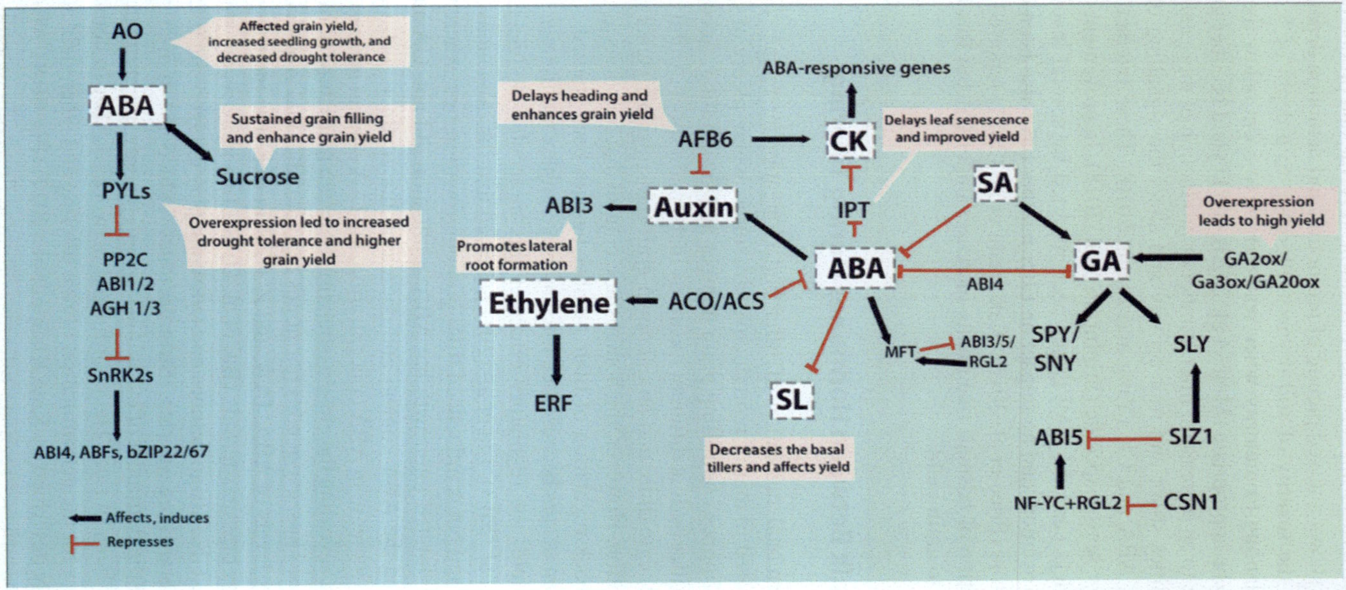

Fig. 10.8 The crosstalk between ABA and other phytohormones strigolactones (SLs), salicylic acid (SA), cytokinin (CK), auxin, ethylene, and gibberellic acid (GA)] influencing plant growth and yield. *ABA* abscisic acid, *PP2C* protein phosphatase 2C, *PYL* pyrabactin resistance-like, *SnRK2s* SNF1-related protein kinase 2, *ABI* ABA insensitive, *ABF* ABA responsive element binding factor, *bZIP* Basic-domain leucine zipper, *ERF* ethylene response factor, *ACO* ACC oxidase, *ACS* ACC synthase. (Adapted from Kavi Kishor et al. 2022 with permission)

10.14 Advantages of Restricted Tillering in Ideal Plant Architecture of Rice

Rice yields increased significantly in the late 1960s, and the rate of increase continued thereafter for two to three decades because of varietal improvement and accompanying efficient management practices. However, the yields stabilized and reached a plateau thereafter. Further efforts to increase yield did not result in visible gains, and propositions were made to change the plant architecture to protect against the undesirable traits of semi-dwarf high-yielding rice plants. High-tillering ability was one of the attributes marginalized in pursuance because late tillers do not contribute to yield. The lack of priority for development (Jaffuel and Dauzat 2005) in a high-tillering rice plant discriminates biomass production from late-formed tillers and suppresses growth. A study conducted on physiological aspects of hormone dynamics revealed the presence of high ethylene concentrations in the panicle boot of later initiated tillers in high-tillering rice plants to the detriment of grain filling, and the application of cytokinin minimized the level of ethylene to abolish the deleterious effects on assimilate partitioning (Kariali and Mohapatra 2007). In this work, the diversity spectrum between the development of different classes of tillers was wider in the high-tillering rice cultivar than in the other cultivars, and the late tillers were discriminated against on account of this gradient. The application of cytokinin reduced the developmental gap between the early-tiller and late-tiller strains. Compared to those of the high-tillering cultivar, no such differences in grain yield occurred between tillers of the low-tillering rice cultivars. This process generated a few tillers, and their development was relatively more synchronous owing to the homogeneous sharing of resources, including root-derived cytokinin and nutrients.

The study of the hormonal dynamics of tillers emphasized a new plant-type cultivar with a reduced number of tillers, in which uniform maturation is essential for achieving new heights in yield potential. The development of an ideal plant architecture conceptually relies on the purpose of enhancing rice yields. The plant architecture critically linked to grain yield is determined by plant height, tiller number and angle, panicle morphology, the position of the organs on the plants, and a plethora of genes controlling these agronomic traits in rice. *Ideal Plant Architecture 1*(*IPA1*) is one of the genes that has been identified through a map-based cloning approach; this gene determines the shoot architecture of rice for increased grain yield, and its beneficial traits include fewer unproductive tillers, stronger culms, and larger panicles. This semi-dominant quantitative trait locus is traced to chromosome 8, and it profoundly affects rice plant architecture to increase grain yield (Jiao et al. 2010). In plants, this QTL encodes the OsSPL14 (SQUAMOSA PROMOTER-BINDING PROTEIN-LIKE 14) protein and is regulated by the microRNA *OsmiR156*. In the *ipa1* mutant, a point mutation involving one nucleotide substitution, C to A, located at the recognition site for *OsmiR156* disturbs *IPA1*mRNA degradation, leading to the accumulation of OsSPL14 and the generation of an ideal rice plant with a reduced tiller number and greater grain yield (Miura et al. 2010). Additional research by a Chinese group (Lu et al. 2013) revealed that 1067 and 2185 genes are associated with *IPA1*-binding sites in shoot apices and young panicles, respectively. *IPA1* directly binds to the promoter of rice *TEOSINTE BRANCHED 1*(*OsTB1*), a negative regulator of axillary tiller buds. The activated expression of *OsTB1* suppresses tillering. The combined module also positively and directly regulates *DENSE AND ERECT PANICLE 1*, a gene that regulates panicle architecture. Furthermore, Duan et al. (2019) identified rice *SHORT INTERNODES1* (*OsSH1*), which is expressed in axillary buds and young panicles and encodes a plant-specific transcription factor of the SH1 family. *OsSH1* modulates the transcriptional activity of *IPA1*. It physically interacts with *IPA1* to repress its transcriptional activity by affecting DNA-binding activity to the promoters of *OsTB1* resulting in increased tiller number and diminished panicle size. Taken together, these findings reveal the joint influence of these genes on tiller number, plant height, and panicle length.

10.15 Regulation of *IPA1* Expression

The regulation activity of *IPA1 has* been identified by Chinese researchers at Dr. Jiang Li's laboratory and his collaborators through the discovery of an *IPA1*-interacting protein, called IPI1, a Ring Finger E3 Ligase. IPI1 has an antagonistic interaction with *IPA1*; it promotes *IPA1* degradation in the panicles but stabilizes it in the shoot apex. The antagonistic response is linked to the presence of different types of polyubiquitin chains, i.e. K48-linked in panicles and K-63-linked in the shoot apex. The *ipi1* loss-of-function mutants exhibit more tillers, larger panicles, and increased plant yield. Tissue-specific ubiquitination by IPI1 accounts for the regulation of *IPA1* protein levels and plant architecture. It is possible that the knockout of IPI1 can increase the tiller number and panicle size to achieve high grain yields (Wang et al. 2017; Wang and Wang 2017) (Fig. 10.9).

In another venture, Zhang et al. (2017) found a few super rice cultivars with the *IPA* trait possessing *ipa1-1D* and *ipa1-2D* alleles; compared to the former, the latter is a weak allele for upregulation of *IPA1* expression, which contributes to a large effect on panicle size but a marginal effect on tiller number. Both alleles hold great potential for improving rice yield. Zhang et al. (2017) proposed that tiller number reduction could be achieved in both the *ipa1-1D* and *ipa1-2D* alleles by integrating them with other yield-related *QTLs*. To determine the optimal *IPA1* dosage for achieving a higher yield potential, a model has been suggested indicating various combinations of the *IPA1*, *ipa1-1D*, and *ipa1-2D* alleles (Fig. 10.10).

10.16 Other Genes Associated with Tiller Number

In addition to *OsSPL14*, several other genes regulate plant height, grains per panicle, and the number of tillers (Table 10.6). *OsPAGN1* (*PLANT ARCHITECTURE and GRAIN NUMBER 1*) is one of these genes. In addition, plants with the recessive allele *pagn1* produce a greater number of grains and increase plant height and tiller numbers (Yan et al. 2022). The tiller angle controlling QTL/gene (*TAC1*), which controls tiller angle, is a key gene, as a rational tiller angle is an important characteristic in determining plant architecture to achieve high yields. An upright tiller and compact panicle are efficient at utilizing solar energy and bearing more grain per

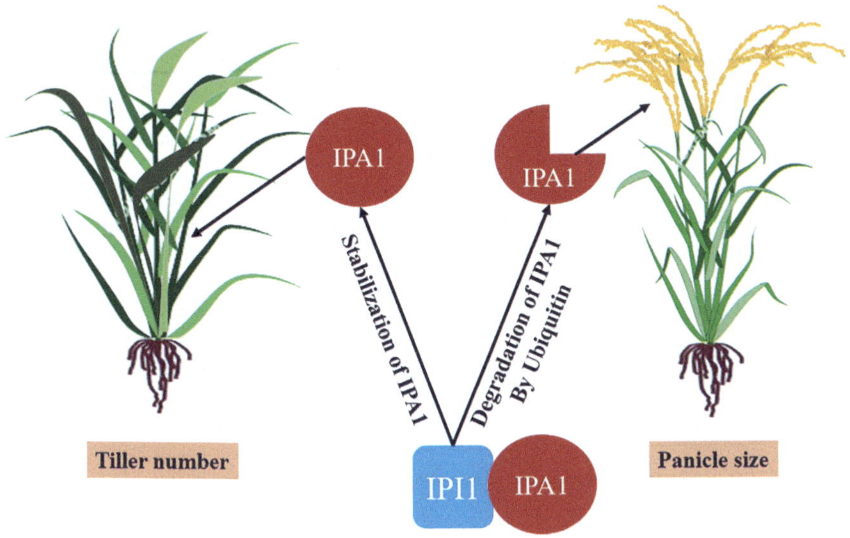

Fig. 10.9 Antagonistic interaction between *IPA1* and *IPI1*. *IPI1* regulates rice plant architecture by controlling tiller number and panicle size by modulating the degradation of IPA1 proteins in the panicles and stabilization of the proteins in the shoot apex in a tissue-specific manner. (Adapted from Wang and Wang 2017)

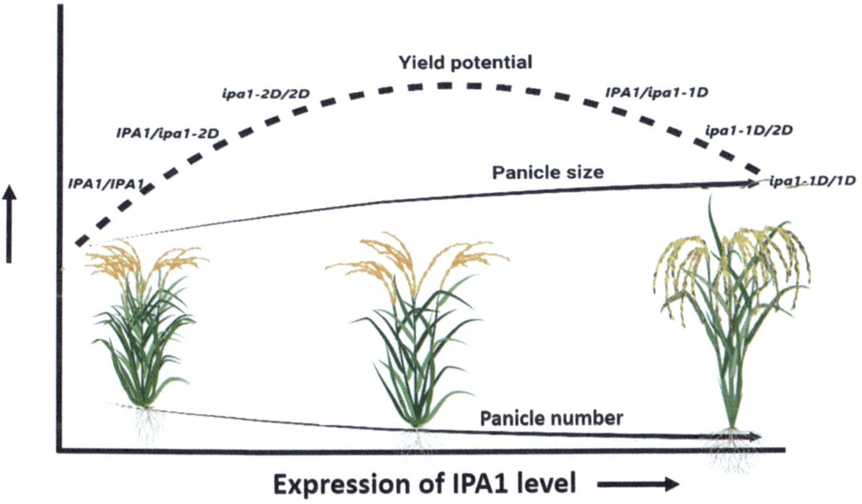

Fig. 10.10 A model of yield potential shaped by different combinations of *IPA1* alleles (*IPA1*, *ipa1-1D*, and *ipa1-2D*). (Adapted from Wang and Wang 2017)

panicle, respectively. Plants with the recessive allele *tac1* were found to produce almost straight tillers and compact panicles (Gao et al. 2021). Several miRNAs have been found to control different genes associated with tiller morphology, and the number of miRNAs has also been shown to be highly important for determining grain yield (Table 10.6).

Table 10.6 Genes associated with reduced tiller formation in rice

Genes/miRNA	Genes expression/mutant/ target genes	Phenotype
OsSPL14$^{ipa1/ipa1}$	Recessive alleles of *Ideal Plant Architecture 1* (*IPA1*)	Reduce effective tiller, increase spikelet and grain number, and greater yield
OsSPL14$^{IPA1/IPA1}$	Dominant alleles of *Ideal Plant Architecture 1* (*IPA1*)	Profuse tillering, spikelet, and grain number decrease, less yield
Plant architecture and grain number 1 (*PAGN1*)	Recessive alleles (*pagn1*) or knock-out of *PAGN1*	More tiller, greater spikelets, and more yield
Tiller Angle Controlling (*TAC1*) QTL	Recessive alleles (*tac1*)	Erect tiller (nearer to 90°), compact panicle, and enhanced yield
OsNAC2	Over-expression	Thicker stem, long panicle, greater grain number per panicle
miR156	Target gene *SPL14/SPL7*	The unproductive tiller number decreases
miR160	Target gene *ARF18*	Tiller number decreases
miR393	Target gene *TIR1*, *AFB2*	Tiller number increases

10.17 Interaction Between IPA1 and Strigolactones

As a pleiotropic regulator, *IPA1* regulates many agronomic traits of rice plants, such as tiller number, plant height, and panicle morphology. In a recent review, several miRNAs were shown to play important roles in the regulatory process for the achievement of ideal plant architecture by manipulating yield-related traits, such as fewer unproductive tillers, more panicle branches, and heavier grains, with much emphasis given to the regulation of *miRNA156* (Kaur et al. 2020). *miRNA156* regulates *IPA1* expression, but a point mutation in the *miRNA156* recognition site disturbs the miRNA-regulated degradation of the *IPA1*mRNA resulting in reduced tiller number and increased plant height and panicle branching. Conversely, CRSPR/Cas9-generated *ipa1* loss of function mutants exhibit opposite phenotypes (Li et al. 2016). This impact on plant morphology is reminiscent of the dwarf and high-tillering phenology exhibited by strigolactone-deficient or strigolactone-signalling mutant plants. The similarities in the responses open a doorway for strong interactions/cooperation between *IPA1* and strigolactones in the molecular mechanism of signalling. To suppress tillering, *IPA1* binds to the promoter region of *OsTB1*(Rice *TEOSINTE BRANCHED1*). In *Arabidopsis*, *BRC1 (BRANCHED1)* is orthologous to *OsTB1,* and the gene is responsive to strigolactone; the strigolactone-induced inhibition of axillary bud outgrowth occurs through the regulation of *BRC1* transcription. In general, strigolactones inhibit shoot branching by suppressing the outgrowth of axillary buds. *D53* (*DWARF53*) is the key repressor in the strigolactone signalling pathway, which regulates the downstream transcriptional network of the strigolactone response. Song et al. (2017) reported that the *IPA1* gene functions as a downstream component of *D53* signalling in regulating tiller number and strigolactone-induced gene expression. In this process, *IPA1* directly binds to the *D53*

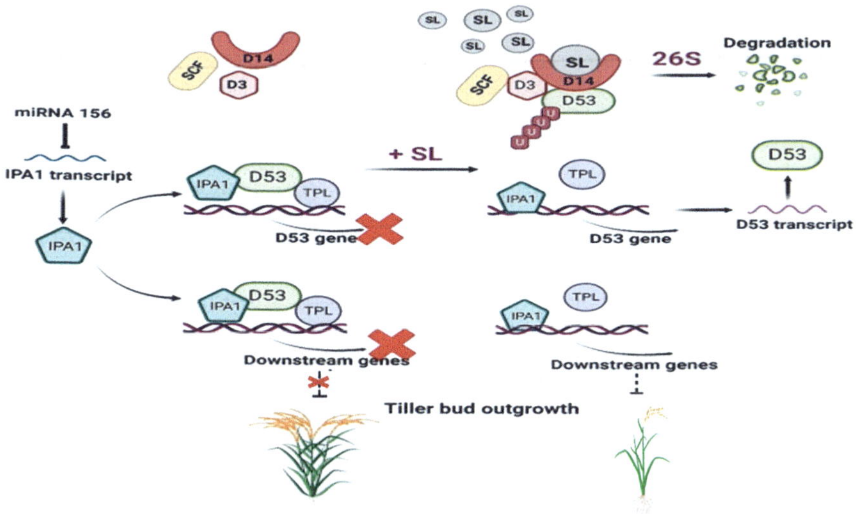

Fig. 10.11 Proposed model of the *IPA1*-mediated SL signalling pathway. In the absence of SLs, the D53 protein binds to IPA1, and together with TPL/TPR proteins, this protein represses the transcriptional activity of *IPA1*. In the presence of SLs, the perception of SL leads to the degradation of D53 by the proteasome system, which in turn releases the repression of *IPA1*-regulated gene expression and leads to an SL response. (Adapted from Song et al. 2017)

promoter and determines the feedback regulation of strigolactone-induced *D53* expression. Overall, these findings confirm that *IPA1* is a downstream transcription factor that acts in concert with D53 to regulate tiller development in rice. Song et al. (2017) proposed a schematic model for the IPA1-mediated strigolactone signalling pathway (Fig. 10.11).

References

Ahmad SA, Husain H, Ali H, Ahmad A (2005) Transplanted fine rice (*Oryza sativa* L.) productivity as affected by plant density and irrigation regimes. Int J Agric Biol 7:445–447

Badshah MA, Tu N, Zou Y, Ibrahim M, Wang K (2014) Yield and tillering response of super hybrid rice Liangyoupeijiu to tillage and establishment methods. Crop J 2(1):79–86

Bai S, Smith SM, Li J (2018) Rice plant architecture: molecular basis and application in breeding. In: Sasaki T, Ashikari M (eds) Rice genomics, genetics and breeding. Springer, Singapore. https://doi.org/10.1007/978-981-10-7461-5_8

Blom CWPM (1999) Adaptations to flooding stress: from plant community to molecule. Plant Biol 1:261–273

Chandler RF (1969) Plant morphology and stand geometry about nitrogen (Chapter 12). In: Eastin JD, Haskin FA, Sullivan CY, van Bavel CHM (eds) Physiological aspects of crop yield. https://doi.org/10.2135/1969.physiologicalaspects.c30

Chang TT, Adair CR, Johnston TH (1982) The conservation and use of rice genetic resources. Adv Agron 35:37–91

Chen WM (2001) Wild rice biological characteristics and integrated control. In: Proceedings of the 18th Asia Pacific Weed Science Society conference, Beijing, China, May 20

Cheng H, Shu K, Zhu T, Wang L, Liu X, Cai W, Qi Z, Feng S (2022) Effects of alternate wetting and drying irrigation on yield, water and nitrogen use, and greenhouse gas emissions in rice paddy fields. J Clean Prod 349:131487

De Datta SK (1981) Principle and practices of rice production. Wiley, New York

Dingkuhn M, Kropff M (1996) Rice. In: Zamski E, Schaffer AA (eds) Photo-assimilate distribution in plants and crops: source–sink relationships. Marcel Dekker Inc., New York, pp 519–547

Duan E, Wang Y, Li X, Lin Q, Zhang T, Wang Y, Zhou C, Zhang H, Jiang L, Wang J, Lei C, Zhang X, Guo X, Wang H, Wan J (2019) OsSHI1 regulates plant architecture through modulating the transcriptional activity of IPA1 in rice. Plant Cell 31:1026–1042

Dun EA, Ferguson BJ, Beveridge CA (2006) Apical dominance and shoot branching. Divergent opinions or divergent mechanisms. Plant Physiol 142:812–819

Gao J, Liang H, Huang J, Qing D, Wu H, Zhou W, Chen W, Pan Y, Dai G, Gao L, Deng G (2021) Development of the PARMS marker of the *TAC1* gene and its utilization in rice plant architecture breeding. Euphytica 217:49

Guo S, Xu Y, Liu H, Mao Z, Chang C, Ma Y, Zhang Q, Meng Z, Chong K (2013) The interaction between OsMADS57 and OsTB1 modulates rice tillering via *DWARF14*. Nat Commun 4:1566. https://doi.org/10.1038/ncomms2542

Huke RE, Huke EH (1990) Rice: then and now. International Rice Research Institute, Philippines

Jaffuel S, Dauzat J (2005) Synchronism of leaf and tiller emergence relative to the position and main stem development stage in a rice cultivar. Ann Bot 95:401–412

Jiao Y, Wang Y, Xue D, Wang J, Yan M, Liu G, Dong G, Zeng D, Lu Z, Zhu X, Qian Q, Li J (2010) Regulation of *OsSPL14* by OsmiR156 defines ideal plant architecture in rice. Nat Genet 42:541–544. https://doi.org/10.1038/ng.591

Kakehashi T, Kikuta M, Menge DM, Gichuhi EW, Samejima H, Makihara D (2021) Alternate wetting and drying (AWD) mitigates the decline in grain filling of Basmati 370 due to low temperature in tropical highlands. Agronomy 11:2345. https://doi.org/10.3390/agronomy11112345

Kariali E (2014) Environmental factors undermine the genetic expression of tiller dynamics in wild rice *Oryza nivara* and *Oryza rufipogon*. Am J Plant Sci 5:18

Kariali E, Mohapatra PK (2007) Hormonal regulation of tiller dynamics in differentially tillering rice cultivars. Plant Growth Regul 53:215–223

Kariali E, Kuanar SR, Mohapatra PK (2008) Individual tiller dynamics of two wild *Oryza* species in contrasting habitats. Plant Prod Sci 11:355–360

Kariali E, Patel S, Panigrahi R, Mohapatra PK (2010) Change of adaptive plasticity of *Oryza rufipogon* Griff. in response to variation in growth habitats. Indian J Plant Physiol 15:288–292

Kariali E, Sarangi S, Panigrahi R, Panda BB, Mohapatra PK (2012) Variation in senescence pattern of different classes of rice tillers and its effect on panicle biomass growth and grain yield. Am J Plant Sci 3:1047–1057

Kaur R, Bhunia RK, Rajam MV (2020) MicroRNAs as potential targets for improving rice yield via plant architecture modulation: recent studies and future perspectives. J Biosci 45:116

Kavi Kishor PV, Tiozon RN Jr, Fernie AR, Sreenivasulu N (2022) Abscisic acid and its role in the modulation of plant growth, development, and yield stability. Trends Plant Sci 27:1283. https://doi.org/10.1016/j.tplants.2022.08.013

Khush GS (1997) Origin, dispersal, cultivation and variation of rice. Plant Mol Biol 35:25–34

Kim H, Jung J, Singh N, Greenberg A, Doyle JF, Tyagi W, Chung JW, Kimball J, Hamilton RS, McCouch SR (2016) Population dynamics among six major groups of the *Oryza rufipogon* species complex, wild relative of cultivated Asian rice. Rice 9:56. https://doi.org/10.1186/s12284-016-0119-0

Li M, Li X, Zhou Z, Wu P, Fang M, Pan X, Lin Q, Luo W, Wu G, Li H (2016) Reassessment of the four yield-related genes *Gn1a*, *DEP1*, *GS3*, and *IPA1* in rice using a CRISPR/Cas9 system. Front Plant Sci 7:377

Liu Y, Ding Y, Wang Q, Li G, Xu J, Liu Z, Wang S (2011) Effect of plant growth regulators on the growth of rice tiller bud and changes of endogenous hormones. Acta Agron Sin 37:670–676

Lu Z, Yu H, Xiong G, Wang J, Liu G, Jing Y, Meng X, Hu X, Fu X, Wang Y, Li J (2013) Genome-wide binding analysis of the transcription activator IDEAL PLANT ARCHITECTURE1 reveals a complex network regulating rice plant architecture. Plant Cell 25:3743–3759

Miura K, Ikeda M, Matsubara A, Song XJ, Ito M, Asano K, Matsuoka M, Kitano H, Ashikari M (2010) OsSPL14 promotes panicle branching and higher grain productivity in rice. Nat Genet 42:545–549

Mohanan KV, Mini CB (2008) Relative contribution of rice tillers of different status towards yield. Int J Plant Breed Genet 2:9–12

Mohapatra PK, Kariali E (2008) Time of emergence determines the pattern of dominance of rice tillers. Aust J Crop Sci 1:53–62

Mohapatra PK, Panda BB, Kariali E (2011) Plasticity of tiller dynamics in wild rice *Oryza rufipogon* Griff.: a strategy for resilience in suboptimal environments. Int J Agron 2011:543237, 9 pages. https://doi.org/10.1155/2011/543237

Moldenhauer KAK, Gibbons JH (2003) Rice morphology and development. In: Smith CW, Dilday RH (eds) Rice: origin, history, technology, and production. Wiley, New Jersey, pp 103–128

Mote K, Rao VP, Anitha V (2020) Alternate wetting and drying. Indian Farm 70:6–9

Norton GJ, Shafaei M, Travis AJ, Deacon CM, Danku J, Pond D, Cochrane N, Lockhart K, Salt D, Zhang H, Dodd IC, Hossain M, Islam MR, Price AH (2017) Impact of alternate wetting and drying on rice physiology, grain production, and grain quality. Field Crop Res 205:1–13

Oka HI (1991) Genetic diversity of wild and cultivated rice in rice biotechnology. In: Khush GS, Toenniessen GH (eds) Biotechnology in agriculture no. 6. CAB International, Wallingford, pp 55–81

Peng S, Khush GS, Cassman KG (1994) Evolution of the new plant ideotype for increased yield potential. In: Cassman KG (ed) Breaking the yield barrier. International Rice Research Institute, Philippines, pp 5–20

Price AH, Norton GJ, Salt DE, Ebenhoeh O, Meharg AA, Meharg C, Islam MR, Sarma RN, Dasgupta T, Ismail AM, McNally KL, Zhang H, Dodd IC, Davies WJ (2013) Alternate wetting and drying irrigation for rice in Bangladesh: is it sustainable and has plant breeding something to offer? Food Energy Sec 2:120–129

Rao SA, Phetpaseut V, Bounphanousay C, Jackson MT (1997) Spontaneous interspecific hybrids in *Oryza* in Lao PDR. Int Rice Res Notes 22:4–5

Ren M, Huang M, Qiu H, Chun Y, Li L, Kumar A, Fang J, Zhao J, He H, Li X (2021) Genome-Wide Association Study of the genetic basis of effective tiller number in rice. Rice 14:56. https://doi.org/10.1186/s12284-021-00495-8

Sharma SD, Shastry SVS (1965) Taxonomic studies in genus *Oryza* L. III. *O. rufipogon* Griff. sensu strict and *O. nivara* Sharma et. Shastry nom. nov. Indian J Genet Plant Breed 25:157–167

Song X, Lu Z, Yu H, Saho G, Xiong J, Meng X, Jing Y, Liu G, Xiong G, Duan J, Yao X, Liu C, Li H, Wang Y, Li J (2017) IPA1 functions as a downstream transcription factor repressed by D53 in strigolactone signalling in rice. Cell Res 27:1128–1141

Sriphirom P, Chidthaisong A, Towprayoon S (2018) Rice cultivation to cope with drought situation by alternate wet and dry (AWD) water management system: a case study of Ratchaburi Province, Thailand. Chem Eng Trans 63:139–144

Sultan SE (2000) Phenotypic plasticity for plant development, function, and life history. Trends Plant Sci 5:537–542

Sultan SE (2004) Promising directions in plant phenotypic plasticity. Perspect Plant Ecol Evol System 6:227–233

Takeoka T (1962) Taxonomic studies of *Oryza*. II. Several species complexes. Bot Mag Tokyo 75:455–461

Tzvelev NN (1989) The system of grasses (*Poaceae*) and their evolution. Bot Rev 55:141–203

Vaughan DA (1994) The wild relatives of rice: a genetic resources handbook. International Rice Research Institute, Philippines

Vigueira CC, Qi X, Song BK, Li LF, Caicedo AL, Jia Y, Olsen KM (2019) Call of the wild rice: *Oryza rufipogon* shapes weedy rice evolution in Southeast Asia. Evol Appl 12:93–104

Wang B, Wang H (2017) IPA1: a new "green revolution" gene? Mol Plant 10:779–781

Wang F, Cheng F, Zhang G (2007) Difference in grain yield and quality among tillers in rice genotypes differing in tillering capacity. Rice Sci 14:135–140
Wang J, Yu H, Xiong G, Lu Z, Jiao Y, Meng X, Liu G, Chen X, Wang Y, Li J (2017) Tissue-specific ubiquitination by IPA1 INTERACTING PROTEIN 1 modulates IPA1 protein levels to regulate plant architecture in rice. Plant Cell 29(4):697–707. https://doi.org/10.1105/tpc.16.00879
Watson L, Clifford HT, Dallwitz MJ (1985) The classification of *Poaceae*: subfamilies and super tribes. Aust J Bot 33:433–484
Yan P, Zhu Y, Wang Y, Ma F, Lan D, Niu F, Dong S, Zhang X, Hu J, Liu S, Guo T, Xin X, Zhang S, Yang J, Cao L, Luo X (2022) A new RING Finger Protein, Plant Architecture and Grain Number 1, affects plant architecture and grain yield in rice. Int J Mol Sci 23:824
Yang JC, Zhang JH (2010) Crop management techniques to enhance harvest index in rice. J Exp Bot 61:3177–3189
Yoshida S (1973) Effects of temperature on growth of the rice plant (*Oryza Sativa* L.) in a controlled environment. Soil Sci Plant Nutr 19:299–310
Yoshida S (1981) Fundamentals rice crop science. International Rice Research Institute, Philippines
Zha M, Zhao Y, Wang Y, Chen B, Tan Z (2022) Strigolactones and cytokinin interaction in buds in the control of rice tillering. Front Plant Sci 13:837136. https://doi.org/10.3389/fpls.2022.837136
Zhang L, Yu H, Ma B, Liu G, Wang J, Wang J, Gao R, Li J, Liu J, Xu J, Zhang Y, Li Q, Huang X, Xu J, Li J, Qian Q, Han B, He Z, Li J (2017) A natural tandem array alleviates epigenetic repression of IPA1 and leads to superior yielding rice. Nat Commun 8:14789. https://doi.org/10.1038/ncomms14789

The manufacturer's authorised representative in the EU is Springer Nature Customer Service Centre GmbH, Europaplatz 3, 69115 Heidelberg, Germany. If you have any concerns regarding our products, please contact ProductSafety@springernature.com

Printed and bound by CPI Group (UK) Ltd, Croydon, CR0 4YY

01/10/2026

02225930-0009